NUCLEAR MAGNETIC RESONANCE SPECTROSCOPY IN ENVIRONMENTAL CHEMISTRY

NUCLEAR MAGNETIC RESONANCE SPECTROSCOPY IN ENVIRONMENTAL CHEMISTRY

Edited by

MARK A. NANNY

ROGER A. MINEAR

JERRY A. LEENHEER

New York   Oxford
Oxford Unversity Press
1997

Oxford University Press

Oxford   New York
Athens   Auckland   Bangkok   Bogota   Bombay   Buenos Aires
Calcutta   Cape Town   Dar es Salaam   Delhi   Florence   Hong Kong
Istanbul   Karachi   Kuala Lumpur   Madras   Madrid   Melbourne
Mexico City   Nairobi   Paris   Singapore   Taipei   Tokyo   Toronto

and associated companies in
Berlin   Ibadan

Copyright © 1997 by Oxford University Pres, Inc.

Published by Oxford University Press, Inc.
198 Madison Avenue, New York, New York 10016

Oxford is a registered trademark of Oxford University Press

Library of Congress Cataloging-in-Publication Data
Nanny, Mark A.
Nuclear magnetic resonance spectroscopy in environmental
chemistry
edited by Mark A. Nanny, Roger A. Minear, Jerry A. Leenheer.
    p.   cm.
Includes bibliographical references and index.
ISBN 0-19-509751-3
1. Environmental chemistry.   2. Nuclear magnetic resonance
spectroscopy.   I. Minear, R. A.   II. Leenheer, J. A.   III. Title.
TD193-N36   1997
628–dc20   95-47215

1   3   5   7   9   8   6   4   2

Printed in the United States of America
on acid-free paper

Preface

Nuclear magnetic resonance (NMR) spectroscopy has been an important analytical and qualitative tool in chemistry, material science, and organic geochemistry for quite some time, and now is becoming recognized as a valuable tool in environmental science and technology. Because it is a non-destructive and element-specific probe, it is useful for examining a myriad of environmental reactions occurring in a variety of complex and heterogeneous matrices. The intent of this book is to show how NMR is becoming such a useful tool, and how it is being applied to a wide variety of environmental science and technology topics. Research involving numerous NMR-active nuclei (^{1}H, ^{2}H, ^{13}C, ^{15}N, ^{19}F, ^{23}Na, ^{27}Al, ^{31}P, and ^{133}Cs) in highly heterogeneous and complex samples of environmental significance is presented. Also included is a wealth of information pertaining to sample preparation, instrumental techniques, and data interpretation.

A major reason why NMR has not been considered useful for environmental research is that NMR spectroscopy is not sensitive enough for most of the chemical concentrations that occur in many environmental reactions. In addition, environmental reactions usually occur in heterogeneous matrices which often hinder and reduce the resolution and sensitivity of NMR spectroscopy. Likewise, many isolation and concentration methods used to overcome low concentrations are not feasible or introduce unacceptable sample alterations. Despite the fact that these problems still exist and in many instances are difficult to overcome, the introduction of stronger magnets, the development of new and better software and hardware, new sample processing and experimental methods, and most importantly, increased accessibility to NMR instrumentation have permitted environmental NMR studies to gain a foothold and grow. The research presented in this book illustrates how these problems have been addressed and how the advances listed above have furthered environmental studies. This book provides a survey of the current application of NMR to research problems in environmental science and technology, and also demonstrates the diversity and breadth of environmental research problems that NMR spectroscopy can address.

The book is divided into three parts, of which the first, on Contaminant Interactions, presents NMR research examining interactions of various molecules, such as pollutants with humic materials. The second part, Solution and Condensed Phase Characterization, presents studies examining the chemistry of various environmentally important species in a variety of phases: solution, colloidal, and condensed phase. Nutrient and Natural Organic Matter Cycling in the Environment, the final part, includes research exploring phosphorus and nitrogen cycling in the environment, in addition to new research involving the characterization of natural organic carbon material, especially with difficult samples, i.e., low-carbon sediments.

To complement each part, an overview chapter is included, providing a discussion as to why and how NMR spectroscopy has advanced research in the areas concerned. In addition, an introductory chapter is included that reviews the basic principles of solution and solid-state NMR spectroscopy for those unfamiliar with NMR. Finally, the future needs and directions of NMR research in environmental science and technology are presented in the closing chapter, which results from a panel discussion involving several of the chapter authors. In this chapter, they recommend areas in which environmental NMR spectroscopy can be the most advantageous technique and, in addition, discuss the difficulties an environmental scientist faces in implementing NMR in a research project.

With its focus on NMR studies in environmental research, it is hoped that this book will be of use to NMR spectroscopists interested in analyzing environmental samples, and requiring a basic body of knowledge regarding spectroscopy of environmental samples. Likewise, it is hoped that environmental scientists will use this book as a guide to incorporating NMR into their research. Hopefully, it will serve as a benchmark for current environmental NMR research, as well as a stepping-stone to spur future work in the same area.

M.A.N
R.A.M
J.A.L

Contents

Contributors

S. J. Anderson
Department of Crop & Soil Sciences
Michigan State University
East Lansing, MI

P. M. Bertsch
Division of Biogeochemistry
University of Georgia
Savannah River Ecology Laboratory
Aiken, SC

J. M. Bortiatynski
Fuel Science Program
The Pennsylvania State University
University Park, PA

Jean Yves Bottero
Laboratorie Environnement &
    Minéralurgic
UA 235 CNRS
Vandoeuvre, France

J. A. Clark
Western Research Institute
365 North 9th Street
Laramie, WY

L. M. Condron
Department of Soil Science
Lincoln University
Canterbury
New Zealand

B. Conyers
Department of Chemistry &
    Biochemistry
Old Dominion University
Norfolk, VA

J. D. Cox
Western Research Institute
365 North 9th Street
Laramie, WY

W. L. Earl
Chemical Sciences & Technology
    Division
Los Alamos National Laboratory
Los Alamos, NM

E. Forrer
Department of Chemistry &
    Biochemistry
Old Dominion University
Norfolk, VA

E. Frossard
Institute for Plant Sciences
Swiss Federal Institute of Technology
CH-8092 Zurich, Switzerland

R. Fründ
Institut für Biophysik & Physikalische
    Biochimie
Universität Regensburg
Regensburg, Germany

P. G. Hatcher
Fuel Science Program
The Pennsylvania State University
University Park, PA

J. D. Hem
US Geological Survey
Menlo Park, CA

B. E. Herbert
Department of Geology & Geophysics
Texas A&M University
College Station, TX

C. T. Johnston
Crop, Soil & Environmental Sciences
Purdue University
West Lafayette, IN

H. Knicker
Lehrstuhl für Bodenkunde
TU München
D-85350 Freising, Germany

L. S. Kotlyar
Institute for Environmental Chemistry
National Research Council of Canada
Ottawa, Ontario

A. Labouriau
Chemical Sciences & Technology
    Division
Los Alamos National Laboratory
Los Alamos, NM

D. C. Lane
Western Research Institute
365 North 9th Street
Laramie, WY

L. A. Leenheer
US Geological Survey
Denver, CO

E. P. Locke
Department of Chemistry &
    Biochemistry
Old Dominion University
Norfolk, VA

H.-D. Lüdemann
Institut für Biophysik & Physikalische
    Biochemie
Universität Regensburg
Regensburg, Germany

G. E. Maciel
Department of Chemistry
Colorado State University
Fort Collins, CO

Armand Masion
Laboratoire Environnement &
    Minéralurgie
UA 235 CNRS
Vandoeuvre, France

F. P. Miknis
Western Research Institute
365 North 9th Street
Laramie, WY

R. D. Minard
Department of Chemistry
The Pennsylvania State University
University Park, PA

R. A. Minear
Department of Civil Engineering
University of Illinois at Urbana—
    Champaign
Urbana, IL

J. L. Morel
ENSAIA-INRA
Vandoeuvre-lès-Nancy, France

M. A. Nanny
School of Civil Engineering and
    Environmental Science
University of Oklahoma
Norman, OK

D. A. Netzel
Western Research Institute
365 North 9th Street
Laramie, WY

R. H. Newman
Industrial Research Limited
Lower Hutt, New Zealand

T. I. Noyes
US Geological Survey
Denver, CO

J. A. Ripmeester
Steacie Institute for Molecular Sciences
National Research Council of Canada
Ottawa, Ontario

J. F. Rovani
Western Research Institute
365 North 9th Street
Laramie, WY

R. Schutte
Syncrude Research
Edmonton, Alberta

F. E. Scully, Jr.
Department of Chemistry &
    Biochemistry
Old Dominion University, Norfolk, VA

B. D. Sparks
Institute for Environmental Chemistry
National Research Council of Canada
Ottawa, Ontario

P. Tekely
Laboratoire de Méthodologie
  RMN
Université de Nancy
Vandoeuvre, France

F. Thomas
Laboratoire Environnement &
  Minéralurgie
UA 235 CNRS
Vandoeuvre, France

K. A. Thorn
US Geological Survey
Arvada, CO

D. V. Vivit
US Geological Survey
Menlo Park, CA

R. L. Wershaw
US Geological Survey
Denver, CO

NUCLEAR MAGNETIC RESONANCE SPECTROSCOPY IN ENVIRONMENTAL CHEMISTRY

Introduction

Nuclear Magnetic Resonance Spectroscopy

Basic Theory and Background

HEIKE KNICKER & MARK A. NANNY

Nuclear magnetic resonance (NMR) spectroscopy is one of the most powerful experimental methods available for atomic and molecular level structure elucidation. It is a powerful technique in that it is a noninvasive probe that can be used to identify individual compounds, aid in determining structures of large macromolecules, such as proteins, and examine the kinetics of certain reactions. NMR spectroscopy takes advantage of the magnetic properties of the observed nucleus that are influenced not only by its chemical environment, but also by physical interactions with its environment. Both can be examined by measuring specific NMR parameters such as coupling constants, relaxation times, or changes in chemical shifts. As NMR techniques and instrumentation advance, NMR spectroscopy is becoming more important in the environmental sciences, tackling problems and questions that previously were difficult to answer. For example, sensitivity enhancement techniques increase the ability to examine a sample without chemical or physical pretreatment. A sample examined in this manner is in its original state and is unaffected by chemical or physical reactions caused by the pretreatment procedure.

Despite its increasing popularity and numerous advantages, NMR spectroscopy can be a mysterious, and at times daunting, technique. The purpose of this chapter is to provide an overview of basic NMR theory and background for the uninitiated. It is hoped that it will provide enough information to those unfamiliar with NMR and its terminology for them to find the remaining chapters understandable and interesting. Those who desire a greater understanding are referred to the many textbooks on solution-state NMR,[1–3] solid-state NMR,[4] and the application of NMR to geochemistry, soil chemistry, oils and coals,[5,6] and carbonaceous solids.[7]

The advance that led to NMR spectroscopy came in 1939 with resonance experiments by Rabi and coworkers,[8] who demonstrated the property of

3

nuclear spin. In 1945, the research groups of Bloch[9] and Purcell[10] independently obtained the first nuclear resonance signals. For this they won the 1952 Nobel prize. The first application of NMR spectroscopy in the field of humic substance research was ^{1}H NMR of liquids.[11-13] González-Vila et al.[14] were the first to apply ^{13}C solution-state NMR to natural humic acids. Eventually, improvements in instrumentation and data processing made quantitative analysis of such spectra feasible. The crucial breakthrough, however, was the development of the Fourier Transformation technique by Ernst and Anderson.[15] This advance permitted the recording of NMR spectra of samples that were of low concentration. Hence, obtaining liquid-state natural-abundance ^{13}C NMR spectra became feasible.

Development of the high resolution solid-state cross-polarization magic angle spinning (CPMAS) technique by Schaefer and Stejskal[16] opened the door for the examination of solid-state samples. The first solid-state ^{13}C NMR spectra of humic substances were published only about 25 years ago.[17-20] With this development solid-state NMR became recognized as a valuable technique in soil and environmental science. Since then it has become a routine method.

Basic NMR Theory

The Resonance Experiment

Nuclei spin around their axis with a spin angular momentum L, which is represented by the spin angular momentum vector $I\hbar$. I is the vector representation of the nuclear spin I and $\hbar$ is Planck's constant h divided by 2π. The movement of a spinning nucleus can be considered as a positive electric current flowing in a loop. Because flowing current generates a magnetic field, the nucleus can be considered to be a magnetic dipole containing a magnetic moment μ, which is given by:

$$\mu = \gamma L \tag{1}$$

which can also be expressed as:

$$\mu = \gamma \hbar I \tag{2}$$

where γ is the gyromagnetic ratio, a nuclear constant which is unique to each elemental nucleus. From equation (2) it is obvious that nuclei with $I \neq 0$ possess a magnetic moment and therefore can be used in NMR experiments. It has been found experimentally that the situation $I \neq 0$ exists for nuclei where either the atomic mass number or the atomic number is odd.

When a spinning nucleus, with $I \neq 0$, is placed in a static magnetic field B_0, which for this case will be defined as parallel to the z-axis, the magnetic moment μ aligns with the magnetic field, B_0. This is similar to a simple bar magnet aligning itself with an external magnetic field. In the B_0 magnetic field, the nucleus feels torque which causes it to precess around the z-axis of the B_0 field. This is similar to a spinning top precessing about the Earth's gravitational field. The frequency of this precession is the Larmor frequency, w_0. Its magnitude is the product of B_0 and the gyromagnetic ratio γ, and is specific for each nucleus.

$$\omega_0 = -\gamma B_0 \tag{3}$$

Quantum mechanical rules state that an object with spin, i.e., one with angular momentum, has a discrete number of spin states. The basic principles of quantum mechanics state that the energy of a spin in a magnetic field must be quantized. The spin states and energy levels which are allowed to a nucleus are described by the magnetic spin quantum number m_I where

$$m_I = I, I - 1 \ldots - I \tag{4}$$

$$\text{with } 2I + 1 \text{ values for } m_I \tag{5}$$

It follows that $2I + 1$ different energy levels of the spins are possible, each with the energy

$$E = -\gamma \hbar m_I B_0 \tag{6}$$

Thus, when an assembly of nuclei with $I = 1/2$ is placed in a magnetic field B_0, the nuclei with $m_I = +1/2$ will align themselves so that their spin magnetic moment vectors are parallel to the B_0 field and those with $m_I = -1/2$ are aligned antiparallel to B_0. The nuclei that are antiparallel to B_0 will have slightly higher energy than those that are aligned parallel to B_0. The difference in energy between the $m_I = 1/2$ level and the $m_i = -1/2$ level is called nuclear Zeeman splitting and is $\Delta E = -\gamma \hbar B_0 \Delta m_I$. This is illustrated with an energy diagram in Figure 1.

At thermal equilibrium the population distribution of an assembly of nuclei with $I = 1/2$ is described by the Boltzmann distribution law:

$$N_\beta / N_\alpha = \exp(-\Delta E/kT) = \exp(-\gamma \hbar B_0/kT) \tag{7}$$

N_α represents the number of spins at the lower energy level where $m = +1/2$, and N_β the number of spins in the upper energy level where $m = -1/2$, k is the Boltzmann constant, and T is the temperature in kelvin units. The energy difference $\Delta E = E_\beta - E_\alpha$ between the two levels can be calculated from equation (7) and is equal to

$$\Delta E = -\gamma \hbar \Delta m_I B_0 \tag{8}$$

Bohr's frequency law states that

$$\Delta E = h\nu = \hbar \omega \tag{9}$$

where ν is the resonance frequency in hertz and ω the resonance frequency in radians/s. From this the nuclear magnetic resonance condition follows:

$$\omega_0 = -\gamma B_0 \Delta m_I \tag{10}$$

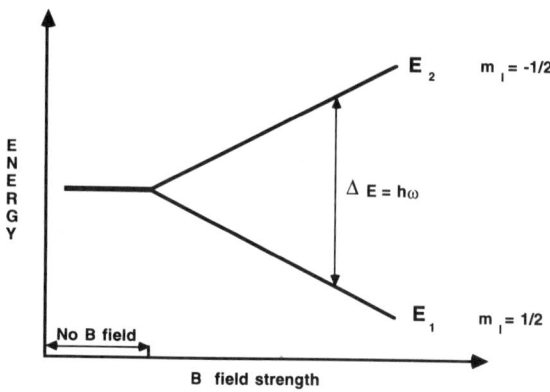

Figure 1. Zeeman splitting of the $m = 1/2$ system in a magnetic field.

where Δm_I is the difference between the ground state or the lowest energy state and the excited state. Quantum mechanical rules state that the only transitions allowed are those in which $\Delta m = 1$.

Equation (10) contains the basic essence of NMR. It states that for a spin transition between energy levels in a magnetic field B_0 to occur, an energy quantum of ω_0 is necessary. A transition from a lower energy level to a higher energy level is detected as an absorption of energy, while a transition from a higher energy level to a lower energy level is detected as an emission of energy. The net amount of all transitions occurring in a sample is detected and identified in an NMR spectrum as a resonance signal occurring at the specific resonance frequency of the nucleus, ω_0. The signals detected in pulse Fourier Transform NMR (FT-NMR) may be considered to be emission signals. These emission signals are Fourier-transformed to yield the spectrum.

The energy required to induce a transition is supplied as a pulse of a second oscillating electromagnetic radiofrequency field B_1 with the frequency ν. B_1 is perpendicular to B_0. If the frequency ν of the pulse satisfies the resonance condition in equation (10), the nuclei absorb energy and the spins are excited to a higher energy state. Equation (10) shows that the magnetic field strength and the resonance frequency are directly proportional. The greater the field strength B_0, the higher the resonance frequency ω_0. Examining equation (8), it can also be seen that the application of higher magnetic fields increases the energy difference ΔE, and therefore the amount of absorbed energy also increases. In practice, this results in greater intensity and higher sensitivity of the resonance signal in the NMR experiment. After the pulse, the nuclei begin to relax and emit energy at a specific Larmor frequency that is detected as an emission signal and Fourier-Transformed to yield the spectrum. The intensity of the emission signal and the spectrum is proportional to the total energy emitted and, therefore, to the number of excited nuclei. The signal intensity thus contains quantitative information about the numbers of nuclear spins excited to the higher energy.

Continuous, rather than pulsed, irradiation results in continuous absorption of the energy by the nuclei. After extended, continuous irradiation, the number of transitions to the upper energy level is the same as to the lower level, resulting in an equal distribution of spins among the two energy states. The energy absorbed by the nuclei in the lower energy state becomes equal to the energy emitted during the relaxation of the nuclei in the higher energy state. Therefore, a net absorption and resonance signal is not detected. At this point, the system is said to be saturated. After termination of the radiofrequency (rf) irradiation B_1, the spin system relaxes to its original thermal equilibrium by losing energy to the environment. There are two types of relaxation which apply to simple pulse FT-NMR, spin–spin relaxation and spin–lattice relaxation. Spin–spin relaxation is an entropic process, in which the coherent phase of the excited spin system is randomized in the x–y plane at a rate governed by the spin–spin relaxation time constant T_2. During spin–lattice relaxation, however, energy is transferred from the excited spin system to the surrounding lattice. This process is governed by the spin–lattice relaxation constant T_1.

T_1 and T_2 relaxation play an important role in practical NMR spectroscopy. To increase the signal to noise (S/N) ratio of a spectrum, several single spectra are accumulated and added. In such an approach, the noise averages out, while the signal is augmented. In order to achieve the maximal signal intensity and to obtain quantitative data, saturation has to be avoided. This means that the spin system has to return completely to its thermal equilibrium before a new rf pulse is applied. If there is not enough time between B_1 pulses, some of the nuclei will still be in the excited state when the pulse occurs. These previously excited nuclei will not be detected. Unless sufficient delay times are allowed, more and more nuclei will remain in the excited state after each pulse until the energy levels have the equal spin population. At this point, a signal will not be detected and the system is saturated. A general rule is that the delay time between two rf pulses should be $5 \times T_1$. T_1 can range from values as short as a few milliseconds to as long as days, and are dependent upon the mobility of the nucleus and upon the type of functional group or molecule in which the nucleus resides. In both liquid- and solid-state NMR unpaired electrons in paramagnetic materials offer extremely efficent relaxation pathways; this greatly reduces T_1. In solids, where the molecular motion can be very slow, the efficiency of energy donation to the environment by the spin system during relaxation is very low. This can result in very long relaxation times.

There are two methods for performing an NMR experiment. One is by continuous-wave NMR (CW-NMR). In this experiment, the detection frequency is fixed and the magnetic field is slowly increased while each nucleus undergoes a transition, when its specific Larmor frequency is generated. This provides a typical NMR spectrum of frequency (x-axis) vs. signal intensity

(y-axis). The higher the resolution desired, the more slowly the frequency must be swept. This may take up to several minutes. If the sample concentration is low, repeated scans are necessary. Thus, CW-NMR experiments are severely limited by concentration constraints; for this reason, CW-NMR is rarely used anymore. Instead, FT-NMR spectroscopy is now used. It allows rapid repeated pulsing of the sample so that a single NMR scan can be obtained for most cases, within a few seconds. This is achieved by exciting all nuclei with a single, very intense, B_1 pulse having a wide frequency range. Thus, if samples have low concentrations, they can be repeatedly pulsed and the individual emission signals are added together. This permits a much higher level of sensitivity to be obtained as the NMR signals are constructively added and the noise, for all practical considerations, is destructively canceled.

FT-NMR experiments are performed by subjecting the nuclei to a very intense B_1 pulse of fixed frequency that excites a wide frequency range (typically 50 kHz to 1 MHz) centered at the fixed frequency, causing all nuclei that have a ω_0 near the pulse frequency to undergo a transition. After the rf pulse, the emitted signal is detected as an oscillating current. As the excited nuclei return or relax to their equilibrium state, the amplitude of the detected oscillating current decreases with the relaxation time constants T_1 and T_2. The relaxation decay pattern is called the free induction decay (FID). Nuclei with different frequencies will decay at different rates and the FID contains this information in the time domain, i.e., the x-axis is time and the y-axis is intensity. A Fourier Transformation is then performed upon the FID, transforming the time domain information to the frequency domain information, i.e., the NMR spectrum. Because all the nuclei are excited at once in an FT-NMR experiment, the length of time required for a single scan is usually only a few seconds. The pulse itself is on the order of 10 μs and collection of the FID is usually on the order of milliseconds to a few seconds. The longest portion of an FT-NMR experiment is the length of time required between pulses if more than one pulse is used. Still, compared to a CW-NMR experiment, the amount of time for an FT-NMR experiment can be a thousand times shorter.

Interactions in an NMR Experiment

So far, only the Zeeman interaction of nuclei with an external static field, B_0, and an applied additional radiofrequency (rf) field, B_1, has been discussed, giving the impression that all nuclei of one element resonate at one specific frequency. If this were the case, NMR would not be useful for much more than elemental analysis. Fortunately, the situation is much more complex, permitting NMR to differentiate between chemical functional groups. In a macroscopic sample the local magnetic field, B_{loc}, of an individual nucleus is influenced by interactions with its environment. These interactions slightly change the B_{loc} of the individual nucleus and cause it to resonate or undergo a transition at a modified ω_0. This change in ω_0 provides valuable information regarding the nucleus and its interactions which, in turn, provide chemical information. The possible interactions are:

1. chemical shift;
2. spin–spin (scalar) interactions;
3. dipole–dipole interactions;
4. quadrupole interactions.

Consideration of these interactions and their influence on the behavior of the nuclear spins has led to the development of many advanced experiments, which give additional information on molecular structure and dynamics.

Chemical Shift

Nuclei in a compound are surrounded by electrons. Interaction of these electrons with the external magnetic field B_0 reduces the local magnetic field B_{loc} felt by the nucleus. The magnitude of the effective magnetic field B_{eff} (the magnetic field perceived by the nucleus) depends upon B_{loc}, hence the specific electron density around the observed nucleus. Expressed with the dimensionless constant σ, it can be written:

$$B_{eff} = B_0 - \sigma B_0 = B_0 \, (1 - \sigma) \tag{11}$$

where σ is the chemical shielding, usually expressed in parts per million (ppm). Chemically nonequivalent nuclei have different electron densities and therefore are exposed to different magnetic fields. The nuclei precess with different frequencies and therefore produce resonance lines which are separated in the spectrum. Because the Larmor frequency of a nucleus is dependent upon the magnitude of the static magnetic field (equation (3)), it is difficult to compare NMR spectra which are of the same substance, but obtained with different magnetic field strengths. To solve this problem, the resonance frequencies of nuclei are measured with respect to some arbitrary standard, usually tetramethylsilane (TMS) for ^{1}H and ^{13}C nuclei. Unfortunately for nuclei less routinely used in NMR spectroscopy, such as ^{15}N, ^{31}P and some others, no common standard exists. In these cases, a comparison of spectra is still possible by converting the chemical shifts obtained with one reference to the chemical shift obtained with another reference, using conversion factors.

The difference in resonance frequency with respect to a reference is termed the chemical shift (δ) and is defined with the frequency of the examined substance (ν_s) and the frequency of a reference substance (ν_{ref}).

$$\delta = [(\nu_s - \nu_{ref})/\nu_{ref}] \times 10^6 \tag{12}$$

Chemical shifts are about 10^{-3} to 10^{-6} of the applied magnetic field. Therefore, they are usually given in parts per million (ppm), which accounts for the 10^6 term in equation (12).

In molecules, the electron distribution is anisotropic. The actual value of the chemical shielding to the effective magnetic field B_{eff} is, therefore, dependent upon the specific orientation of the molecules relative to the B_0 field. For molecules in rapid motion, such as those in liquid, the anisotropy averages to zero. On the other hand, the molecular motion is slow in the solid state. Because of the immobility of the molecules, the chemical shielding of each nucleus is not averaged to its isotropic value. Each nucleus in a different orientation relative to B_0 is affected by a different B_{loc} and therefore precesses with a slightly different frequency. For a powder, where all such orientations occur, a broad, average resonance line is observed in the spectrum. These broad resonance lines can have the magnitude of several kilohertz, making interpretation of the solid-state spectra more difficult.

Scalar Coupling or Spin–Spin Coupling

In a macroscopic sample, the nuclear spins are coupled to each other by their surrounding electrons. This coupling is transmitted by interactions of the electrons of bonded nuclei and is known as spin–spin coupling. It is sometimes further categorized as indirect or scalar coupling to distinguish it from the direct dipolar interaction. The spin–spin coupling can occur for both heteronuclear and homonuclear interactions. The modification of B_{loc} of the interacting nuclei by the spin–spin coupling causes splitting of the nucleus Zeeman energy levels of the spin system and can be seen in a spectrum by the splitting of the signal into multiplets.

The multiplicity and intensity of the signals correspond to the possible numbers of spin states of the coupling spins. The frequency difference between the multiplet peaks indicates the coupling constant. This constant is specific for the nuclei involved, the number of intervening bonds, and the nature of the bonds. It is another NMR parameter that can be used for structural analysis of chemical compounds. In NMR experiments with nuclei with a low natural abundance, such as ^{13}C or ^{15}N, the homonuclear coupling is less important. It is a rare occurrence that two NMR-active nuclei of the same kind are bonded and influence each other. In experiments where labeled compounds are used, this may not be the case and the coupling can be observed. On the other hand, the heteronuclear spin–spin coupling between protons and the observed nucleus is a common phenomenon. In large organic molecules this coupling can produce complicated spectra. The splitting of the signal into multiplets as a result of both homonuclear and heteronuclear spin–spin coupling can be eliminated by radiation with an additional B_2 field close to the resonance frequency of the coupling nucleus. This irradiation causes fast transitions of spins between their energy levels and results in an equal population of the relevant spin energy levels (saturation). The observed nuclei only feel an average interaction of the coupled nuclei. In the spectrum only one resonance line at the resonance frequency of the observed nucleus is detected and the spectrum is said to be decoupled. This decoupling causes a change in the population of the relevant energy levels and results in an enhancement of the intensity of the observed signal. This effect is known as the Nuclear Overhauser Enhancement (NOE). Because this enhancement is not equal for nuclei in different physical and chemical environments, it is obvious that such decoupled spectra cannot be analyzed quantitatively.

Dipole–Dipole Interaction

At the molecular level the magnetic dipole field of a nucleus with spin I_1 affects the local magnetic field of its neighbor nucleus with spin I_2, resulting in an increase or decrease of the local magnetic field B_{loc} of both nuclei. The spins precess with a slightly different Larmor frequency. This causes splitting of the signal which cannot be resolved by the instrument software and thus there is broadening of the resonance line. The increase in the field is dependent upon the magnetic moments of both nuclei, their distance from each other, and their orientation relative to the static external magnetic field B_0. Due to the Brownian molecular motion in liquids, the orientation-dependent interactions are averaged, which results in narrow signals. In solids, the rigid molecules stay at a fixed angle in every possible orientation. This results in different local magnetic fields and different Larmor frequencies of the nuclei for each orientation and therefore in broad resonance lines in solid-state NMR spectra. As in the case of spin–spin coupling, this interaction can be suppressed by decoupling the unobserved nuclei.

Quadrupole Interaction

Nuclei with $I > 1/2$ have a nonspherical electric charge distribution around the nucleus. The electric field gradients of the electron density interact with this quadrupole moment. This results in an orientation-dependent spectrum for solids. Solid-state ^{14}N and ^{17}O NMR suffer enormous line broadening from this interaction. This effect is so large that, even in the liquid state, ^{14}N line widths of 100 to 1000 Hz are encountered. Therefore, for the examination of nitrogen-containing organic material the much less abundant, and therefore less sensitive, ^{15}N isotope has to be used.

Solid-state NMR

The NMR spectra of whole soils or insoluble soil fractions can be obtained with solid-state NMR. As compared to liquids, the mobility of molecules in the solid state is much lower. In a solid-state NMR experiment, the orientation-dependent interactions, such as dipolar coupling, quadrupolar interaction and chemical shift anisotropy, cannot be averaged over time, resulting in broad resonance lines. The slow molecular motion in solids also offers few relaxation pathways, resulting in long spin–lattice relaxation times T_1. Application of the CPMAS (cross-polarization magic angle spinning) technique,[16] which is a combination of the magic angle spinning technique[21] and cross-polarization high power decoupling methods,[22,23] yields high resolution spectra with enhanced signal intensity for solid samples.

Magic Angle Spinning

Dipolar, quadrupolar, and chemical shift interactions are orientation-dependent. These anisotropic interactions are all proportional to the term $3\cos^2\beta - 1$, where β is the angle between the rotor and the magnetic field. This term vanishes in the case of a solid if β is made equal to the magic angle of $54°44'$. Spinning the sample with sufficient speed at this magic angle imitates molecular motion and removes the homonuclear dipolar coupling, quadrupole interactions, and chemical shift anisotropy. For the complete removal of any of these anisotropic interactions, the magnitude of the spinning frequency must be at least as high as the line broadening caused by the interaction itself. In the case of chemical shift anisotropy, this line broadening is dependent upon the applied static field and can have values of up to 500 ppm. This requires spinning speeds of several kilohertz. Because the chemical shift anisotropy is proportional to the stationary magnetic field, it is obvious that for higher magnetic fields increased spinning speeds must be used. In the case of insufficient spinning speed, spinning sidebands appear in the spectrum. They are observed on each side of the isotropic signal at a frequency distance equal to the spinning frequency. Their appearance can cause quantification problems because they include the signal intensity of the main signal and can obscure smaller resonance signals occurring in their chemical shift region. The present upper limit of spinning speeds is approximately 20 kHz, which is still insufficient for some nuclei in high magnetic fields (i.e., for ^{13}C at $B_0 > 7$ tesla (T)). Even though an NMR spectroscopist prefers higher magnetic fields to increase the sensitivity of an NMR experiment, the application of higher fields for solid-state NMR is not always an advantage. When low sensitivity nuclei, such as ^{15}N, are examined, high magnetic fields are desirable. A compromise must be made in these situations and interpretation of the spectra must include consideration of the sidebands. Elegant experiments have been designed, such as TOSS (total sideband suppression), to remove the sidebands.[24] Applying this pulse sequence, it must be kept in mind that intensity losses may occur.

High-power Decoupling

In solids, heteronuclear dipolar coupling is largely caused by coupling to protons. For the complete removal of the heteronuclear dipolar interaction, unrealistic spinning speeds of approximately 60 kHz have to be achieved. Therefore, high-power decoupling is employed. A high-frequency ^1H field is applied perpendicularly to the static magnetic field B_0. If the ^1H nuclei are irradiated close to their resonance frequency, fast transitions between the energy levels of the proton spin system occur and the observed nuclei "feels" only an average proton field. The coupling with the observed nuclei is removed. In comparison with the decoupling in liquids, a higher decoupling power is needed for solid-state NMR. ^{1}H amplifiers with a frequency range of several kilohertz and a power of several hundred watts are used for such applications. In general, the NOE is suppressed by an application of an inverse decoupler.

Cross Polarization

In a solid-state NMR experiment for nuclei with a low natural abundance, such as ^{13}C or ^{15}N, long spin–lattice relaxation times are expected. This can increase the measurement time necessary to achieve a tolerable S/N ratio, sometimes to days. By applying cross polarization, such long measurement times can be avoided, while at the same time signal intensity enhancement is achieved.

This technique is based on an energy or magnetization transfer from a nuclear spin system with a high relative abundance and fast spin–lattice relaxation time to a nuclear spin system X with low natural abundance and long T_1. The X nuclei are then observed after the magnetization transfer. The principles of the cross polarization technique will be explained below for a sample where the magnetization is transferred from a proton spin system to an X (^{13}C or ^{15}N) spin system. However, it is important to point out that cross polarization is also possible for other spin system combinations.

In a sample containing ^{1}H and X nuclei, the energy levels of the spin systems are split as a function of the strength of the external magnetic field $\boldsymbol{B}_0$ and their specific gyromagnetic ratios. The energy levels of the ^{1}H and X spin systems are different because of their different gyromagnetic ratios; therefore, transitions between energy levels for both systems occur at different Larmor frequencies. Based upon the principles of quantum mechanics, any transfer of the population difference (polarization) from one spin system to another is only possible if the energy difference between the two spin systems is the same for each. In a cross-polarization experiment this is a achieved by simultanous irradiation of the ^{1}H and X spin systems with variable magnetic fields during a contact time t. Hereby the fields B_{1H} and B_{1X} have to be adjusted in such a way that the energy differences between the energy levels of both systems are equal, fullfilling the Hartmann–Hahn condition.

$$\Delta E_H = \gamma_H B_{1H} = \gamma_X B_{1X} = \Delta E_X \qquad (13)$$

In a cross-polarization experiment, usually the less abundant X spin system is polarized by the more highly abundant protons. The cross-polarization dynamics are described by two parameters, the polarization transfer time T_{XH} and the proton spin–lattice relaxation time in the rotating frame $T_{1\rho H}$. T_{XH} expresses the time needed for the cross polarization. $T_{1\rho H}$ represents the relaxation of the protons during the cross-polarization experiment.

With increasing contact time t, the X-magnetization and therefore the signal intensity of the observed X increases with the first-order rate constant $1/T_{XH}$. After the optimal contact time, the X-magnetization and therefore the X signal intensity decrease exponentionally because of loss of magnetization due to $T_{1\rho H}$. The efficiency of the magnetization transfer is dependent upon a number of factors, such as molecular motion and the distance of the X nucleus from the protons. In general, T_{XH} is shorter for XH_2 and XH than for XH_3, which, in turn, is shorter than for nonprotonated X nuclei. Considering this fact, it becomes obvious that in a cross-polarization experiment the maximal signal intensity of the X nuclei can only be achieved if the polarization transfer is complete before $T_{1\rho H}$ has started. This is true if $T_{1\rho H} \gg T_{XH}$. If T_{XH} approaches $T_{1\rho H}$, the signal intensity is not any longer proportional to the spin concentration and a quantitative analysis of the spectrum is not possible. When $T_{1\rho H} \ll T_{XH}$ the signal is completely suppressed. More detailed information about quantification problems in solid-state NMR is available in a text written by Wilson.[5]

The signal enhancement which can be achieved in a cross-polarization experiment is dependent upon the ratios of the gyromagnetic ratios of the participating spin systems. The experimental repetition rate in a cross-polarization experiment is limited by the ^{1}H spin–lattice relaxation, which is faster than that of less abundant X nuclei such as ^{13}C and ^{15}N.

References

1. Harris, R. K., *Nuclear Magnetic Resnance Spectroscopy—A Physicochemical View*, Pitman, Marshfield, MA, 1983.
2. Derome, A. E., *Modern NMR Techniques for Chemistry Research*, Pergamon, New York, NY, 1987.
3. Sanders, J. K. M., and Hunter, B. K., *Modern NMR Spectroscopy: A Guide for Chemists*, Oxford University Press, New York, NY, 1993.
4. Stejskal, E. O., *High Resolution NMR in the Solid State: Fundamentals of CP/MAS*, Oxford University Press, New York, NY, 1994.
5. Wilson, M. A., *N.M.R. Techniques and Applications in Geochemistry and Soil Chemistry*, Pergamon, New York, NY, 1987.
6. Wershaw, R. L., and Mikita, M. A., eds, *NMR of Humic Substances and Coal—Techniques, Problems, and Solutions*, Lewis Publishers, Chelsea, MI, 1987.
7. Botto, R. E., and Sanada Yuzo, (eds), *Magnetic Resonance of Carbonaceous Solids*, ACS Adv. Chem. Ser. No. 229, American Chemical Society, Washington DC, 1993.
8. Rabi, I., Millman, S., Kusch, P., and Zacharias, J. P., The molecular beam resonance method for measuring nuclear magnetic moments. The magnetic moments of $_3$Li^6, $_3$Li^7 and $_9$F^{19}. *Phys. Rev.* 55, 526–535, 1939.
9. Bloch, F., Hansen, W. W., and Packard, M. E., Nuclear induction. *Phys. Rev.* 69, 127–127, 1946.
10. Purcell, E. M., Torrey, H. C., and Pound, R. V., Resonance absorption by nuclear magnetic moments in a solid. *Phys. Rev.* 69, 37–38, 1946.

11. Schnitzer, M., and Barton, D. H. R., A new experimental approach to the humic acid problem. *Nature (London)* 198, 217–219, 1963.
12. Neyroud, J. A., and Schnitzer, M., The chemistry of high molecular weight fulvic acid fractions. *Can. J. Chem.* 52, 4123–4132, 1972.
13. Lüdemann, H.-D., Lentz, H., and Liechmann, W., Protonenresonanzspektroskopie von Ligninen und Huminsäuren bei 100 Megahertz. *Erdöl, Kohle, Erdgas, Petrochem. Brennst. Chem.* 26, 506–509, 1973.
14. González-Vila, F. J., Lentz, H., and Lüdemann, H.-D., FT-C-13 nuclear magnetic resonance spectra of natural humic substances. *Biochem. Biophys. Res. Commun.* 72, 1063–1070, 1976.
15. Ernst, R. R., and Anderson, W. A., Application of Fourier transform spectroscopy to magnetic resonance. *Rev. Sci. Instrum.* 37, 93–102, 1966.
16. Schaefer, J., and Stejskal, E. O., Carbon-13 nuclear magnetic resonance of polymers spinning at magic angle. *J. Am. Chem. Soc.* 98, 1031–1032, 1976.
17. Mikinis, M. A., Bartuska, V. J., and Maciel, G. E., Cross-polarization magic angle spinning ^{13}C NMR spectra of oil shales. *Org. Geochem.* 1, 169–176, 1979.
18. Newman, R. H., Tate, K. R., Barron, P. F., and Wilson, M. A., Towards a direct, non-destructive method of characterising soil humic substances using ^{13}C NMR. *J. Soil Sci.* 31, 623–631, 1980.
19. Hatcher, P. G., Breger, I. A., and Mattingly, M. A., Structural characteristics of fulvic acids from continental shelf sediments. *Nature (London)* 285, 560–562, 1980.
20. Hatcher, P. G., Rowan, R., and Mattingly, M. A., ^1H and ^{13}C NMR of marine humic acids. *Org. Geochem.* 2, 77–85, 1980.
21. Andrew, E. R., The narrowing of NMR spectra of solids by high speed specimen rotation and the resolution of solids. *Prog. Nucl. Magn. Reson. Spectrosc.* 8, 1–39, 1971.
22. Hartmann, S. R., and Hahn, E. L., Nuclear double resonance in the rotating frame. *Phys. Rev.* 128, 2042–2053, 1962.
23. Pines, A., Gibby, G. G., and Waugh, J. S., Proton-enhanced NMR of dilute spins in solids. *J. Chem. Phys.* 59, 569–590, 1973.
24. Dixon, W. T., Spinning-sideband-free and spinning-sideband-only NMR spectra in spinning samples. *J. Chem. Phys.* 77, 1800–1809, 1982.

CONTAMINANT INTERACTIONS

1.

Sorption Processes in the Environment

Nuclear Magnetic Resonance Spectroscopy as a New Analytical Method

MARK A. NANNY

The transport and fate of chemicals in the environment comprise one of the most pertinent issues in environmental chemistry. Physical and chemical interactions between the chemical of interest (sorbate) and the various components present in soil, water, and sediment (sorbent) can dramatically influence the transport and fate of the chemical. For example, organic material such as humic acids or surfactants, when bound to the surface of soil, sediment, and clay particles, can enhance sorption of chemicals to the particle surface, immobilizing the chemicals and possibly protecting them from degradation.[1,2] The term "sorption," as used in this chapter, refers to any physical or chemical association between the sorbate and sorbent. Thus, this definition includes molecular associations ranging from hydrophobic partitioning processes to covalent binding. It has also been observed that sorbents can act as a "buffer system" for the sorbate by taking up the sorbate when it is present at high concentrations and then slowly releasing it back into solution during periods of low or zero sorbate concentration.[3] This type of behavior is important when the environmental presence of the chemical is episodic, such as in seasonal application of pesticides and herbicides. A dramatic example of this is presented by Steinberg et al.,[3] who detected the volatile, soil fumigant 1,2-dibromoethane in agricultural topsoils up to 19 years after the last application.

On the other hand, soluble organic matter such as dissolved humic and fulvic acids can dramatically increase the apparent solubility of many hydrophobic chemicals in soil and sediment pore water.[4–7] In this manner, the mobility of the chemical is increased. Thus, the nature and presence of organic material can strongly influence the transport properties of many chemicals in the environment.[8]

19

It is well known that sorption processes are a function of numerous parameters: the chemical characteristics of the sorbate (e.g., hydrophobicity, polarity), and the identity and chemical characteristics of the mineral phase[9] and its organic coating.[10–13] Other parameters important to sorption processes are the porosity of the sorbent particles, the presence of dissolved organic matter, and the solution pH and ionic strength. Understanding the influence of sorption processes upon the transport and fate of chemicals in soils and sediments is central to evaluating numerous environmental situations ranging from pesticide contamination of groundwater[14] to bioassimilation of xenobiotic chemicals.[15] Despite the importance of sorption processes in the environment, mechanisms of interaction are poorly understood. Most of the information regarding sorption is based upon macroscopic observations and measurements. The heterogeneity and complexity of environmental systems hinder detailed, analytical studies of sorption processes on a microscopic scale. It is in this area that nuclear magnetic resonance (NMR) spectroscopy has potential as an analytical tool, overcoming many of these problems and providing information on a molecular scale which is typically unattainable with conventional sorption experiments.

Numerous experimental designs are used for sorption studies, the most common of which are the batch process,[7,16] column,[7] and HPLC methodologies.[17–22] In general, these experiments determine fractionation by measuring the concentration of dissolved and sorbed compound. Many experiments utilize radioactive isotopes and use radiotracer methods to measure sorption.[23] Other studies depend upon solvent extraction or purge and trap methods before measuring solute concentrations with gas chromatography.[24] The caveats of extraction techniques concern the potential for inducing a chemical change in the system which can modify the partition equilibrium, or the possible incomplete recovery of sorbed residues. A final problem is the possible formation of reaction by-products that are hydrophilic, nonvolatile, or have formed covalent bonds with macromolecular material such as humic substances. These by-products usually are not amenable to typical analytical methods such as gas chromatography.

Clearly, there exists a strong need for noninvasive, analytical methods which can examine interactions on a microscopic level between a chemical and soil, sediment, or humic materials. Unfortunately, molecular interactions between a specific chemical and soil, sediment, or clay particles are difficult to study because of the great heterogeneity and diversity of organic matter, mineral material, and environmental conditions. NMR spectroscopy is an analytical method that is demonstrating usefulness in this area. It is a nondestructive, noninvasive probe that, in specific situations, can follow molecular interactions in highly heterogeneous environments, both covalent and noncovalent. It is beginning to be used to examine sorption processes occurring in both the liquid and solid state; results from both are presented in the following chapters of this part.

NMR Spectroscopy and Sorption Studies

NMR, both liquid- and solid-state, has been used for several decades in a variety of environmental studies,[25,26] especially for the characterization of humic and fulvic materials. Even so, its value for sorption studies is just now being recognized (Chapters 2–5, this text). Its noninvasive nature and ability to distinguish between different functional groups and compounds make NMR spectroscopy a powerful tool. The major difficulty in using it as an analytical tool for environmental studies usually has been its lack of sensitivity. NMR characterization of natural organic matter often requires some isolation and concentration procedures before analysis. In most situations where sorption processes are of interest, the concentration of sorbate is usually much too low for NMR analysis of unconcentrated samples, and use of NMR has been limited. An additional problem with using NMR spectroscopy for such studies is the inability to distinguish between the signals of the sorbate and the sorbent, especially when the sorbent concentration greatly exceeds that of the sorbate.

Advances in NMR instrumentation such as increases in magnetic field strength, Fourier Transform processing, and new pulse sequences have helped address sensitivity problems and have increased the application of NMR to environmental studies as a whole. But more importantly, sorption studies have been made possible recently through the use of sorbate molecules labeled with NMR-sensitive nuclei at or near chemically active sites. This method involves labeling molecules with NMR-sensitive nuclei in combination with standard NMR techniques. The NMR-sensitive nuclei labels serve as a flag to monitor chemical changes at or near the labeled sites in the reacting molecule.[27–32] Nuclei typically used are ^{13}C and ^{15}N, and more recently ^{2}H and ^{19}F.

For example, since the natural abundance of ^{13}C is only 1.1%, signals from specifically ^{13}C-labeled sites will be approximately 100 times more intense than signals from naturally abundant ^{13}C nuclei. The latter signals will be of such low intensity that they will be lost in baseline noise. Thus, the chemistry occurring at or near the labeled sites can be exclusively monitored by ^{13}C NMR.

The use of NMR for sorption studies can provide three types of data: changes in chemical shift positions, line broadening changes, and changes in the spin–lattice relaxation time (T_1). These are all functions of the media surrounding the sorbate. Therefore, any changes due to associations through partitioning or adsorption will cause a change in the NMR spectrum by changing one or more of these parameters. Detailed explanations of these effects are presented in the Introduction of this book and in the opening sections of the chapters in this Part. Because the chemical shift is a function of the electronic density surrounding the nucleus, any change in this density will result in a change in the position of the NMR signal. Alterations in electron density occur anytime changes in covalent or ionic bonding occur.

Therefore, if sorbing molecules form covalent bonds with the sorbent, changes in the spectral signals occur. Solvent polarity and hydrogen bonding can also influence the electronic density surrounding the nuclei, and thus changes in NMR chemical shifts can be used to examine these phenomena. Changes in line broadening result from changes in the correlation times, which are indicative of the rotational motion of the sorbate. While in solution, the sorptive compound, which is usually a small molecule, can tumble freely, but once it is associated with macromolecules such as humic acid, it becomes restricted in its motion and, in turn, its correlation time increases. Changes in the T_1 relaxation time are a function of dipole–dipole relaxation, which can be also related to the tumbling motion of the molecule in solution. Interaction with macromolecular material causes a small molecule to decrease its tumbling frequency and, in turn, experience a reduction in its T_1 relaxation time.

Employing ^{13}C NMR and examining changes in chemical shift values, Bortiatynski et al. (Chapter 2, this text) have examined the formation of covalent bonds between ^{13}C-labeled phenol and soluble humic material in the presence of horseradish peroxidase. They have carried this research one step beyond their previous studies with ^{13}C-labeled phenol[33] by elegantly examining and quantifying the noncovalent association between the ^{13}C-labeled phenol and the soluble humic material. This was achieved by measuring changes in the T_1 relaxation time of the ^{13}C nucleus as the labeled phenol interacts with the humic material. Likewise, using ^{1}H and ^{19}F NMR, Anderson (Chapter 3, this text) utilized changes in chemical shifts and line widths, to examine the noncovalent interactions in solution between various pesticides and natural macromolecules such as soluble humic acids. Herbert and Bertsch (Chapter 4, this text) used ^{2}H and ^{19}F NMR, to study the noncovalent interactions of nonionic molecules with solvents of varying polarity. They then examined the interactions of the same nonionic compounds with surfactants and humic acids and compared these results with those from the solvent polarity studies to obtain an indication of the association. This provided a better understanding and greater insight into solvent, surfactant, and humic acid interactions with nonionic molecules. Netzel et al. (Chapter 5, this text) use changes in the chemical shifts of solid-state ^{13}C NMR spectra to examine the sorption of pyridine, pentachlorophenol, naphthalene, and 1,1,2,2-tetrachloroethane on two types of coal fly ash. Since unlabeled materials were used, detection of low sorbate concentrations was achieved by means of a large-volume probe. The collective results of these studies demonstrate that NMR is a valuable and powerful tool for examining sorption processes in both the solution and the solid state.

Future Research

NMR has the potential to be a very useful and powerful technique to study sorption processes. Besides chemical shift changes, which are indicative of covalent bond breaking and formation, changes in the line broadening, T_1 relaxation times, and spin–spin coupling can provide information regarding partitioning and adsorption processes. The chemistry occurring on a microscopic level during sorption, including the kinetics and mechanisms of binding, remains poorly understood. NMR has the potential to examine such processes. Besides providing basic fundamental information regarding sorption mechanisms and processes, the use of NMR for sorption studies can have direct applications.

One topic which is currently of interest and is tied into the interactions between the chemical and its environment is the issue of bioavailability.[34,35] The increasing use of bioremediation as a mitigation strategy has led to the awareness that many other factors besides just the biodegradability of a chemical are important to understanding the rates and mechanisms of degradation. When a chemical becomes sorbed to a soil or sediment particle, its bioavailability can be dramatically changed. Chemicals that are sorbed for extended periods tend to become even more unavailable to biotransformation; this process is often referred to as aging. Thus aging influences the extent of biodegradation and, in turn, the feasibility of bioremediation strategies for contaminated sites. Not only can NMR examine the sorption behavior of specific pollutants, but it can also follow the formation and fate of reaction by-products.[36]

Other bioavailability issues in which NMR can be useful include examining the sorption and other similar association interactions between dissolved pollutants and soluble macromolecular compounds such as humic matter, and how these interactions effect the biotoxicity of the pollutants.[37] Benson and Long[38] have shown that various pesticides and herbicides can exhibit synergistic or antagonistic toxic effects arising from sorption interactions with dissolved humic material. Solution-state NMR has the potential to characterize the mechanisms of association between such pollutants and dissolved humic matter, which, in turn, could provide insight into the issue of biotoxicity.

An additional research topic of interest in which NMR studies will be valuable is the influence of surfactants upon the sorption mechanism. Much research has been done using batch processes and column experiments to explore the interaction of surfactants with sorbents and sorbates.[39–41] In many cases, the surfactant concentration strongly influences the sorption behavior of the solute. Usually, at surfactant concentrations below the critical micelle concentration (CMC), the surfactant forms a monolayer on the sorbate surface and enhances solute adsorption. At surfactant concentrations above the CMC, the micelles interact with the solute and prevent it from associating with the sorbent. This can have the effect of solubilizing the sorbate and enhancing mobility. One remediation strategy involves removing organic contaminants from soil or sediments by washing them with a soluble surfactant. NMR studies would be quite useful in this area; for example, labeling solute or surfactant molecules with NMR-sensitive nuclei could examine these interactions on a molecular level.

References

1. Hsu, T. S., and Bartha, R., *Soil Sci.* 116, 1974.
2. Chiou, C. T., Porter, P. E., and Schmedding, D. W., *Environ. Sci. Technol.* 17(4), 227–231, 1983.
3. Steinberg, S. M., Pignatello, J. J., and Sawhney, B. L., *Environ. Sci. Technol.* 21(12), 1201–1208, 1987.
4. Ballard, T. M., *Soil Sci. Am. Proc.*, 35, 145–147, 1971.
5. Caron, G., Suffet, I. H., and Belton, T., *Chemosphere* 14(8), 993–1000, 1985.
6. Webster, G. R. B., Muldrew, D. H., Graham, J. J., Sarna, L. P., and Muir, D. C. G., *Chemosphere* 15(9–12), 1279–1386, 1986.
7. Johnson-Logan, L. R., Broshears, R. E., and Klaine, S. J., *Environ. Sci. Technol.* 26(11), 2234–2239, 1992.
8. McCarthy, J. F., Williams, T. M., Liang, L., Jardine, P. M., Jolley, L. W., Tayor, D. L., Palumbo, A. V., and Cooper, L. W., *Environ. Sci. Technol.* 27(4), 667–676, 1993.
9. Schwandt, H., Kogel-Knabner, I., Stanjek, H., and Totsche, K., *Sci. Tot. Environ.* 123/124, 121–123, 1992.
10. Lambert, S. M., Porter, P. E., and Schieferstein, R. H., *Weeds* 13, 185–190, 1965.
11. Mingelgrin, U., and Gerstl, Z., *J. Environ. Qual.* 12, 1–11, 1983.
12. Garbarini, D. R., and Lion, L. W., *Environ. Sci. Technol.* 20(12), 1263–1269, 1986.
13. Grathwohl, P., *Environ. Sci. Technol.* 24(11), 1687–1693, 1990.
14. Foster, S. S. D., Chilton, P. J., and Stuart, M. E., *J. IWEM* 5, 186–193, 1991.
15. Neilson, A. H., *Organic Chemicals in the Aquatic Environment: Distribution, Persistence, and Toxicity*, CRC Press, Boca Raton, FL, 1994, pp. 53–108.
16. Ball, W. P., and Roberts, P. V., *Environ. Sci. Technol.*, 25(7), 1223–1236, 1991.
17. Landrum, P. F., Nihart, S. R., Eadle, B. J., and Gardner, W. S., *Environ. Sci. Technol.* 18(3), 187–192, 1984.
18. Vowles, P. D., and Mantoura, R. F. C., *Chemosphere*, 16(1), 109–116, 1987.
19. Szabo, G., Farkas, G., and Bulman, R. A., *Chemosphere* 24(4), 403–412, 1992.
20. Szabo, G., and Bulman, R. A., *J. Liq. Chromatogr.* 17(12), 2593, 1994.
21. Kordel, W., *Chemosphere* 27(12), 2341, 1993.
22. Kordel, W., *Sci. Tot. Environ.* 162(2/3), 119, 1995.
23. Sarkar, J. M., Malcolm, R. L., and Bollag, J.-M., *Soil Sci. Soc. Am. J.* 52, 1988.
24. Alford-Stevens, A. L., Eichelberger, J. W., and Budde, W. L., *Environ. Sci. Technol.* 22, 304–312, 1988.
25. Wershaw, R. L., and Mitkita, M. A., eds, *NMR of Humic Substances and Coal—Techniques, Problems, and Solutions*, Lewis Publishers, Chelsea, MI, 1987.
26. Wilson, M. A., *NMR Techniques and Applications in Geochemistry and Soil Chemistry*, Pergamon, Sydney, Australia, 1987.
27. Zelibor, Jr., J. L., Romankiw, L., Hatcher, P. G., and Colwell, R. R., *Appl. Environ. Microbiol.* 54(4), 1051–1060, 1988.
28. Baldock, J. A., Oades, J. M., Vassallo, A. M., and Wilson, M. A., *Environ. Sci. Technol.* 24(4), 527–530, 1990.
29. Thorn, K. A., Arterburn, J. B., and Mikita, M. A., *Environ. Sci. Technol.* 26(1), 107–116, 1992.
30. Haider, K., Spiteller, M., Wais, A., and Fild, M., *Int. J. Environ. Anal. Chem.* 53(2), 125–137, 1993.
31. Hatcher, P. G., Bortiatynski, J. M., Minard, R. D., Dec, J., and Bollag, J.-M., *Environ. Sci. Technol.* 27(10), 2098–2103, 1993.

32. Knicker, H., and Lüdermann, H.-D., *Org. Geochem.* 23, 329–341, 1995.
33. Bortiatynski, J. M., Hatcher, P. G., Minard, R. D., Dec, J., and Bollag, J.-M., in *Humic Substances in the Global Environment and Implications on Human Health*, Senesi, N., and Miano, T. M. (eds), Elsevier Scientific, New York, 1994, pp. 1091–1099.
34. Weissenfels, W. D., Klewer, H.-J., and Langhoff, J., *Appl. Microbiol. Biotechnol.* 36, 689–696, 1992.
35. Alexander, M., *Biodegradation and Bioremediation*, Academic Press, New York, NY, 1994.
36. Nanny, M. A., Bortiatynski, J. M., Tien, M., and Hatcher, P. G., *Environ. Toxicol. Chem.* 15, 1857–1864, 1996.
37. Dell'Angola, G., Ferrari, G., and Nardi, S., *Pestic. Biochem. Physiol.* 15, 101–104, 1981.
38. Benson, W. H., and Long, S. F., *Ecotoxicol. Environ. Safety* 21, 301–307, 1991.
39. Adeel, Z., and Luthy, R. G., *Environ. Sci. Technol.* 29(4), 1032–1042, 1995.
40. Deitsch, J. J., and Smith, J. A., *Environ. Sci. Technol.* 29(4), 1069–1080, 1995.
41. Sun, S., Instep, W. P., and Boyd, S. A., *Environ. Sci. Technol.* 29(4), 903–913, 1995.

2.

The Development of ^{13}C Labeling and ^{13}C NMR Spectroscopy Techniques to Study the Interaction of Pollutants with Humic Substances

JACQUELINE M. BORTIATYNSKI, PATRICK G. HATCHER,
& ROBERT D. MINARD

Modern agricultural practices have contributed to the accumulation of herbicides, pesticides and their decomposition products in the soil. These pollutants are known to interact with soil organic matter to form covalent and/or noncovalent bonding associations. The covalent bonds are thought to result from addition or oxidative coupling reactions, some of which may be catalyzed by oxidoreductive enzymes.[1-4] Noncovalent associations include such interactions as ion exchange, hydrogen bonding, protonation, charge transfer, ligand exchange, coordination through metal ions, van der Waals forces, and hydrophobic bonding.[5-8]

The association of pollutants with soil organic matter is an area of study that is of extreme interest for two reasons. First, dissolved organic matter present in lakes and streams is known to enhance the solubility of pollutants, which poses a real threat to the quality of fresh water supplies.[9-15] Therefore, if we are to predict the movement of pollutants in the water table we need to have a mechanistic understanding of their interactions with dissolved humic materials. Second, early studies had indicated that some pollutants chemically bind to humic materials, thus reducing the risk of further transport and dispersion.[2,16-19] If this chemical binding of the pollutants is irreversible, then this process may serve as a natural means for their detoxification. Regardless of the type of association, the first task in any mechanistic study is to characterize the reaction products structurally.

In the case of noncovalent binding mechanisms, studies have focused on the physical characteristics of the process and not on the structure of the associated pollutant. Association studies are used to determine the sorption kinetics and transport of pollutants as well as their association constants. These types of studies utilize various techniques such as batch sorption, gas-purge desorption, column adsorption, and miscible displacement.[16,20-25]

All of these techniques are only capable of providing quantitative information on the amount of pollutant sorbed by a substrate.

The study of the covalent binding of pollutants to humic substances has utilized ^{14}C labeling in addition to various spectrometric techniques such as ultraviolet (UV) difference, fluorescence polarization and infrared (IR) spectroscopy. These studies have all inferred that covalent binding does indeed occur, but direct structural evidence to support such a conclusion could not be provided. Recently, ^{13}C and ^{15}N NMR in combination with site-specific NMR labeling has been used to examine the binding of [13]C- and ^{15}N-labeled pollutants to humic substances. The resulting spectra provide the structural information that is necessary to assign specific types of covalent binding interactions. The success of these recent studies has prompted us to examine the association of pollutants with soil organic matter using this powerful analytical tool.

In this chapter we will summarize the various techniques that have been utilized to study the interaction of pollutants with soil organic matter, and the role that NMR and site-specific NMR labeling have played in providing valuable structural information concerning covalent binding processes. In addition, we will present the preliminary results of a new study in which we have utilized spin–lattice relaxation times T_1 as a probe to examine the association of C-1 ^{13}C-labeled phenol with a soil humic acid at neutral pH. A dramatic change in T_1 as a function of humic acid concentration should be indicative of the amount of associated phenol. This experimental method may be an important means of determining association constants.

Background Information

Association Techniques

As mentioned above, a number of techniques are used to study the association of pollutants with soil organic matter. One such method is batch sorption, which is typically used to determine sorption kinetics.[26] A slurry or solution containing the pollutant and the sorbent is agitated and allowed to equilibrate. Following equilibration, the mixture is phase-separated by acidification and centrifugation or just by centrifugation. The supernatant is then analyzed by such methods as gas chromatography, high-pressure liquid chromatography, or ^{14}C radiometry to determine the concentration of pollutant remaining in the liquid phase.

The sorption kinetics of volatile pollutants are investigated using a gas-purge technique developed by Karickhoff.[23,26–28] This technique is generally used to examine dissociation processes, since the solution containing a preequilibrated mixture of sorbent and solute is swept with a gas (air or nitrogen) which induces continuous evaporation of the solute. The gas stream is then trapped and the mass flux of the solute is determined. A modified version of this technique has been developed by Brusseau et al. which allows both association and dissociation processes to be observed.[23] Their modifications involved additional valves so that the system can operate in a closed manner, which allows them to observe association processes also. In the modified closed system, a known amount of solute is injected into the system and then the gas flow is monitored continuously by gas chromatography until an equilibrium state is reached, at which point the concentration of associated pollutant is determined.

Column adsorption studies are used to examine the transport of pollutants in solutions containing dissolved humic matter. [29,30] Once packed with a sorbent (soil, sand), the chosen pollutant is loaded at the head of the column. Then the column is continuously leached with a solution containing dissolved organic matter at a predetermined rate, and the effluent is examined for the presence of the pollutant. When the effluent is found to be free from any pollutant, the column is sectioned and the concentration of the pollutant that is associated with the sorbent is determined.

Miscible displacement studies are another means of examining solute transport and sorption kinetics.[23,24,26] This method, like the column adsorption studies, is used to examine the movement of a pollutant through a column packed with a given sorbent. One major difference between the two techniques is that with the miscible displacement method the pollutant is not loaded on the column prior to leaching. Instead it is dissolved in a stock solution and then it is pumped through the column packed with the sorbent material. The concentration of pollutant in the effluent is analyzed by chromatographic or other spectroscopic methods. A recent improvement by Brusseau et al. uses a flow-through UV detector system that prevents the loss of volatile pollutants during the analytical process.[23]

It is clear from the brief descriptions given above that these techniques (miscible displacement, column adsorption, gas-purge desorption, and batch sorption), which are currently used to study the association of pollutants with soil or soil organic matter, are designed to determine the concentration and rate of binding. Although the data that have been obtained from such studies constitute an integral part of our current understanding of the association process, structural information is still lacking. Without an adequate description of the reaction products, it is impossible to identify reaction pathways and assign the type of interaction. In fact, a distinct possibility exists that both associative and covalent binding mechanisms may be operative in some instances (for example, in the soil).[16] It is extremely important to distinguish among binding mechanisms, since the mobility and toxicity of covalently bound pollutants are significantly different from those of pollutants bound associatively.

Covalent Binding Techniques

From ^{14}C radiolabeling studies conducted to examine the decomposition of agricultural waste products, it has been established that even after exhaustive extraction of contaminated soils some pollutants remain bound to humic substances.[31–36] It is thought that naturally occurring enzymes in the soil catalyze the oxidative coupling of some pollutants to soil organic matter.[2,16,37–41] Once a xenobiotic is covalently bound to humic substances, it becomes an intrinsic and irretrievable part of the soil organic matter.[2,3,42–44] For this reason, agricultural scientists have viewed the covalent binding of pollutants to humic materials as a natural means of removing xenobiotics from the soil.[2,16] The real concerns with this type of binding are the subsequent release of the pollutant and the generation of new products that are more toxic than the parent compound.[45,46] One such example is the formation of dioxins and chlorinated furans from the addition of horseradish peroxidase to water contaminated with chlorinated phenols.[47]

Even if the covalent binding of pollutants to humic materials was not considered as a potential mitigation strategy for the detoxification of soil, the movement and release of xenobiotics are processes that must be understood. Like its associative counterpart, the covalent binding of pollutants to soil organic matter can serve as a means of transport for the attached pollutant into the water table. Incorporation of the pollutant into the soil organic matter may be a means of detoxification; however, any subsequent release of the xenobiotic would result in its dispersion.

A number of analytical techniques have been used to examine the covalent binding of pollutants to organic matter.[37–42,48–51] [14]C labeling studies, by far the most common techniques, have been used to quantify the amount of radiolabel that remains attached to soil organic matter under various conditions.[52] The ^{14}C labeling studies are convenient and effective in tracing the path of the pollutant from the initial point of contact to its mineralization to [14]CO$_2$, but the mere presence of the label cannot be used to determine the structure or form of the intermediate or final products that are produced in the covalent binding process.

Spectrometric techniques such as UV difference, fluorescence polarization and IR spectroscopy have also been used to examine the covalent binding of pollutants to humic substances.[48–51] Although these techniques are commonly used to determine the presence of organic functional groups they are not successful for the examination of covalent binding interactions, due to the complexity of the humic acid matrix.

In an effort to reduce the complexity induced by the macromolecular structure of humic materials, studies were carried out to examine the enzymatic covalent binding of pollutants to individual compounds thought to be models for humic acid structures. Bollag and coworkers have oxidatively coupled compounds such guaiacol, catechol, and syringic acid, as well as substituted phenols, with pollutants such as chlorinated phenols and anilines, to determine the propensity of these pollutants to couple with compounds containing functional groups characteristic of those in humic acids.[2,52,53] From these model studies, C–O, C–C, C–N, and N–N coupled dimers and trimers were isolated using high-pressure liquid chromatography and then structurally characterized by mass spectrometry and ^{1}H NMR.[54-57] In addition, these studies demonstrated how pH, enzyme activity, reaction time, and temperature affect the yield of reaction products.[58,59] Therefore, the model studies have provided a wealth of information concerning the reaction conditions and the types of products that are produced from the enzymatic coupling of pollutants to the model compounds.

NMR Techniques

NMR is one of the most powerful analytical tools for the structural elucidation of organic molecules in complex systems. ^{1}H and ^{13}C NMR have been used extensively to study the structure of humic materials.[60-62] ^{13}C NMR, in particular, with its large chemical shift range, has provided the most comprehensive functional group assignments. Although solid-state ^{13}C NMR provides the most convenient and unobtrusive means of examining the structure of humic substances, it does not provide the resolution that is available from high resolution NMR of liquids. Preston and Blackwell have demonstrated that quantitative ^{13}C NMR spectra (within $\pm 10\%$) of humic solutions in 0.5 M NaOD can be obtained using an inverse-gated pulse sequence with a 45° pulse and a delay of approximately 2 s between pulses.[63]

When compared to ^{1}H NMR, ^{13}C NMR is over 100 times less sensitive and has a natural abundance of only 1.0%. Since ^{13}C NMR suffers from this inherent insensitivity, generally the experiment times are longer and the sample sizes are larger than those needed for ^{1}H NMR. This insensitivity can be surmounted, however, with the use of site-specific ^{13}C labeling. The 100% enrichment of a specific carbon with ^{13}C will result in a 100-fold signal enhancement for that carbon in the NMR spectrum.[64] In complex reactions where a number of different products are formed, site-specific labeling not only increases the sensitivity of the experiment, but it also simplifies the NMR spectrum. The simplification of the spectra results from the fact that those signals which are due to ^{13}C atoms at natural abundance levels are significantly reduced in size or they only appear as noise in the baseline, in comparison with those peaks which arise from the ^{13}C-labeled carbons.[65]

Unlike ^{13}C NMR, ^{15}N NMR has seldom been utilized to study the structure of humic acids. ^{15}N NMR is less sensitive (approximately 3.8×10^{-6} times) and has longer relaxation times than ^{1}H NMR, but it also has a negative gyromagnetic ratio which results in a negative nuclear Overhauser effect (NOE).[66] Inverse-gated ^{15}N NMR is used to eliminate the effect of the negative NOE and obtain quantitative spectra, while polarization transfer techniques such as Insensitive Nuclei Enhanced by Polarization Transfer (INEPT) and Distortionless Enhanced Polarization Transfer (DEPT) in combination with ^{15}N labeling are used to increase the sensitivity of the spectra.

^{13}C and ^{15}N labeling in combination with standard NMR techniques has been used to identify and study the chemistry of functional groups that are present in humic substances.[66-72] These studies have used derivatization techniques to introduce the NMR label. The types of functional groups that are present in the humic substances are inferred on the basis of the NMR identification of the labeled products. Generally the reactions include the methylation or acylation of hydroxyl groups and the transformation of carboxyl carbons to form oximes. In addition, the derivatization of soils with ^{13}C-labeled silylating agents has been used to enhance the extractability of bound organic residues.[72] The identification of reaction products from the various derivatization reactions is simplified due to the signal enhancement which is observed for the labeled carbons in the NMR spectrum.

Methylation reactions with ^{13}C-labeled diazomethane and methyl iodide have been used to distinguish between hydroxyl functionalities.[67-69,73] Diazomethane in the absence of a Lewis acid catalyst is known to methylate all carboxylic acids, phenols, and enols of β-diketones, while methyl iodide is known to methylate the remaining hydroxyl functionalities such as carbohydrates, alcohols, and hindered phenols. The chemical shift of the ^{13}C-labeled carbon of the methyl group is used to identify the type of carbon–oxygen bond that has been produced in the reaction. In addition to methylation, hydroxyl groups can also be derivatized with an acylating reagent such as ^{13}C-labeled sodium acetate.[70] Carbonyl functionalities can be identified by derivatization with ^{15}N-labeled hydroxylamine.[66,71] The carboxyl carbons of ketones, quinones, esters, and lactones have been derivatized with hydroxylamine to form oximes which were identified using standard ^{15}N NMR spectroscopy.

Both the ^{13}C and the ^{15}N derivatization studies have provided valuable structural information concerning humic acid functional group composition and reactivity. These studies also demonstrated the potential utility of NMR and site-specific NMR labeling to study complex processes involving humic acids. Thorn and Mikita recognized this potential and used ^{15}N labeling and standard ^{15}N NMR spectroscopy to study ammonia fixation by humic substances. They followed the reaction of ^{15}N-labeled ammonium hydroxide with Suwannee River fulvic acid, a peat humic acid, and a Leonardite humic acid using high resolution ^{15}N NMR techniques such as INEPT and DEPT.[74] By this analytical approach, they were able to identify a number of heterocyclic compounds as the primary products of ammonia fixation reactions by humic substances. Similar heterocyclic products were found in an additional study in which ^{15}N-labeled chloramine was bound to Suwannee River fulvic acid.[75]

The binding of ^{15}N-labeled aniline to humic substances in the presence and absence of oxidoreductive enzymes was also examined using ^{15}N NMR techniques. The ^{15}N-labeled products formed from the oxidative coupling and the nucleophilic addition of aromatic amines to humic matter were identified, and it was found that aniline covalently binds to humic substances in the presence and absence of the enzyme.[76]

Another process that has been developed by Hatcher et al. using ^{13}C labeling and ^{13}C NMR is the enzyme-catalyzed binding of 2,4-dichlorophenol to a Minnesota peat humic acid in the presence of horseradish peroxidase.[65] They were the first to demonstrate that covalent bonds are formed between a pollutant labeled with ^{13}C and humic acids. Chlorophenols are known to be toxic and can be generated from a number of processes such as the degradation of pesticides and wood products. Therefore they are excellent candidates for binding studies to examine a type of pollutant which is able to form C–O and C–C bonds with humic acid functional groups. The model studies of Minard et al., and others, have provided the information that was needed to predict target sites of reactivity for the oxidative coupling of phenols to humic acid functional groups.[78] Based on the results of these model studies, the C-1, C-2, and C-6 sites were chosen by Hatcher et al. to carry the ^{13}C label.[65] A complete discussion of the enzyme-catalyzed binding of 1-^{13}C- and 2,6-^{13}C-labeled 2,4-dichlorophenol to a Minnesota peat humic acid appears elsewhere, but a synopsis of the results will be presented here to show the utility of such a method.

Hatcher et al.[65] and Bortiatynski et al.[77] showed that 1-^{13}C- as well as 2,6-^{13}C-labeled 2,4-dichlorophenol did not form covalent bonds with humic acids in the absence of enzyme. The ^{13}C NMR spectra obtained of these pollutants in the presence of humic acids were essentially those of the ^{13}C-labeled 2,4-dichlorophenol—one peak in the case of the 1-^{13}C-labeled phenol representing the C-1 carbon, and two sets of doublets for the 2,6-^{13}C-labeled phenol. When enzyme was added to induce covalent binding, a multitude of new peaks were observed, representing sites at or near the labeled carbons which had formed covalent bonds with humic acid or with the 2,4-dichlorophenol itself.

Figure 2.1 is the inverse-gated ^{13}C NMR spectrum of 1-^{13}C-labeled 2,4-dichlorophenol which was enzymatically bound to the Minnesota peat humic acid and then dissolved in 0.5 M NaOD. Figure 2.2 is the corresponding ^{13}C NMR spectrum of the bound 2,6-^{13}C-labeled 2,4-dichlorophenol. The complexity of the spectra and the dispersion of ^{13}C chemical shifts reflect the numerous sites that are available for binding in the humic acid matrix. The macromolecular structure of humic acid provides a number of unique chemical environments. Thus the ^{13}C-labeled carbons that are at or near the site of the covalent linkage (carbon–carbon, carbon–oxygen ether, and carbon–oxygen ester) are subjected to many different electronic effects; this results in a dispersion of the ^{13}C chemical shifts.[79]

In an effort to simplify the spectral assignments, the signals in the NMR spectrum have been grouped into regions and labeled (A through J). The signals in the NMR spectrum in Figure 2.1 result from the C-1-^{13}C-labeled carbon, while the signals in Figure 2.2 result from the 2,6-^{13}C-labeled carbons. The ^{13}C NMR signals of greatest intensity were thought to represent preferred bonding interactions. The interactions that have been assigned to the largest peaks in both spectra are carbon–carbon bonding linkages formed at the C-4 and C-6 carbons of the 2,4-dichlorophenol to aromatic and aliphatic carbons of the humic acid. The signals for these carbon–carbon bond-

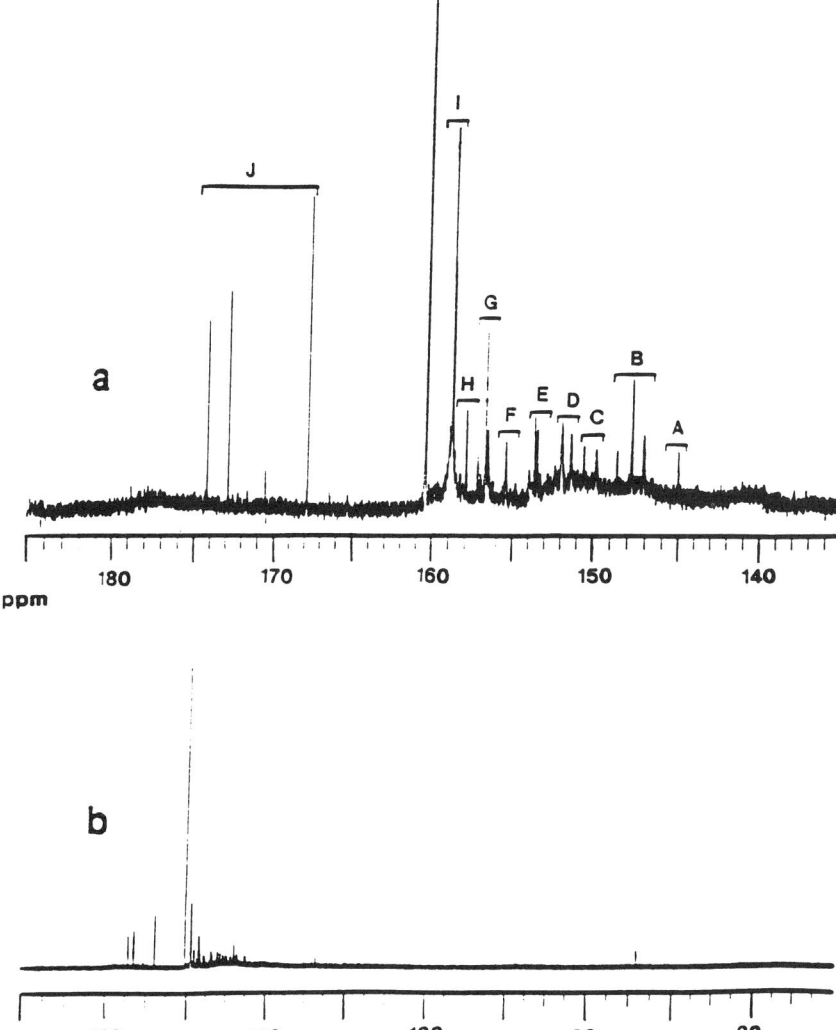

Figure 2.1   The inverse-gated ^{13}C NMR spectrum of 1-^{13}C-labeled 2,4-dichlorophe-
nol enzymatically bound to a Minnesota peat humic acid: (a) expanded region, 130–
200 ppm; (b) full spectrum, 0–200 ppm.

ing interactions are found in region I in Figure 2.1 and regions A, C, D, H,
and I in Figure 2.2. In addition, both spectra indicate the presence of a dimer
moiety that is formed from the polymerization of 2,4-dichlorophenol with
itself, which results in a carbon–carbon linkage from the C-6 carbon of one
monomer to the C-3 carbon of the other monomer. The chemical shifts of the
labeled carbons as well as the ^{13}C–^{13}C coupling patterns (two doublets in

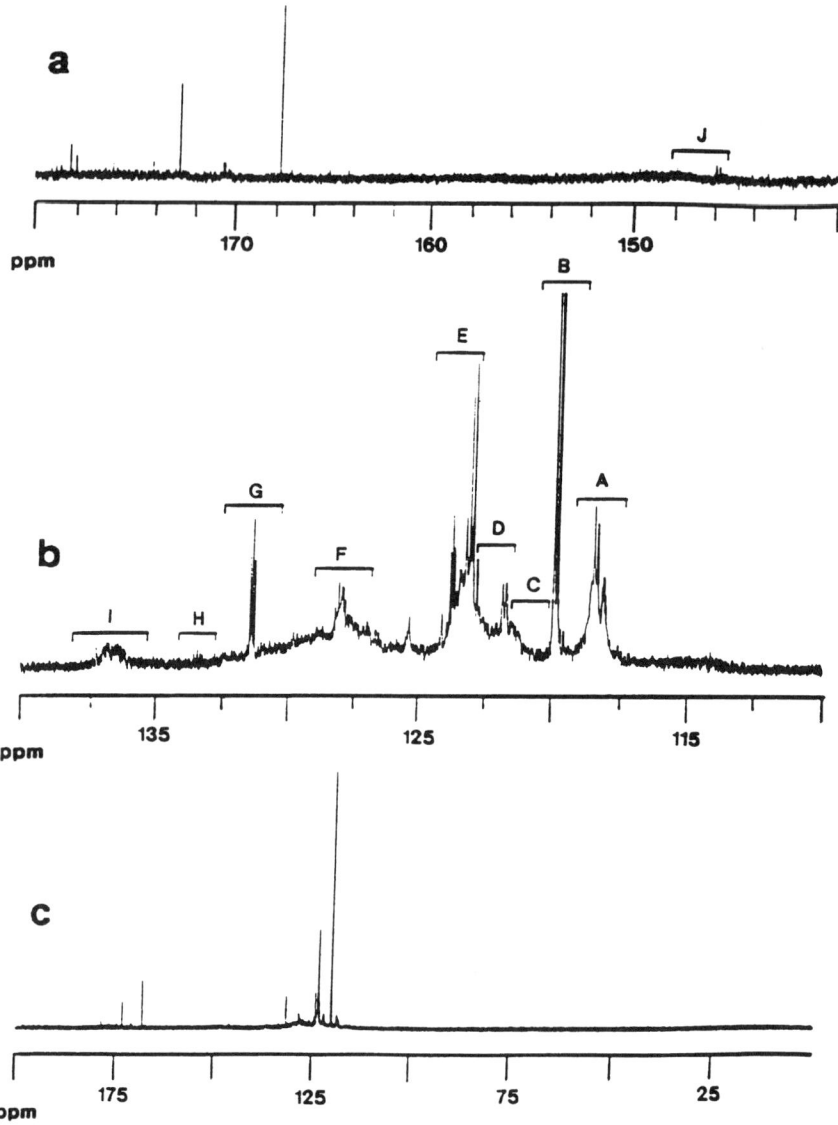

Figure 2.2 The inverse-gated ^{13}C NMR spectrum of 2,6-^{13}C-labeled 2,4-dichlorophenol enzymatically bound to a Minnesota peat humic acid: (a) expanded region, 140–180 ppm; (b) expanded region, 110–140 ppm; (c) full spectrum, 0–200 ppm.

regions B and E and two doublets of doublets in regions E and G) in Figure 2.1 and the large singlet at 160 ppm in Figure 2.2 highly suggest this polymerized product.

The presence of the dimer moiety raised concern that all of the signals in Figures 2.1 and 2.2 may result from self-polymerization and not actual covalent bonding interactions with humic acids. The enzyme-catalyzed polymerization of 2,4-dichlorophenol proceeds quite differently compared to the reaction with humic acid, since the polymerization is accompanied by the immediate precipitation of the polymerized products and yields a significantly different ^{13}C NMR spectrum.

The success that has been achieved in applying ^{13}C and ^{15}N labeling in combination with NMR to examine covalent binding processes has urged us to find a similar protocol for the examination of the noncovalent association of pollutants with humic materials. Lee et al.[24,80,81] and Schellenberg et al.[22] have examined the association of phenols and chlorinated phenols with soils and sediments using miscible displacement and batch techniques. We chose to use a modified batch equilibrium technique in which ^{13}C-labeled 2,4-dichlorophenol (either 1-^{13}C or 2,6-^{13}C) was allowed to react with humic acids at neutral and basic pH in a 5 mm NMR tube. The ^{13}C NMR spectra obtained from these experiments contained no additional signals other than those of the labeled 2,4-dichlorophenol, which highly suggests that covalent bonding interactions between the 2,4-dichlorophenol and the humic acid were not present in these solutions. It is apparent that noncovalent interactions are not capable of inducing chemical shift dispersions, and we must employ other NMR techniques such as the measurement of spin–lattice relaxation times (T_1) to detect such interactions.

The remainder of this chapter is dedicated to the discussion of a preliminary study on the use of ^{13}C NMR relaxation times for labeled carbons in pollutants to follow the course of their noncovalent interaction with humic substances. The basis for these studies derives from the fact that the T_1 values of labeled carbons in solution will be affected by noncovalent associations with humic substances.

Experimental Section

Materials

1-^{13}C-labeled (99%) phenol was purchased from Cambridge Isotope Laboratories, Woburn, MA, and used without further purification. An Armadale humic acid from the Bh horizon of Armadale soil in Prince Edward Island, Canada (kindly provided by M. Schnitzer), was chosen for the study, since it is very soluble at neutral pH and it has been well characterized.[82] This humic acid was isolated by the methods used by the International Humic Substances Society. A pH 7.2 universal buffer (0.2 M acetic acid, 0.2 M boric acid, 0.2 M phosphoric acid in 1 M NaOH) was used in these experiments.

Preparation of Samples

The humic acid stock solution was prepared by dissolving 20 mg of Armadale humic acid in a solution containing 1.6 mL universal buffer, pH 7.2, and 8.4 mL of deionized water. The samples for the T_1 experiments were prepared by adding 2 mg of 1-^{13}C-labeled phenol in 200 μL D_2O to a 5 mm NMR tube, followed by the addition of an appropriate volume of humic acid stock solution and universal buffer (ratio of 1:7 buffer to water including D_2O) to produce a final volume of 1 mL. Some samples were degassed using a three-cycle freeze–thaw vacuum line technique, and then permanently sealed. The remaining samples were not prepared in this manner when it was established that the degassing was unnecessary for accurate T_1 measurements.

Preparation of Combustion Sample

A sample of Armadale humic acid (2.57 mg) was placed in a quartz tube, weighed and placed in a oven at 412 °C for 7 h. The quartz tube was reweighed, and the remaining ash was transferred to a 5 mm NMR tube containing 2 mg of 1-^{13}C-labeled phenol, 200 μL D_2O, 160 μL universal buffer, and 640 μL water.

Preparation of Covalently Bound Pollutant Samples

A synopsis of the procedure that has been used to bind pollutants enzymatically to soil organic matter is presented below. A more detailed description of this procedure appears in Hatcher et al.[65] and Bortiatynski et. al.[77] The humic acid is dissolved in 0.5 M NaOH and then treated with ion exchange resin to produce a neutral stock solution. An aliquot of the humic acid stock solution and the desired ^{13}C-labeled phenol is added to an Erlenmeyer flask. The concentration of the ^{13}C-labeled phenol is based on the minimum amount of label that is need to produce reasonable NMR signals in an 8 h period. We have typically found that 3 to 4 mg is the minimum quantity of labeled material. Horseradish peroxidase and H_2O_2 or any other chosen oxidoreductive enzyme is then added to the humic acid/phenol mixture, and the solution is allowed to sit over night. The solution is then acidified with 6 M HCl and the humic acid is isolated by centrifugation. The precipated humic acid is washed with acidified water and then dissolved in 0.5 M NaOD for NMR analysis.

NMR Experiments

All ^{13}C NMR data that were collected for the sole purpose of chemical shift analysis were obtained on a Bruker AM500 NMR spectrometer using a ^{13}C inverse-gated pulse sequence. The ^{13}C T_1 experiments were carried out using a standard inversion recovery sequence with gated decoupling at 273 °C on a

Bruker WH 360 MHz NMR spectrometer. The experimental parameters were (1) carbon resonance frequency 90.56 MHz, (2) sweep width 21700 Hz, (3) recycle delay at least 5 T_1, (4) 12 τ values per experiment, with 16 scans per variable delay. The data were processed with 1 Hz line broadening, and the resulting T_1 values were obtained using a three-parameter fit.

Results and Discussion

The sample containing 1-^{13}C-labeled phenol (referred to hereafter as ^{13}C-phenol) in the absence of humic acid at neutral pH produces a single peak at 155.1 ppm in the ^{13}C NMR spectrum, shown in Figure 2.3(b). Another sample which contains both the ^{13}C-phenol and 1.3 mg of Armadale humic acid produces an identical NMR spectrum, as shown in Figure 2.3(a). The fact that identical spectra (Figures 2.3(a), 2.3(b)) are obtained for both samples is important for two reasons. First, the sensitivity enhancement that is gained due to the presence of the ^{13}C label is apparent, since the signals for the remaining unlabeled carbons in the ^{13}C-phenol and the humic acid are very weak in comparison and become noise in the baseline. Second, the facts that the chemical shifts of the ^{13}C-labeled carbons are identical and that there are no other peaks in the spectrum indicate that the ^{13}C-phenol has not formed covalent bonds to the humic acid.

When covalent binding interactions between phenols and humic acids are induced by oxidoreductive enzymes, the resulting ^{13}C NMR spectra contain a dispersion of chemical shifts representing the numerous types of binding interactions shown above for 1-[13]C- or 2,6-^{13}C-labeled 2,4-dichlorophenol (Figures 2.1 and 2.2). In an effort to demonstrate that a dispersion of chemical shifts also occurs when ^{13}C-phenol is allowed to react under similar conditions, we show in Figure 2.4 a spectrum of ^{13}C-phenol covalently bonded to a humic acid. The spectrum in Figure 2.4 does indeed contain a dispersion of chemical shifts which primarily consists of two broad sets of signals in the ranges 157 to 158 ppm and 164 to 165.5 ppm. The fact that the signals are very broad in these regions is indicative of the variety of chemical environments that are in close proximity to each type of covalent bonding interaction.[85]

The tentative assignments of the signals found in the two broad regions in the ^{13}C NMR spectrum in Figure 2.4 are found in Table 2.1. Like the spectra in Figures 2.1 and 2.2, the signals for the covalently bound ^{13}C-phenol are assigned using calculated ^{13}C chemical shifts for the C-1 carbon of the ^{13}C-phenol which are representative of C–C and C–O bonding interactions with the humic acid. Bonding at or near the site of the ^{13}C label results in detectable changes in the ^{13}C chemical shift. For example, covalent bonding at the phenolic carbon (C-1) to form C–O ether linkages results in higher ^{13}C chemical shift values for the C-1 carbon (158 to 160 ppm) than that found for ^{13}C-phenol (155 ppm). On the other hand, C–O bonds formed at the C-1

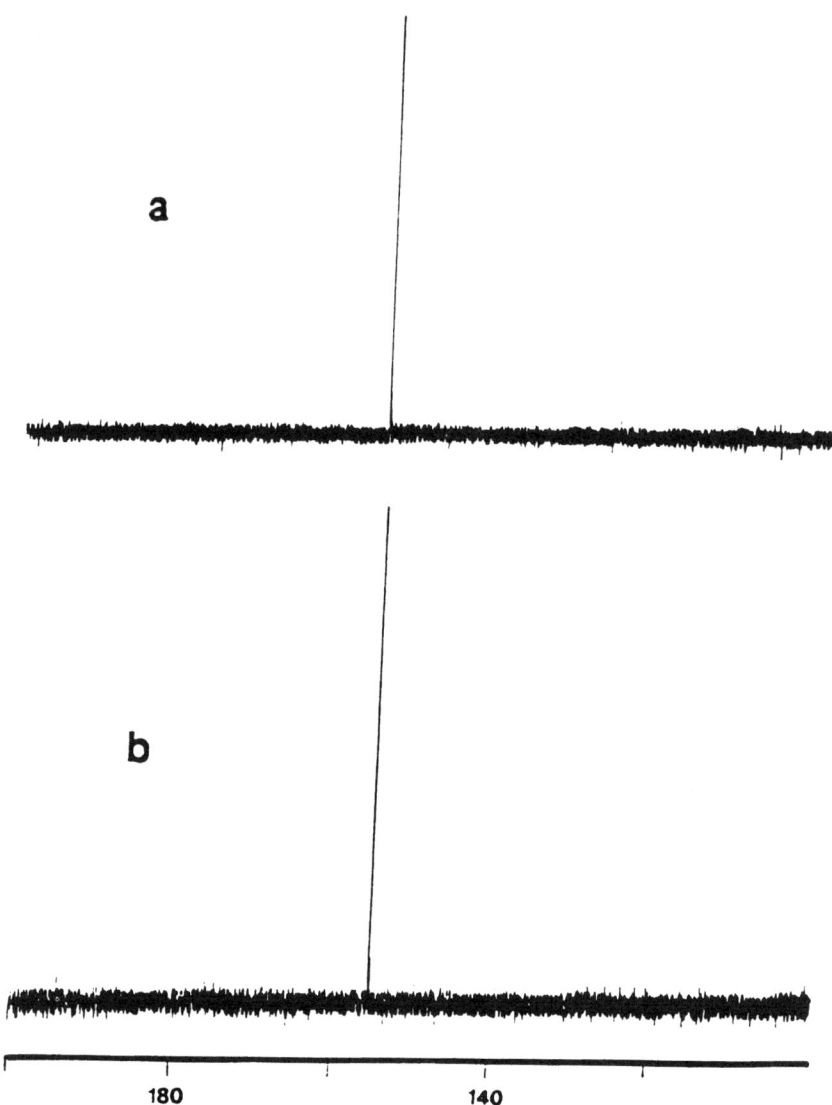

Figure 2.3 The inverse-gated ^{13}C NMR spectrum of ^{13}C-phenol obtained (a) in the presence of Armadale humic acid, and (b) in the absence of Armadale humic acid.

carbon, and which result in ester linkages, produce C-1 ^{13}C chemical shifts at lower values (150 to 151 ppm) than that found for ^{13}C-phenol.

Lee et al.[22,80,81] and Schellenberg et al.[24] have shown that phenols form noncovalent bonding associations with humic substances when no enzymatic reagent is present. However, the spectrum shown in Figure 2.3(a) demonstrates that either not enough phenol is associated with the humic substances

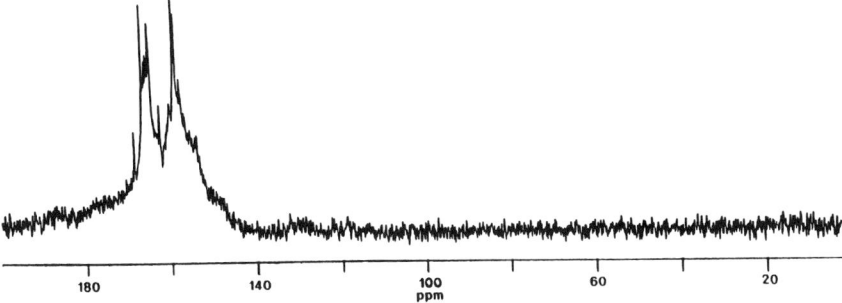

Figure 2.4 The inverse-gated ^{13}C NMR spectrum of ^{13}C-phenol enzymatically bound to Suwannee River humic acid.

or these associations are not strong enough to induce a significant change in the ^{13}C chemical shift of the C-1 carbon of ^{13}C-phenol. A cursory calculation based on a partition constant determined by Schellenberg et al. for 2,4-dichlorophenol at neutral pH with natural sediments predicts that between 1.7 and 1.9 mg of phenol will associate with 1.3 mg of humic material.[22] Although the ^{13}C chemical shifts are not affected enough by noncovalent binding processes to be an effective probe, ^{13}C spin–lattice relaxation times (T_1), are an alternative means of examining these associations.

Table 2.1 Tentative Assignment of the Covalent Bonding Interactions of ^{13}C-phenol with Suwannee River Humic Acid

Site of Binding on ^{13}C-phenol	Type of Covalent Bonding to Humic Acid	C_1 Chemical Shift (ppm)	Found in ^{13}C NMR Spectrum	Compound Representative of the Bonding Interaction
C—1	O$^-$	165.56	Yes	^{13}C-phneolate (noncovalently bonded)
C—1	OCH$_3$	159.9	Yes	Anisole
C—1	OC$_6$H$_6$	157.7	Yes	Diphenyl ether
C—1	OCOCH$_3$	150.9	No	Phenyl acetate
C—2	C$_4$H$_9$	165.0	Yes	2-Butylphenolate
C—4	C$_4$H$_9$	162.8	Yes	4-Butylphenolate
C—2	C$_6$H$_5$	163.7	Yes	2-Phenylphenolate
C—4	C$_6$H$_5$	164.0	Yes	4-Phenylphenolate
C—2	OCH$_3$	151.2	No	2-Methoxyphenolate
C—4	OCH$_3$	157.8	Yes	4-Methoxyphenolate
C—2	OC$_6$H$_5$	156.6	Yes	2-Pehnoxyphenolate
C—4	OC$_6$H$_5$	160.3	Yes	4-Phenoxyphenolate
C—2	OCOCH$_3$	151.2	No	2-Acetylphenolate
C—4	OCOCH$_3$	162.4	Yes	4-Acetylphenolate

Spin–lattice relaxation is the result of energy transfer from the excited nucleus to the fluctuating localized electronic or magnetic fields that are found in the surrounding lattice.[83] These fluctuating fields are present due to molecular motion and the resulting energy transfer occurs by a number of mechanisms. The predominant relaxation pathway for carbon nuclei is through dipole–dipole interactions with other magnetic nuclei or unpaired electrons. When the dipole–dipole relaxation mechanism is the dominant relaxation pathway, the spin–lattice relaxation time T_1 is a reflection of the rotation or reorientation of the molecule in the solution. This type of relaxation is most effective when the rate of reorientation (correlation time τ_c) is in the region of 10^7 or 10^8 Hz. Large molecules generally tumble more slowly in solution and have more time to interact with the lattice; thus, they require a shorter time for their nuclear spins to return to the equilibrium state. The opposite effect takes place with small molecules, which tend to tumble quickly and spend less of their time interacting with the lattice, resulting in longer relaxation times. Therefore, small molecules that have become associated with a macromolecule such as humic acid should exhibit shorter ^{13}C relaxation times than those contained in the unassociated species. For example, phenols in aqueous solution typically have C-1 spin–lattice relaxation times that are significantly longer (approximately ten times) than those of the corresponding phenolic carbons or other carbon atoms of humic acids.[63,84] This large difference in spin–lattice relaxation times provides an excellent means of distinguishing whether a particular tagged molecule, 1-^{13}C-phenol in this case, is associating with humic acids.

Table 2.2 contains spin–lattice relaxation data (T_1) for the C-1 carbon of ^{13}C-phenol at 25 °C in the presence of Armadale humic acid. The T_1 for the C-1 carbon of ^{13}C-phenol alone is relatively long, as expected, since this carbon has no directly attached protons, and phenol is contained in a relatively small molecule. The addition of the Armadale humic acid causes a substantial decrease in the T_1 of the C-1 carbon of the ^{13}C-phenol, which is represented as the change in relaxation time ΔT_1 found in Table 2.2. When the concentration of added humic acid reaches 1.3 mg, the T_1 of the C-1 carbon approaches that of a carbon atom in a soil humic acid (0.2 to 2.4 s).[86]

The calculation of the spin–lattice relaxation times (T_1) assumes a dipole–dipole relaxation mechanism for the C-1 carbon of ^{13}C-phenol. If this assumption is true, the change in the dipole–dipole relaxation must be induced by the ^{13}C-phenol's association with the humic acid. As more of the ^{13}C-phenol becomes associated with the humic acid, the T_1 value should continue to decrease. When most of the ^{13}C-phenol is associated with the humic acid the T_1 of the C-1 carbon approaches that of a similar type of carbon species in the humic acid macromolecular structure. Although the spin–lattice relaxation data presented in Table 2.1 reflect the predicted trend for association between the ^{13}C-phenol and the humic acid, a decrease in T_1 can also result from an increase in the solvent viscosity or relaxation mechanisms other than the preferred dipole–dipole mechanism. If any other

Table 2.2 T_1 (sec) 25 °C of the C-1 Carbon of 1-^{13}C-phenol in a Water Solution Containing Various Concentrations of Armadale Humic Acid

Concn. of ^{13}-labeled Phenol (mg/mL)	Concn. of Added Armadale Humic Acid (mg/mL)	Ratio of Armadale Humic Acid/^{13}C-phenol	T_1 (s)	Standard Deviation (s)	$\Delta T_1{}^a$ (s)	Concn. of Associated ^{13}C-phenol (mg/mL)
2	0.000	0.000	35.64	0.03	0.00	0.00
2	0.130	0.065	12.84	0.01	23.57	1.34
2	0.325	0.160	6.64	0.01	29.56	1.71
2	0.650	0.325	3.74	0.01	32.17	1.88
2	0.975	0.488	2.56	0.01	33.08	1.95
2	1.300	0.650	1.77	0.02	33.92	2.00
2	0.050^b	0.025	23.59	0.01	12.05	—

^{a} $\Delta T_1 = 35.64 - T_1$ (observed) (s).
^{b} The sample weight after combustion.

mechanism is contributing to the change in T_1, then the amount of ^{13}C-phenol that is associated with the humic acid is not a direct reflection of these values.

Changes in solvent viscosity tend to slow down the tumbling rate of the solute and decrease relaxation times. Since the amount of humic material that was added to the solutions increased over the series of experiments, an increase in viscosity may have accompanied this addition. Based on the viscosity values reported by Chen and Schnitzer a 0.2% increase in the viscosity over that of water can be expected for a humic acid solution containing 10 mg/mL at neutral pH.[87] The most concentrated humic acid solution used in this study contained only 1.3 mg/mL. Therefore the increase in viscosity that is possible for this series of samples is not sufficient to affect the relaxation measurements significantly.

In addition to dipole–dipole interactions, other relaxation mechanisms can, in certain cases, contribute to the relaxation of carbon nuclei. One of the most efficient relaxation pathways of excited nuclei is through interactions with paramagnetic materials, e.g., relaxation induced by the presence of the unpaired electrons of oxygen or of metal ions, or by the presence of free radicals. The removal of any oxygen from the sample is one precaution that should be observed. The T_1 values for the C-1 carbon of some of the ^{13}C-phenol samples were obtained before and after dissolved oxygen removal. When compared to the 1-^{13}C-phenol T_1 values that were obtained for the samples prior to degassing, the removal of oxygen only resulted in a 5% change in the spin–lattice relaxation time, a negligible change. The degassing of samples was time consuming, and the sample tubes were very prone to cracking. In light of the small error that is introduced due to the presence of the oxygen and the problems encountered during the degassing process, we felt it was unnecessary to deoxygenate the remaining samples. Any discussion concerning the T_1 values obtained for the ^{13}C-phenol samples includes relaxation data from samples without oxygen removal.

In addition to dissolved oxygen, mineral matter trapped in the humic acid can also contain paramagnetic metal ions. Soil humic acids generally contain Fe^{2+} which is bound by the organic residues, and when released forms Fe^{3+}, a paramagnetic metal ion.[88] In an effort to release Fe^{3+} and any other bound metals, a 2.6 mg sample of Armadale humic acid (twice the amount used in the most concentrated sample containing humic acid) was combusted in a quartz tube at 412 °C for 7 h. After combustion, the sample was weighed (0.05 mg), and it was found that the original sample contained 1.9% mineral matter. It is important to point out that the ash did not completely dissolve in the D_2O. The fact that the pyrolyzed solid was insoluble raises serious questions concerning this experiment and a more appropriate examination of the effect of humic-bound mineral matter on the relaxation times will be carried out in future studies. The presence of the residue ash in a sample with ^{13}C-phenol was found to reduce the T_1 of the C-1 carbon by approximately 12 s or 4.6 s/mg humic acid. Considering the fact that such a small effect is observed and it is unlikely that oxidized forms of mineral matter would be present in the humic acids, we can neglect this effect on the values of the T_1.

Free radicals also produce paramagnetic centers that will affect the rate of relaxation of the phenolic carbons. Since the experiments were carried out at neutral pH, the concentration of radicals produced by the humic materials and the phenol itself were reduced but not eliminated. Electron spin resonance experiments on these solutions will be carried out to determine the concentration of free radicals in these solutions.

Although the spin lattice relaxation of the C-1 carbon is only slightly affected by the presence of dissolved oxygen and mineral matter, the observed T_1 values are still not a direct measure of the amount of associated ^{13}C-phenol, with two exceptions. The first exception is when no humic acid is added to the solution. In this case, there is no association, and the observed T_1 value directly reflects the status of the ^{13}C-phenol. The second exception occurs when all of the ^{13}C-phenol is associated with the humic acid. In this case, the observed T_1 has reached a steady state and does not change in the presence of additional humic acid.

In all other cases except the two presented above, the amount of associated ^{13}C-phenol can only be obtained indirectly, due to the fact that the signals for the free and associated species are not separated in the ^{13}C NMR spectrum. The insufficient sensitivity of the ^{13}C chemical shifts to separate these species results in an observed T_1 that reflects contributions from the free as well as the associated ^{13}C-labeled carbons of the ^{13}C-phenol. Equation (1) defines the observed spin–lattice relaxation time, T_1, as the weighted sum of the relaxation times for the C-1 carbons of the associated (T_{1A}) and free (T_{1F}) ^{13}C-phenol.

$$T_{1(\text{obs})} = T_{1A}A + T_{1F}F \qquad (1)$$

where A and F are the amounts of associated and free ^{13}C-phenol, respectively, as fractions of the total ^{13}C-phenol. The relaxation time for the associated species (T_{1A}) in this case is assumed to be 1.77 s. This T_1 value is found in Table 2.2 for the sample containing 1.3 mg of added Armadale humic acid. As discussed above, the T_1 value for the C-1 carbon of associated ^{13}C-phenol is anticipated to be in the range of 0.2 to 2.4 s. The spin–lattice relaxation time for T_{1F}, also found in Table 2.2, is 35.64 s. Equation (2) expresses the relationship between the fractions of associated (A) and free (F) ^{13}C-phenol:

$$A + F = 1 \tag{2}$$

When equation (2) is solved in terms of the amount of free (F) ^{13}C-phenol and then inserted into equation (1), the amount of associated ^{13}C-phenol can be calculated from equation (3):

$$A = (T_{1obs} - 35.64)/ - 33.87 \tag{3}$$

The last column in Table 2.2 contains the calculated amount of associated ^{13}C-phenol for each sample based on the observed T_1. When the amount of associated ^{13}C-phenol is plotted versus the observed spin–lattice relaxation time, a linear relationship is found as shown in Figure 2.5. Therefore there is a linear relationship between the concentration of the humic acid and the number of binding sites available to the ^{13}C-phenol. Additional information concerning the reaction surface of the humic acid would be useful in order to examine the association process further.

One way of examining surface chemical interactions is by determining the fractal geometry of the reaction surface. If a molecular system is described by a fractal dimension (D) of 1 then the surface is open and one-dimensional. When a fractal dimension value is 2 then the system has a closed, planar surface. Rice and Lin have found that humic acids in the solid state are described by a surface fractal (average $D = 2.3$) which characterizes them as having a closed, highly irregular surface, while the solution-state species are described by a surface fractal (average $D = 2.3$) or by a mass fractal $(D = 1.6)$ which characterizes an open, less compact surface.[89] The fractal dimension is dependent upon the type of humic acid that is considered.

Pfeifer and Avnir have shown that fractals can be obtained to describe the formation of monolayers at reaction surfaces using the power law shown in equation (4):

$$t_{1(c-1)}(HA) \propto HA^{-D} \tag{4}$$

where D represents the fractal dimension and $T_{1(c-1)}(HA)$ is the spin–lattice relaxation time T_1 as a function of the amount of added humic acid, HA, in milligrams.[90] A plot of the spin–lattice relaxation time versus the concentration c of humic acid from Table 2.2 produces a curve which is best represented by equation (5), as illustrated by Figure 2.6. When the data in Table 2.2 are plotted as the log of the concentration of added humic acid versus the log of the $c - 1$ relaxation time, T_1, the trend is linear and is defined by

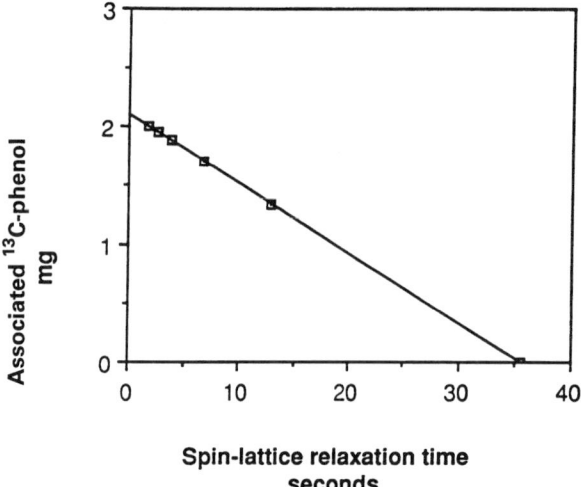

Figure 2.5 A plot of the amount of associated ^{13}C-phenol versus the spin–lattice relaxation time of the C-1 carbon of ^{13}C-phenol.

equation (6), with a correlation coefficient of 0.992. By representing the data in this form, information concerning the geometry of the humic acid surface can be predicted. The slope of the line from the fit of the data yields a fractal dimension of approximately 1.2 for the humic acid surface. As discussed previously, this suggests that it is a mass fractal describing a reaction surface intermediate between a line and a plane.

$$T_{1(c-1)}(\text{HA}) = \text{HA}^{-D} \tag{5}$$

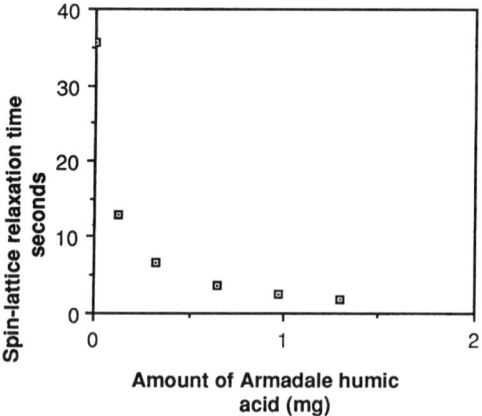

Figure 2.6 A plot of the amount of added Armadale humic acid versus the spin–lattice relaxation time, T_1, of the C-1 carbon of ^{13}C-phenol.

$$\log T_{1(c-1)} = -D \log \text{HA} \qquad (6)$$

Conclusions

The interaction of pollutants such as phenols and their halogenated counterparts with humic acids is an area of study that has only just begun to benefit from the stuctural information that is made available by NMR in combination with site-specific NMR labeling. This report emphasizes the need for such information and demonstrates how this technique can be applied to gain structural data concerning covalent and noncovalent bonding interactions. For instance, the covalent bonding of ^{13}C-phenol to Armadale humic acid produces a ^{13}C NMR spectrum which displays a dispersion of ^{13}C chemical shifts for the labeled C-1 carbon that can be assigned to specific types of bonding interactions. Noncovalent associative interactions, on the other hand, are not powerful enough to effect a change in the ^{13}C chemical shift of the labeled carbon; however, spin–lattice relaxation times T_1 constitute a sensitive probe to examine these weaker interactions. Other analytical methods were not capable of producing such structural information, so the benefit of this NMR technique is obvious.

We have also presented data on a new application of ^{13}C NMR and NMR labeling in which T_1 relaxation times were used to measure the concentration of the associated ^{13}C-phenol in solutions containing various amounts of Armadale humic acid. A linear relationship exists between the amount of associated ^{13}C-phenol and the observed T_1. In addition, the T_1 data were used to determine a fractal dimension for the humic acid surface of approximately 1.2, which suggests that the reaction surface is intermediate between a plane and a line. The application of NMR to study environmental problems is a challenging area of research, since information that was once not attainable can now be sought, if not obtained. This technique does have its limitations, however, and the most critical of these is sensitivity. With site-specific labeling we and others have been pushing the limits of the sensitivity of this technique and we will continue to explore new applications of this useful structural tool.

References

1. Hsu, T. S., and Bartha, R., Biodegradation of chloroaniline humus complexes in soil and in culture solution. *Soil Sci.* 118, 213, 1974.
2. Bollag, J.-M., Cross-coupling of humus constituents and xenobiotic substances. In *Aquatic and Terrestrial Humic Materials*, Christman, R.F., and Gjessing, E. T. (eds.), Ann Arbor Science, Ann Arbor, MI, 1983, Chapter 6.
3. Berry, D. F., and Boyd, S. A., Decontamination of soil through enhanced formation of bound residues. *Environ. Sci. Technol.* 19, 1132, 1985.
4. Thorn, K. A., and Mikita, M. A., Ammonia fixation by humic substances: a nitrogen-15 and carbon-13 NMR study. *Sci. Total Environ.* 113, 67, 1992.

5. Hayes, M. H. B., Adsorption of triazine herbicides on soil organic matter, including a short review on soil organic matter chemistry. *Residue Rev.* 32, 131, 1970.
6. Khan, S. U., Adsorption of pesticide by humic substances: a review. *Environ. Lett.* 3, 1, 1972.
7. Stevenson, F.J., Organic matter reactions involving herbicides in soil. *J. Environ. Qual.* 1, 333, 1972.
8. Khan, S. U., The interaction of organic matter with pesticides. In *Soil Organic Matter*, Schnitzer, M., and Khan, S. U. (eds), Elsevier Scientific, New York, NY, 1978, Chapter 4.
9. Gjessing, E. T., and Berglind, L., Adsorption of PAH to aquatic humus. *Arch. Hydrobiol.* 92, 24, 1981.
10. Carlburg, G. E., and Martinsen, K., Adsorption/complexation of organic micropollutants to aquatic humus: influence of aquatic humus with time on organic pollutants and comparison of two analytical methods for analyzing organic pollutants in humus water. *Sci. Tot. Environ.* 25, 245, 1982.
11. Carter, C. W., and Suffet, I. W., Interactions between dissolved humic and fulvic acids and pollutants in aquatic environment. In *Fate of Chemicals in the Environment*, Swann, R. L., and Eschenroeder, A. (eds), ACS Symp. Ser. No. 259, American Chemical Society, Washington DC, 1983, p. 215.
12. Caron, G., Suffet, I. H., and Belton, T., Effect of dissolved organic carbon on the environmental distribution of nonpolar organic compounds. *Chemosphere* 14, 993, 1985.
13. Means, J. C., and Wijayaratne, R. D., Role of natural colloids in the transport of hydrophobic pollutants. *Science*, 215, 968, 1982.
14. Chiou, C. T., Malcolm, R. L., Brinton, T. I., and Kile, D. E., Water solubility enhancement of some organic pollutants and pesticides by dissolved humic and fulvic acids. *Environ. Sci. Technol.* 20, 502, 1986.
15. Chiou, C. T., Kile, D. E., Brinton, T. I., Malcolm, R. L., Leenheer, J. A., and MacCarthy, P., A comparison of water solubility enhancements of organic solutes by aquatic humic materials and commercial humic acids. *Environ. Sci. Technol.* 21, 1231, 1987.
16. Stevenson, F. J., Organic matter reactions involving pesticides in soil. In *Bound and Conjugated Pesticide Residues*, Kaufman, D. D., Still, G. G., Paulson, G. D., and Bandal, S. K. (eds), ACS Symp. Ser. No. 29, American Chemical Society, Washington DC, 1976, p. 180.
17. Bartha, R., Pesticide residues in humus, *Am. Soc. Microb. News* 46, 356, 1980.
18. Martin, J. P., Haider, K., and Linhares, L. F., Decomposition and stabilization of ring-^{14}C-labeled catechol in soil. *Soil Sci. Soc. Am. J.* 43, 100, 1979.
19. Pignatello, J. J., Sorption dynamics of organic compounds in soils and sediments. In *Reactions and Movement of Organic Chemicals in Soils*, Shawney, B. L., and Brown, K. (eds), Spec. Publ. No. 22, Soil Science Society of America, Madison, WI, 1989, p. 45.
20. Chiou, C., Malcolm, R. L., Brinton, T. I., and Kile, D. E., Water solubility enhancement of some organic pollutants and pesticides by dissolved humic and fulvic acids. *Environ. Sci. Technol.* 20, 502, 1986.
21. Chiou, C. T., Kile, D. E., Brinton, T. I., Malcolm, R. L., and Leeneer, J. A., A comparison of water solubility enhancements of organic solutes by aquatic humic materials and commercial humic acids. *Environ. Sci. Technol.* 21, 1231, 1987.

22. Schellenberg, K., Leuenberger, C., and Schwarzenbach, R. P., Sorption of chlorinated phenols by natural sediments and aquifer materials. *Environ. Sci. Technol.* 18, 652, 1984.

23. Brusseau, M. L., Rao, P. S., and Jessup, R. E., Sorption kinetics of organic chemicals: evaluation of gas-purge and miscible-displacement techniques. *Environ. Sci. Technol.* 24, 727, 1990.

24. Lee, L. S., Suresh, P., Rao, R. S. C., and Brusseau, M. L., Nonequilibrium sorption and transport of neutral and ionized chlorophenols. *Environ. Sci. Technol.* 25, 722, 1991.

25. Brusseau, M. L., and Rao, R. S. C., Sorption nonideality during organic contaminant transport in porous media. *CRC Crit. Rev. Environ. Control* 19, 33, 1989.

26. Brusseau, M. L., and Rao, R. S. C., The influence of sorbate–organic matter interactions on sorption nonequilibrium. *Chemosphere* 18, 1691, 1989.

27. Karickhoff, S. W., and Morris, K. R., Sorption dynamics of hydrophobic pollutants in sediment suspensions. *Environ. Toxicol. Chem.* 4, 469, 1985.

28. Karickhoff, S. W., Sorption kinetics of hydrophobic pollutants in natural sediments. In *Contaminants and Sediments*, Vol. 2, Baker, R. A. (ed.), Ann Arbor Science, Ann Arbor, MI, 1980, p. 193.

29. Johnson-Logan, L. R., Broshears, R. E., and Klaine, S. J., Partitioning behavior and the mobility of chlordane in groundwater. *Environ. Sci. Technol.* 26, 2234, 1992.

30. Nkedi-Kizza, P., Suresh, P., Rao, R. S. C., and Hornsby, A. G., Influence of organic cosolvents on leaching of hydrophobic organic chemicals through soils. *Environ. Sci. Technol.* 21, 1107, 1987.

31. Martin, J. P., Haider, K., and Linhares, L. F., Decomposition and stabilization of ring-^{14}C-labeled catechol in soil. *Soil Sci. Soc. Am. J.* 43, 100, 1979.

32. Bartha, R., Fate of herbicide-derived chloroanilines in soil. *J. Agr. Food Chem.* 19, 385, 1971.

33. Wilson, R. G., and Cheng, H. H., Breakdown and movement of 2,4-D in the soil under field conditions. *Weed Sci.* 24, 462, 1976.

34. Cheng, H. H., and Führ, F., Extraction of methabenzthiazuron from the soil. *J. Agric. Food Chem.* 24, 421, 1976.

35. Wilson, Jr., R. G., and Cheng, H. H., Fate of 2,4-D in a Naff silt loam soil. *J. Environ. Qual.* 7, 281, 1978.

36. Meikle, R. W., Regoli, A. J., Kurihara, N. H., and Laskowski, D. A., Classification of bound residues soil organic matter: polymeric nature of residues in humic substance. In *Bound and Conjugated Pesticide Residues*, Kaufman, D. D., Still, G. G., Paulson, G. D., and Bandal, S. K. (eds), ACS Symp. Ser. No. 29, American Chemical Society, Washington DC, 1976, p. 272.

37. Hsu, T. S., and Bartha, R., Biodegradation of chloraniline humus complexes in soil and in culture solution. *Soil Sci.* 118, 213, 1974.

38. Bollag, J.-M., Liu, S.-Y., and Minard, R. D., Cross-coupling of phenolic humus constituents and 2,4-dichlorophenol. *Soil Sci. Soc. Am. J.* 44, 52, 1980.

39. Bollag, J.-M., Minard, R. D., and Liu, S.-Y., Cross-linkage between anilines and phenolic humus constituents. *Environ. Sci. Technol.* 17, 72, 1983.

40. Bollag, J.-M., and Loll, M. J., Incorporation of xenobiotics into soil humus. *Experientia* 39, 1221, 1983.

41. Bollag, J.-M., Liu, S.-Y., and Deune, E. G., Transformation of 2,6-diethylaniline in soil. *Soil Sci.* 143, 56, 1987.

42. Bollag, J.-M., Blattmann, P., and Laanio, T., Adsorption and transformation of four substituted anilines in soil. *J. Agric. Food Chem.* 26, 1302, 1978.
43. McCarthy, J. F., and Jimenez, B. D., Reduction in bioavailability to bluegills of polycyclic aromatic hydrocarbons bound to dissolved humic matter. *Environ. Toxicol. Chem.* 4, 511, 1985.
44. Ogram, A. V., Jessup, R. E., Ou, L. T., and Pao, R. S. C., Effects of sorption on biological degradation rates of (2,4-dichlorophenoxy)acetic acid in soils. *Appl. Environ. Microbiol.* 49, 582, 1985.
45. Bollag, J.-M., Shuttleworth, K. L., and Anderson, D. H., Laccase-mediated detoxification of phenolic compounds. *Appl. Environ. Microbiol.* 54, 3086, 1988.
46. Dec, J., and Bollag, J.-M., Microbial release and degradation of catechol and chlorophenols bound to synthetic humic acid. *Soil Sci. Soc. Am. J.* 52, 1366, 1988.
47. Maloney, S. W., Manem, J., Mallevialle, J., and Fiessinger, F., Transformation of trace organic compounds in drinking water by enzymatic oxidative coupling. *Environ. Sci. Technol.* 20, 249, 1986.
48. Parris, G. E., Covalent binding of aromatic amines to humate. 1. Reactions with carbonyls and quinones. *Environ. Sci. Technol.* 14, 1099, 1980.
49. Olson, B. M., The use of fluorescence spectroscopy to study herbicide–humic acid interactions: preliminary observations. *Can. J. Soil Sci.* 70, 515, 1990.
50. Melcer, M. E., Zalewski, M. S., Brisk, M. A., and Hassett, J. P., Evidence for a charge-transfer interaction between dissolved humic materials and organic molecules: I. Study of the binding interaction between humic materials and chloranil. *Chemosphere* 16, 1115, 1987.
51. Morra, M. J., Corapcioglu, M. O., von Wandruszka, R. M. A., Marshall, D. B., and Topper, K., Fluorescence quenching and polarization studies of naphthalene and 1-naphthol interaction with humic acid. *Soil Sci. Soc. Am. J.* 54, 1283, 1990.
52. Sarkar, J. M., Malcolm, R. L., and Bollag, J.-M., Enzymatic coupling of 2,4-dichlorophenol to stream fulvic acid in the presence of oxidoreductases. *Soil Sci. Soc. Am. J.* 52, 688, 1988.
53. Sjoblad, R. D., and Bollag, J.-M., Oxidative coupling of aromatic compounds by enzymes from soil microorganisms. In *Soil Biochemistry*, Vol. 5, Paul, E. A., and Ladd, J. N. (eds), Marcel Dekker, New York, NY, 1981, Chapter 3.
54. Bollag, J.-M., Enzymes catalyzing oxidative coupling reactions of pollutants. In *Metal Ions in Biological Systems*, Vol. 28, Sigel, H., and Sigel, A. (eds), Marcel Dekker, New York, NY, 1992, Chapter 6.
55. Bollag, J.-M., and Liu, S.-Y., Copolymerization of halogenated phenols and syringic acid. *Pestic. Biochem. Physiol.* 23, 261, 1985.
56. Bollag, J.-M, Liu, S.-Y., and Minard, R. D., Cross-coupling of phenolic humus constituents and 2,4-dichlorophenol. *Soil Sci. Soc. Am. J.* 44, 52, 1980.
57. Bollag, J.-M., Minard, R. D., and Liu, S.-Y., Cross-linkage between anilines and phenolic humus constituents. *Environ. Sci. Technol.* 17, 72, 1983.
58. Dec, J., and Bollag, J.-M., Detoxification of substituted phenols by oxidoreductive enzymes through polymerization reactions. *Arch. Environ. Contam. Toxicol.* 19, 543, 1990.
59. Leonowicz, A., Edgehill, R. U., and Bollag, J.-M., The effect of pH on the transformation of syringic and vanillic acids by the laccases of *Rhizoctonia praticola* and *Trametes versicolor*. *Arch. Microbiol.* 137, 89, 1984.
60. Frye, J. S., Bronnimann, C. E., and Maciel, G. E., Solid-state NMR of humic materials. In *NMR of Humic Substances and Coal—Techniques, Problems, and*

Solutions, Wershaw, R. L., and Mikita, M. A., (eds), Lewis Publishers, Chelsea, MI, 1987, Chapter 3.

61. Preston, C. M., Review of solution NMR of humic substances. In *NMR of Humic Substances and Coal—Techniques, Problems and Solutions*, Wershaw, M. L., and Mikita, M. A. (eds), Lewis Publishers, Chelsea, MI, 1987, Chapter 2.

62. Wilson, M. A., *NMR Techniques and Applications in Geochemistry and Soil Chemistry*, Pergamon, New York, NY, 1987.

63. Preston, C. M., and Blackwell, B. A., Carbon-13 nuclear magnetic resonance for a humic and fulvic acid: signal-to-noise optimization, quantitation, and spin–echo techniques. *Soil Sci.* 139, 88, 1985.

64. Levy, G. C., Lichter, R. L., and Nelson, G. L., *Carbon-13 Nuclear Magnetic Resonance Spectroscopy*, 2nd edn, John Wiley, New York, NY, 1980, Chapter 1.

65. Hatcher, P. G., Bortiatynski, J. M., Minard, R. D., Dec, J., and Bollag, J.-M., Use of high resolution ^{13}C NMR to examine the enzymatic covalent binding of ^{13}C-labeled 2,4-dichlorophenol to humic substances. *Environ. Sci. Technol.* 27, 2098, 1993.

66. Thorn, K. A., Folan, D. W., Arterbrun, J. B., Mikita, M. A., and MacCarthy, P., Application of INEPT nitrogen-15 and silicon-29 nuclear magnetic resonance spectroscopy to derivatized fulvic acids. *Sci. Tot. Environ.* 81/82, 209, 1989.

67. Thorn, K. A., Steelink, C., and Wershaw, R. L., Methylation patterns of aquatic humic substances determined by ^{13}C NMR spectroscopy. *Org. Geochem.* 11, 123, 1987.

68. Mikita, M. A., and Steelink, C., Carbon-13 enriched nuclear magnetic resonance method for the determination of hydroxyl functionality in humic substances. *Anal. Chem.* 53, 1715, 1981.

69. Thorn, K. A., NMR structural investigations of aquatic humic substances. PhD Dissertation, University of Arizona, 1984, p. 44.

70. Preston, C. M., and Ripmeester, J. A., ^{13}C-labeling for NMR studies of soils: CPMAS NMR observation of ^{13}C-acetate transformation in a mineral soil. *Can. J. Soil Sci.* 63, 495, 1983.

71. Thorn, K. A., Arterburn, J. B., and Mikita, M. A., ^{15}N and ^{13}C NMR investigation of hydroxylamine-derivatized humic substances. *Environ. Sci. Technol.* 26, 107, 1992.

72. Haider, K., Spiteller, M., Reichert, K., and Fild, M., Derivatization of humic compounds: an analytical approach for bound organic residues. *Int. J. Environ. Anal. Chem.* 46, 201, 1992.

73. Gonzalez-vila, F. J., Lüdemann, H.-D., and Martin, F., ^{13}C-NMR structural features of soil humic acids and their methylated, hydrolyzed and extracted derivatives. *Geoderma* 31, 3, 1983.

74. Thorn, K. A., and Mikita, M. A., Ammonia fixation by humic substances: a nitrogen-15 and carbon-13 NMR study. *Sci. Tot. Environ.* 113, 67, 1992.

75. Grinwalla, A. S., and Mikita, M. A., Reaction of Suwannee river fulvic acid with chloramine: characterization of products via ^{15}N NMR. *Environ. Sci. Technol.* 26, 1148, 1992.

76. Thorn, K. A., Weber, E. J., Spidle, D. L., and Pettigrew, P. J., Kinetic and N-15 NMR spectroscopic studies of the covalent binding of aniline to humic substances in the presence and absence of phenoloxidase enzymes. In *6th Int. Meeting Int. Humic Substances Soc. Abstracts*, Senesi, N., and Miano, T. M. (eds), Monopoli, Italy, 1992, p. 119.

77. Bortiatynski, J. M., Hatcher, P. G., Minard, R. D., Dec, J., and Bollag, J.-M., Enzyme catalyzed binding of ^{13}C-labeled 2,4-dichlorophenol to humic acid using high resolution ^{13}C NMR. In *Humic Substances in the Global Environment and Implications on Human Health*, Senesi, N., and Miano, T. M. (eds), Elsevier Scientific, New York, 1994, p. 1091.

78. Minard, R. D., Liu, S.-Y., and Bollag, J.-M., Oligomers and quinones from 2,4-dichlorophenol. *J. Agric. Food Chem.* 29, 250, 1981.

79. Silverstein, R. M., Bassler, G. C., and Morrill, T. C., *Spectrometric Identification of Organic Compounds*, 4th edn, John Wiley, New York, NY, 1981, p. 265.

80. Lee, L. S., Bellin, C. A., Pinal, R., and Rao, P. S. C., Cosolvent effects on sorption of organic acids by soils from mixed solvents. *Environ. Sci. Technol.* 27, 165, 1993.

81. Lee, L. S., Rao, P. S. C., Nkedi-Kizza, P., and Delfino, J. J., Influence of solvent and sorbent characteristics on distribution of pentachlorophenol in octanol–water and soil–water systems. *Environ. Sci. Technol.* 24, 654, 1990.

82. Schulten, H.-R., and Schnitzer, M., Structural studies on soil humic acids by Curie-point pyrolysis–gas chromatography/mass spectrometry. *Soil Sci.* 153, 205, 1992.

83. Derome, A. E., *Modern NMR Techniques for Chemistry Research*, Organic Chemistry Series Vol. 6, Baldwin, J. E. (ed), Pergamon, Oxford, 1988, p. 85.

84. Levy, G. C., ^{13}C spin–lattice relaxation in substituted benzenes. *J. Chem. Soc., Chem. Commun.* 47, 1972.

85. Bortiatynski, J. M., Hatcher, P. G., Minard, R. D., Dec, J., and Bollag, J.-M., unpublished results.

86. Thorn, K. A., Folan, D. W., and MacCarthy, P. *Characterization of the International Humic Substances Society Standard and Reference Fulvic and Humic Acids by Solution State Carbon-13 (^{13}C) and Hydrogen-1 (^{1}H) Nuclear Magnetic Resonance Spectrometry.* Water-Resources Investigations Report 89-4196, US Geological Survey, Denver, CO, 1989, p. 26.

87. Chen, Y., and Schnitzer, M., Viscosity measurements on soil humic substances. *Soil Sci. Soc. Am. J.* 40, 866, 1976.

88. Schnitzer, M., Humic substances: chemistry and reactions. In *Soil Organic Matter*, Schnitzer, M., and Khan, S. U. (eds), Elsevier Scientific, New York, NY, 1978, Chapter 1.

89. Rice, J. A., and Lin, J.-S., Fractal nature of humic materials. *Environ. Sci. Technol.* 27, 413, 1993.

90. Pfeifer, P., and Avnir, D., Chemistry in noninteger dimensions between two and three. I. Fractal theory of heterogeneous surfaces. *J. Chem. Phys.* 79, 3358, 1983.

3.

Proton and ^{19}F NMR Spectroscopy of Pesticide Intermolecular Interactions

SHARON J. ANDERSON

Sorption of organic pollutants by soils and sediments is one of the main chemical processes that controls pollutant migration in the environment. Information about the molecular mechanisms by which an organic pollutant interacts with other solution-phase constituents and with solid-phase sorbents would be invaluable for more accurate prediction of pollutant fate and transport and for optimal design and application of remediation procedures. Many current models and remediation strategies are based upon the "partition theory" of organic compound sorption,[1,2] which predicts sorption coefficients from properties such as water solubility[3-6] or octanol–water partition coefficients.[4,5,7] Partition theory is well suited for nonpolar hydrocarbons but may not be appropriate for pesticides with electrophilic or weakly acidic or basic substituents,[2] which may interact with soils or organic matter through specific interactions such as hydrogen bonding or charge-transfer complexes. If a pesticide can form hydrogen bonds or a charge-transfer complex with a sorbent, sorption may be greater than in the absence of specific interactions.

Nuclear magnetic resonance (NMR) spectroscopy is well suited for the study of pesticide–solution or pesticide–sorbent interactions because NMR is an element-specific method that is extremely sensitive to the electron density (shielding) near the nucleus of interest. Consequently, solution-state NMR can distinguish between closely related functional groups and can provide information about intermolecular interactions. All nuclei with nonzero nuclear spin quantum number can be studied by NMR spectroscopy. Of the more than 100 NMR-active nuclei, ^{1}H and ^{19}F are the easiest to study because both have natural abundances near 100% and greater NMR sensitivity than any other nuclei. In addition, both ^{1}H and ^{19}F have zero quadrupolar moments, which means that sharp, well resolved NMR peaks can be obtained, at least in homogeneous solutions. Proton NMR is well suited for

51

elucidating molecular interactions in solution but cannot be used to study interactions between pesticides and heterogeneous sorbents such as soils, humic acid, or even cell extracts, since protons in the sorbent generally produce broad peaks that mask the NMR peaks from the solute or sorbate of interest. In contrast, ^{19}F NMR can be used to study interactions between fluorine-containing molecules and heterogeneous sorbents because the fluorine concentration in most natural sorbents is negligible. Several herbicides contain fluorine, e.g., fluridone (1-methyl-3-phenyl-5-[3-(trifluoromethyl)-phenyl]-4(1H)-pyridinone), trifluralin (2,6-dinitro-N,N-dipropyl-4-(trifluoromethyl)benzenamine), and primisulfuron (3-[4,6-bis(difluoromethoxy)pyrimidin-2-yl]-1-(2-methoxy-carbonylphenylsulfonyl)urea). In one study, a fluorine-substituted analog of atrazine was synthesized so that atrazine self-association could be studied by ^{19}F NMR.[8] In addition, simple fluoroaromatic solutes can be used as model compounds to obtain information about interactions involving a broad range of functional groups because ^{19}F chemical shifts of fluoroaromatic solutes are very sensitive to changes in the electronic environment of other substituents on a benzene ring.

This chapter is divided into three main sections. First, an overview of how molecular interactions affect nuclear shielding and NMR chemical shifts is provided, along with the requirements for obtaining mechanistic information from NMR chemical shifts and linewidths. The next section contains examples of proton NMR studies of pesticide interactions in homogeneous solutions. The final section describes ^{1}H and ^{19}F NMR experiments designed to provide mechanistic information about interactions between organic solutes and natural organic macromolecules.

Using NMR to Determine Intermolecular Interaction Mechanisms

Contributions to Nuclear Shielding and NMR Chemical Shifts

The electron density around a nucleus determines the NMR chemical shift for a particular atom in a molecule. Chemical shifts for a given nucleus are affected by the properties of neighboring atoms with which chemical bonds are formed and by intermolecular interactions with a solvent and other species in a sample. The effects of the surrounding medium on NMR chemical shifts can be defined qualitatively as the resulting change in chemical shift, measured relative to a common reference compound, when a solute is transferred at infinite dilution from one environment (generally a nonpolar, inert reference solvent) to another:[9]

$$\Delta\delta_{\text{medium}} = \delta_{\text{sample}} - \delta_{\text{refsolv}} \qquad (1)$$

where $\Delta\delta_{\text{medium}}$ denotes the contribution of intermolecular interactions to NMR chemical shifts and δ_{sample} and δ_{refsolv} are the chemical shifts in the sample and reference systems, relative to a common reference compound. The effects of the solution or medium on NMR chemical shifts can be divided, at least conceptually, into five components:[10]

$$\Delta\delta_{\text{medium}} = \Delta\delta_b + \Delta\delta_a + \Delta\delta_w + \Delta\delta_E + \Delta\delta_c \tag{2}$$

The term $\Delta\delta_b$ is the contribution of bulk magnetic susceptibility differences to $\Delta\delta_{\text{medium}}$. The magnetic anisotropy term, $\Delta\delta_a$, is important for anisotropic solvents such as rod-shaped CS_2 or planar benzene, which have anisotropic diamagnetic susceptibilities. The magnitudes of $\Delta\delta_a$ and $\Delta\delta_b$ depend only on the shape and magnetic susceptibility of the surrounding medium, not upon solute properties, and thus should be the same for any nucleus in a sample. The contribution of van der Waals dispersion interactions, $\Delta\delta_w$, depends upon solute size as well as on the polarizability and ionization potential of the solvent. Finally, $\Delta\delta_E$ is the effect of electric dipolar interactions on chemical shifts, and $\Delta\delta_c$ is the contribution from hydrogen-bonding, charge transfer, ion-pair, or other complexes.[10]

Importance of Appropriate Chemical Shift Reference

In order to use NMR to study solute–solute complexes or specific sorbate–sorbent interactions, it is necessary to isolate the effects of dipolar or complexation interactions ($\Delta\delta_E$ and $\Delta\delta_c$) and to minimize or eliminate $\Delta\delta_b$, $\Delta\delta_a$, and $\Delta\delta_w$ by use of an appropriate referencing or calibration procedure. The basic principles and relative merits of two different methods, internal and external referencing, will be described below. Regardless of whether internal or external referencing is used, the reference compound must be nonpolar and must not react with either the solvent or other solutes in solution. In order to minimize differences in polarizability and $\Delta\delta_w$, the reference compound should be the same shape, size, and type (aromatic or aliphatic) as the solute of interest,[10,11] though this criterion is not important for proton NMR because van der Waals dispersion forces make only a small contribution to ^1H chemical shifts.[9] For proton NMR, TMS (tetramethylsilane) is used as an inert reference compound for studying both aliphatic and aromatic solutes in organic solvents, and TSP (sodium 3-trimethylsilylpropionate 2,2,3,3-d_4) is used in aqueous solutions.

For fluorine NMR, van der Waals forces can make a relatively large contribution to chemical shifts and $\Delta\delta_{\text{medium}}$,[9] so it is more important that the polarizability of the solute and reference compounds be similar. For fluoroalkanes (including trifluoromethyl-containing aromatic pesticides), freons such as $CFCl_3$ and nonpolar fluoroalkanes such as hexafluorocyclohexane are commonly used as reference compounds. For fluoroaromatic compounds, hexafluorobenzene (C_6F_6) and 1,4-difluorobenzene (p-DFB) are the same size, shape, and type as the solute of interest and are nonpolar, provided that the quadrupole moment of C_6F_6 makes a negligible contribution to $\Delta\delta_E$.[10,11] It is possible that C_6F_6 and p-DFB may form π–π complexes with other aromatic compounds, but no ^{19}F NMR or UV spectroscopic evidence was found for complex formation between C_6F_6 and hexamethyl-benzene, a π-donor.[12] 1,4-Difluorobenzene may be the best fluoroaromatic reference for aqueous samples, because it is much more soluble in water than is C_6F_6.

In addition to these constraints on the solute, the reference solvent must be inert, with a very low dielectric constant and low polarizability.[9] Given these constraints, cyclohexane probably is the ideal reference solvent. Chloroform and CCl_4, which are common solvents in NMR spectroscopy, do not meet these criteria. Carbon tetrachloride produces solvent shifts compared with cyclohexane because (1) the dielectric constant of CCl_4 is about 10% higher than that of cyclohexane, and (2) the polarizable Cl atoms can interact with many solutes and even form weak hydrogen-bonds with protons of certain solutes. Chloroform is an even poorer reference solvent because it forms hydrogen bonds with many polar solutes.[9]

Comparison of Internal and External Referencing Methods

When a reference compound is added directly to a solution of interest (internal referencing; Figure 3.1(a)), the reference and sample solutes experience the same bulk susceptibility (i.e., $\Delta\delta_b = 0$). An internal reference compound that is the same size, shape, and approximate polarizability (aromatic or aliphatic) as the sample solute will also eliminate $\Delta\delta_a$ and minimize $\Delta\delta_w$, thereby isolating polar and complexation contributions to $\Delta\delta_{medium}$. In other words, with an appropriate choice of internal reference compound, medium effects are simply given by

$$\Delta\delta_{medium} = \Delta\delta_E + \Delta\delta_c \qquad (3)$$

When an external referencing procedure is used, the chemical shift of the solute of interest is measured in the sample solution, but the chemical shift of an inert reference solute is measured in a reference solvent such as cyclohexane, not in the sample solution. The sample and reference solutions may be placed in separate NMR tubes and analyzed separately (Figure 3.1(b)), or the sample and reference solutions may be analyzed simultaneously by putting the reference solution into a coaxial capillary tube that is inserted into the sample NMR tube (Figure 3.1(c)). When either of these external reference methods is used, the contribution of $\Delta\delta_b$ to $\Delta\delta_{medium}$ can be calculated from the bulk volume susceptibilities of the reference solvent ($\chi_{refsolv}$) and the sample solution (χ_{sample}). When bulk volume susceptibility is given in SI units, the general equation to find $\Delta\delta_b$ is

$$\Delta\delta_b = \delta_{corrected} - \delta_{observed} = -(1/3 - \alpha)(\chi_{sample} - \chi_{refsolv}) \qquad (4)$$

Downfield chemical shifts (decreased nuclear shielding) in equation (4) correspond to more positive δ values. In equation (4), $\delta_{corrected}$ is the

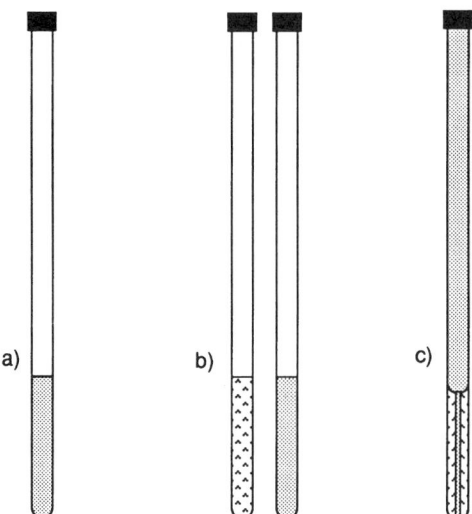

Figure 3.1   Schematic representation of (a) internal referencing, in which the reference compound is added to sample solution; (b) external referencing, in which the solute is in the sample solution and the reference solute is in the reference solvent in a separate tube; (c) external referencing with a capillary tube: the solute in the sample solution is in the annular region of the NMR tube, whereas the reference solute is in the reference solvent in a coaxial capillary tube so that the sample and reference solutions can be analyzed simultaneously.

susceptibility-corrected chemical shift, δ_{observed} is the chemical shift measured in the sample solution, and α is a geometric factor that depends upon the shape of the NMR tube and the orientation of the magnetic field B_0 relative to the sample tube. For spherical NMR tubes, $\alpha = 1/3$ for all orientations of B_0. For cylindrical NMR tubes with B_0 perpendicular to the tube, $\alpha = 1/3$. For cylindrical tubes with B_0 parallel to the tube (the geometry found in FT-NMR spectrometers with superconducting magnets), $\alpha = 0.$[13] For the latter system, equation (4) reduces to

$$\Delta\delta_b = -1/3(\chi_{\text{sample}} - \chi_{\text{refsolv}}) \tag{5}$$

The value of χ in SI units is equal to 4π times the cgs value.

Bulk susceptibility calculations are straightforward if the bulk susceptibility of the sample is known accurately. Bulk susceptibilities of pure solvents can be obtained from published tables, for example in the *CRC Handbook of Chemistry and Physics*, but tabulated data generally are compiled from many different sources and are not always consistent. Furthermore, trace quantities of paramagnetic impurities in the sample or reference solution can produce errors in $\Delta\delta_b$ by causing χ to become less negative than in pure solution.[13] Even the most accurate bulk susceptibility corrections cannot account for $\Delta\delta_a$

and $\Delta\delta_w$, however, so medium effects on externally referenced chemical shifts must be described with equation (6):

$$\Delta\delta_{medium} = \Delta\delta_a + \Delta\delta_w + \Delta\delta_E + \Delta\delta_c \qquad (6)$$

Comparison of equations (3) and (6) shows that internal and external referencing can give different values of $\Delta\delta_{medium}$ and lead to different interpretations of chemical shift data, and that internal referencing is preferable for studies of dipolar and complexation interactions. In spite of its shortcomings, however, external referencing (sometimes even without bulk susceptibility corrections) has been used frequently in NMR studies of intermolecular interactions.

To provide an example of the discrepancy between internal and external referencing results and to compare the suitability of C_6F_6 and p-DFB as ^{19}F reference compounds, ^{19}F chemical shifts of C_6F_6 and p-DFB were measured in five solvents (Table 3.1). First, 0.5 mL of 0.5 mM and 5 mM solutions of C_6F_6 or p-DFB in each solvent were transferred to 5 mm NMR tubes. Then a coaxial capillary tube that contained 5 mM C_6F_6 or p-DFB in cyclohexane was inserted into each NMR tube. Chemical shifts ($\delta_{observed}$) of C_6F_6 and p-DFB in each solvent were measured relative to the peak for that solute in the cyclohexane capillary tube. All spectra were obtained on a Varian VXR 500 MHz NMR spectrometer at 470.268 MHz with a 5 μs (40°) pulse and a 2 s delay. Samples were spun at 20 Hz, and 64 transients were collected for each sample. Proton decoupling was not possible on this spectrometer, so $\delta_{observed}$ for p-DFB represents the center of the doublet.

The data for 0.5 mM solutions are shown in Table 3.1. Values of $\delta_{observed}$ were the same in 0.5 mM and 5 mM solutions, except in water, where $\delta_{observed}$ for p-DFB was 0.1 ppm more negative in 5 mM than 0.5 mM solution (unpublished data). Values of $\Delta\delta_b$ were calculated with equation (5) and susceptibility-corrected chemical shifts ($\delta_{corrected}$) were calculated with equation (4). The $\delta_{corrected}$ values in Table 3.1 are numerically equivalent to the discrepancy between internal referencing ($\delta_{observed}$) and susceptibility-corrected external referencing. Negative values of $\delta_{corrected}$ mean that the signal for the internal reference compound was upfield of the susceptibility-corrected external reference peak and that solute chemical shifts measured by internal referencing would be more downfield (at more positive δ) than shifts calculated by external referencing and bulk susceptibility corrections. The exact source of the difference between internal and external referencing with C_6F_6 and p-DFB is not known, but is more probably attributable to anisotropy and van der Waals interactions than to complex formation if, as discussed above, $\Delta\delta_E$ and $\Delta\delta_c$ are zero for C_6F_6 and p-DFB.

The data in Table 3.1 also show that solute chemical shifts measured in water and methanol will depend upon whether C_6F_6 or p-DFB is used as the fluoroaromatic reference compound. Because p-DFB is more similar to other fluoroaromatic compounds than is C_6F_6, p-DFB is probably a better internal reference compound, at least in hydrogen-bonding solvents. The two

Table 3.1 Bulk Susceptibility Corrections ($\Delta\delta_b$) and Observed and Corrected Chemical Shifts for 0.5 mM Solutions of Hexafluorobenzene and 1,4-Difluorobenzene

Solvent	χ^a	$\Delta\delta_b{}^b$ (ppm)	$\delta_{observed}{}^c$ (ppm)		$\delta_{corrected}{}^d$ (ppm)	
			C_6F_6	p-DFB	C_6F_6	p-DFB
Cyclohexane	−7.829	0	0	0	0	0
Benzene	−7.653	−0.06	−0.31	−0.27	−0.25	−0.21
Acetonitrile	−6.509	−0.44	−1.15	−0.11	−0.71	−0.67
Methanol	−6.572	−0.42	−2.14	−1.78	−1.72	−1.36
Water	−8.998	+0.39	+0.57	+0.92	+0.18	+0.53

[a] Bulk volume susceptibilities (SI units).[13]
[b] $\Delta\delta_b = -\frac{1}{3}(\delta_{sample} - \delta_{cyhex})$.
[c] $\delta_{observed} = \delta_{sample} - \delta_{cyhex}$. Negative value means that observed peak in the solvent is upfield of the peak in cyclohexane.
[d] $\delta_{corrected} = \delta_{observed} - \Delta\delta_b$.

reference compounds gave identical results in benzene, which suggests that π–π interactions are equally important (or unimportant) for these reference compounds, and that using either C_6F_6 or p-DFB should eliminate π–π contributions to ^{19}F chemical shifts of fluoroaromatic solutes.

Mechanistic Information from Chemical Shifts

Mechanistic information about solute–solute and sorbate–sorbent interactions can be obtained from NMR chemical shifts (i.e., $\Delta\delta_{medium}$) if $\Delta\delta_{medium}$ is sufficiently large and is known accurately. The magnitude of $\Delta\delta_{medium}$ increases with increasing orbital overlap between the solute and a complexing species; $\Delta\delta_{medium}$ is greatest when an NMR-active nucleus is in a more anisotropic environment in one medium than another.[10] Because hydrogen bonding causes anisotropic changes in electron density, NMR is a sensitive probe of hydrogen-bonding interactions.

The accuracy with which $\Delta\delta_{medium}$ can be determined depends upon the use of an appropriate reference or calibration method and on the width of the NMR peaks. When changes in $\Delta\delta$ are measured as a function of increasing concentration of a complexation agent, internal and external reference methods generally give identical results, provided that the total concentration of added reagent is not high enough to alter either the magnetic susceptibility or the solvating properties of the solution. In studies of interactions between small solutes and a macromolecular or colloidal sorbent, internal referencing is preferable to external referencing because bulk susceptibilities are difficult to determine accurately and external referencing cannot account for contributions of $\Delta\delta_a$ and $\Delta\delta_w$. Even with internal referencing, though, $\Delta\delta_b$, $\Delta\delta_a$, and $\Delta\delta_w$ may be different for the solute than for the internal reference compound if the colloidal material exhibits very different affinity for the solute than for the reference compound. In such cases, chemical interactions ($\Delta\delta_E$ and $\Delta\delta_c$) will not be the only contributions to $\Delta\delta_{medium}$.

Mechanistic Information From Linewidths

Broad NMR peaks can obscure mechanistic information about chemical interactions, but NMR linewidths themselves can provide complementary information about interaction mechanisms, provided that the magnetic field is homogeneous. Several different types of molecular interactions can contribute to line broadening:

1. fast chemical exchange between different environments;
2. interaction with a continuum of sites;
3. reduced molecular mobility caused by sorption or by increased solution viscosity;
4. interaction with paramagnetic sites.

Carefully designed experiments can often help to rule out one or more of these causes of line broadening. When the chemical exchange rate between two environments is comparable to the NMR frequency, only a single broad peak is observed, not separate peaks for each environment. Similarly, if a heterogeneous sorbent comprises a near-continuum of sorption sites (as is the case for soils and humic acid), separate peaks for each environment generally cannot be resolved, depending upon the exchange rate and the mobility of sorbed species. The contribution of chemical exchange or multiple molecular environments to NMR line broadening can be determined by measuring linewidths (full width at half-height, or $\Delta\nu_{1/2}$) at different spectrometer frequencies. If either chemical exchange or sorption onto heterogeneous sites is the sole cause of line broadening, then $\Delta\nu_{1/2}$ will increase in direct proportion to spectrometer frequency. Any lesser increase in $\Delta\nu_{1/2}$ with increasing frequency would suggest that one or both of these mechanisms contributes to NMR line broadening but is not the sole source.

A decrease in a solute's molecular mobility also causes $\Delta\nu_{1/2}$ to increase. Both sorption by colloidal material and an increase in solution viscosity can reduce a solute's molecular mobility and cause NMR linewidths to increase. Thus, the contribution of solution viscosity to line broadening must be determined separately before an increase in $\Delta\nu_{1/2}$ can be attributed to sorption. For example, the ^{19}F NMR linewidth of 5-fluoro-L-tryptophan was much greater in human serum albumin (HSA) suspension than in aqueous buffer.[14] To determine whether sorption or viscosity caused the increase in $\Delta\nu_{1/2}$, $\Delta\nu_{1/2}$ of 5-fluoro-L-tryptophan was also measured in a glycerol solution that had the same viscosity as the HSA suspension. The 5-fluoro-L-tryptophan linewidth was much greater in HSA than in glycerol, and only slightly greater in glycerol than in aqueous buffer. These results suggested that sorption, not solution viscosity, was the major cause of reduced mobility and line broadening in the HSA suspension.[14]

Interaction with paramagnetic sites also produces extremely broad NMR peaks because of rapid spin–spin relaxation (short T_2). Paramagnetic relaxation is essentially independent of temperature, whereas chemical exchange rates and molecular mobility decrease with decreasing temperature. Thus, $\Delta\nu_{1/2}$ will be independent of temperature if paramagnetic relaxation is the dominant line broadening mechanism, but will increase with decreasing temperature whenchemical exchange or sorption contribute to line broadening.

Relaxation times can provide useful information about molecular mobility and the nature of a sorption site. When several molecular environments contribute to a single NMR peak, however, only an average relaxation time can usually be measured for each NMR peak, not separate relaxation times for each molecular environment. Average spin–spin (transverse) relaxation times can be estimated from $\Delta\nu_{1/2}$ of Lorentzian NMR peaks:

$$T_2 = 1/(\pi\Delta\nu_{1/2}) \qquad (7)$$

The significance of NMR relaxation times and methods for measuring them have been reviewed in detail elsewhere.[15–17]

NMR Studies of Pesticide Molecular Interactions in Solution

Hydrogen Bonding by Atrazine

Proton NMR has shown that atrazine (6-chloro-N-ethyl-N'-(1-methylethyl)-1,3,5-triazine-2,4-diamine) is both a hydrogen-bond donor and a hydrogen-bond acceptor.[18–20] Lone-pair electrons of the alkylamino side-chain N atom are delocalized onto the triazine ring (Figure 3.2(a)), which gives the bond between the triazine ring and the alkylamino N atom partial double-bond character. The partial double-bond character restricts the rotation of the side chain and causes atrazine to exist in four conformational isomers that are related by rotation about this bond. At 288 K, the isopropylamino NH and ethylamino NH of each isomer give separate NMR signals for a total of eight peaks in the NH region.[18] At 335 K, interconversion among conformers is rapid, and the signals from the four isomers coalesce to give a single broad peak for each NH group.[18]

In addition, delocalization of the alkylamino lone-pair electrons causes the ring N atoms to behave as hydrogen-bond acceptors and the alkylamino NH protons to behave as hydrogen-bond donors. The existence of both hydrogen-bond donor and hydrogen-bond acceptor groups on atrazine allows atrazine to form a cyclic dimer by hydrogen-bonding to itself in aprotic solvents (Figure 3.2(b)). The calculated chemical shifts for the hydrogen-bonded dimer are downfield of the monomer peaks by 3.5 and 2.5 ppm for the ethylamino and isopropylamino NH protons, respectively. This result suggests that hydrogen-bonding with the ethylamino NH is stronger than

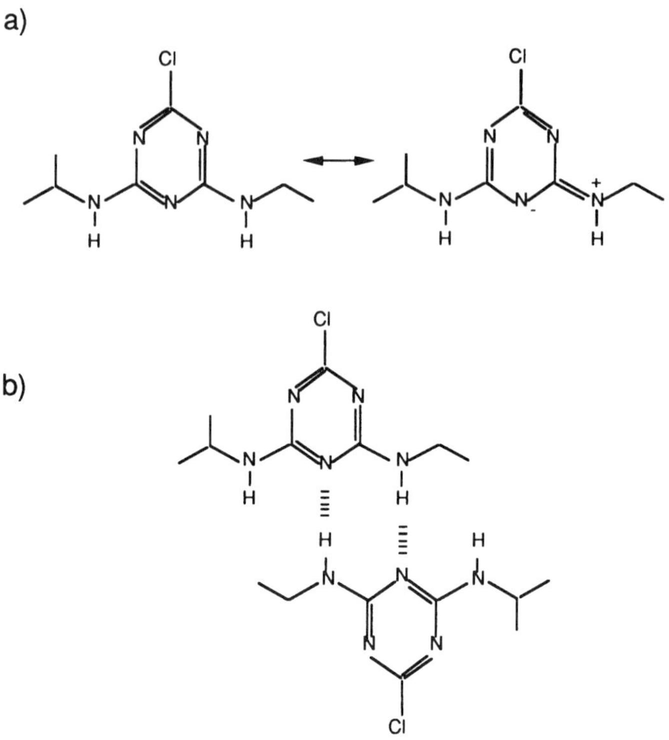

a)

b)

Figure 3.2 (a) Atrazine structure, showing resonance form in which lone-pair electrons from the alkylamino N are delocalized on the triazine ring. Both the ethylamino and the isopropyl N can donate electrons to the triazine ring; (b) atrazine dimer formed by cooperative hydrogen bonding. Modified from ref. 18.

isopropylamino hydrogen-bonding, possibly due to greater steric hindrance for the bulkier isopropyl group.[18]

Atrazine hydrogen-bond formation constants have been calculated from measurements of atrazine NH chemical shifts as a function of the concentration of hydrogen-bond donor or acceptor compounds.[19] The formation constants were used to calculate atrazine's hydrogen-bond donor (α) and hydrogen-bond acceptor (β) parameters using the universal hydrogen-bond donor and acceptor scales proposed by Abraham et al.[21,22] The calculated hydrogen-bond donor and acceptor properties for atrazine ($\alpha = 0.41 \pm 0.02$; $\beta = 0.49 \pm 0.02$) are very similar to the values for acetone ($\alpha = 0.4$; $\beta = 0.5$), which is considered to form hydrogen bonds of moderate strength. Thus, atrazine should form moderately strong hydrogen-bonded complexes with monofunctional donors and acceptors.[19] Such complexes may be sufficiently strong to affect the sorption of atrazine by soil organic matter. Because dipolar and orbital overlap interactions involved in hydrogen-bonding are stronger in solvents with low dielectric constant, atrazine hydrogen-bonding in soils should be strongest within hydrophobic domains of soil organic matter.[20]

Proton NMR also demonstrated that atrazine can form cooperative hydrogen bonds in CCl_4 with compounds such as acetic acid or pyrrolidinone that contain both donor protons and acceptor O or N groups.[20] Formation constants for atrazine hydrogen-bonding to these two compounds were very large, which suggested that simultaneous donor and acceptor interactions reinforce one another (are "cooperative") through orbital overlap and resonance stabilization.[20] Soil organic matter, which contains both hydrogen-bond donor and acceptor groups in a variety of different configurations, may also form cooperative hydrogen bonds with atrazine. Water also may form hydrogen bonds with either atrazine or soil organic matter, but water may not compete well with atrazine for cooperative hydrogen-bonding sites on soil organic matter because water is unable to form cooperative hydrogen bonds with atrazine.[20]

Glyphosate Ion-pairing

The ability of glyphosate (*N*-(phosphonomethyl)glycine mono(isopropylamine) salt; Figure 3.3) to control weeds in agricultural fields is inhibited by common divalent cations such as Ca^{2+}, which are present in high concentrations in many natural water supplies. The mechanism by which Ca^{2+} interacts with glyphosate has been studied using a combination of ^1H, ^{13}C, and ^{31}P NMR experiments.[23] The HOD peak at 4.65 ppm was used as the internal standard for the proton NMR spectra; ^{13}C and ^{31}P spectra were calibrated with an external standard without bulk susceptibility corrections. When Ca^{2+} was added to technical-grade glyphosate (H^+ form) in a 1:1 molar ratio, the proton NMR doublet for methylene protons adjacent to the carboxyl group shifted upfield by about 0.3 ppm; the doublet for methylene protons adjacent to the phosphonate group shifted upfield by about 0.15 ppm. The signal to noise (S/N) ratio was very poor in the ^{13}C NMR spectra, but it was apparent that the peak for the carboxyl carbon shifted downfield, as did the ^{31}P NMR peak for the phosphorus atom of glyphosate. When Ca^{2+} was added to the isopropylamine formulation of glyphosate, the ^{31}P NMR peak again moved upfield; the upfield shift was greater for 4:1 Ca:glyphosate than for 1:1 Ca:glyphosate. The proton NMR peak for methylene protons adjacent to the carboxyl group was masked by other components of the commercial isopropylamine formulation, but the doublet for methylene protons adjacent to the phosphonate group shifted upfield about 0.05 ppm. The results for both technical-grade glyphosate and the isopropylamine formulation led the authors to conclude that Ca^{2+} associates with both the phosphonate and carboxylate groups of glyphosate.[23]

Approximately 70 min after Ca^{2+} was added to technical-grade glyphosate, a second doublet appeared for methylene protons adjacent to

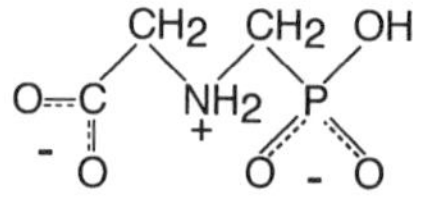

Figure 3.3 Chemical structure of glyphosate anion. Calcium can associate with either the carboxyl or the phosphonate group.

phosphonate.[23] The original doublet gradually disappeared, whereas the second doublet increased in intensity and moved downfield by about 0.05 ppm during the next 13 h. There were no significant changes in the peak for the protons adjacent to the carboxyl group. These results suggested that the original Ca–glyphosate complexes rearranged to thermodynamically more stable complexes as the solutions were allowed to age for about 14 h. The authors proposed that the initial Ca–glyphosate association may involve a combination of monodentate Ca–phosphonate complexes, monodentate Ca–carboxylate complexes, and bidentate phosphonate–Ca–carboxylate complexes. In the latter case, a stable six-membered ring would be formed. At later times, glyphosate may act as a tridentate or tetradentate ligand, with one or two phosphonate oxygen atoms, one carboxylate oxygen, and the nitrogen lone-pair electrons from a single glyphosate molecule all associated with Ca^{2+}.[23] Interaction between glyphosate and Ca^{2+}, which caused decreased plant uptake of glyphosate, can be overcome by addition of ammonium sulfate. In the presence of both Ca^{2+} and NH_4^+, the proton and ^{31}P NMR peaks shifted to reflect partial displacement of Ca^{2+} by NH_4^+.[23]

Solution-phase Complexes Involving DDT

Proton NMR also has yielded mechanistic information about complexes between DDT (1,1,1-trichloro-2,2-bis(p-chlorophenyl)ethane; Figure 3.4(a)) and two types of complexing agents: aromatic π-donors and aliphatic O- and N-containing compounds.[24,25] The CCl_3 group, which is strongly electrophilic, withdraws electron density from the aromatic ring and from the benzhydryl C–H bond (Figure 3.4(a)). Consequently, the aromatic ring of DDT is a π-acceptor and the benzhydryl H atom can form complexes with lone-pair electrons. The benzhydryl proton NMR peak moves downfield when electron donors such as tributyl phosphate, diethyl carbamate, ethyl acetate, or dimethylurea are added to DDT solutions in CCl_4.[25] In contrast, when DDT forms π–π complexes with relatively electron-rich aromatic compounds such as naphthalene and benzene, the chemical shifts of the benzhydryl protons and of the *ortho* and *meta* aromatic protons move upfield. Based on complexation constants calculated from the NMR titration data, π–π complexes are two or three times stronger than complexes between electron donor compounds and the benzhydryl proton.[25]

a)

b)

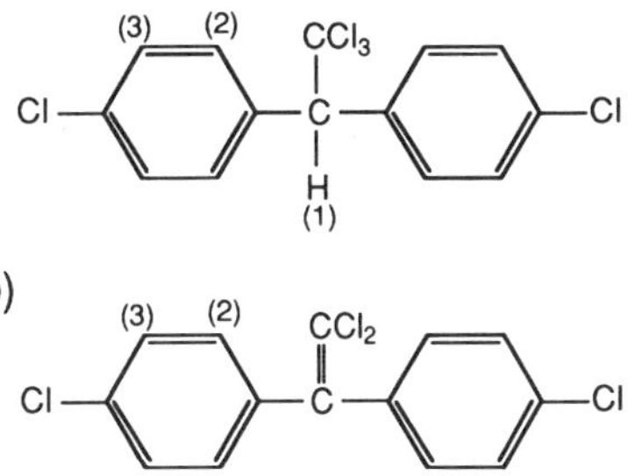

Figure 3.4 (a) Structure of DDT, showing the electrophilic CCl_3 group, the benzhydryl H (1), and the aromatic protons (2 and 3); (b) Structure of DDE (1,1-dichloro-2,2-bis(p-chlorophenyl)ethene), which has no benzhydryl proton.

Interactions Between Organic Compounds and Natural Macromolecules

DDT–Macromolecule Interactions

Information about DDT–protein interactions has been obtained by calculating effective or apparent relaxation rates ($1/T_2$; equation (7)) from NMR linewidths of the aromatic and benzhydryl protons of DDT. As the concentration of bovine and human serum albumin (BSA and HSA) increased from 0 to 4%, relaxation rates for the aromatic *ortho* protons of DDT increased by a factor of five, but only doubled for the benzhydryl proton.[26] The relatively large effect of BSA and HSA on aromatic proton relaxation rates suggests that BSA and HSA interact preferentially with aromatic π electrons of DDT, not with the benzhydryl proton. In addition, the NMR signals for the aromatic protons shifted upfield in the presence of BSA and HSA,[26] which is further evidence[25] that these two proteins interact with DDT by π–π interactions.

Proton NMR has also been used to determine the mechanisms by which p,p'-DDT affects membrane permeability, using lecithin as a simple analog of a phospholipid membrane.[27] When p,p'-DDT was added to lecithin suspensions in CCl_4 or $CDCl_3$, the NMR peaks for the benzhydryl and *ortho* ring protons of DDT shifted downfield, as did the signal for protons adjacent to the phosphate and $N^+(CH_3)_3$ groups of lecithin. This effect of lecithin on DDT chemical shifts[27] was the same as reported above for solution-phase interactions between the benzhydryl H atom of DDT and lone-pair electrons on tributyl phosphate and other electron donor groups.[25] Thus, the NMR results for nonaqueous DDT–lecithin suspensions suggest that the benzhydryl proton of DDT interacts with the phosphate group of lecithin.[27] Additional evidence for complexes between lecithin and the benzhydryl proton of DDT was obtained from proton NMR experiments with DDE (Figure 3.4(b)), which has no benzhydryl H atom. When DDE was added to lecithin suspensions, the NMR signals of DDE and lecithin were unchanged from those in pure solutions, which suggests that DDT and lecithin do not interact unless a benzhydryl proton is present.[27]

No mechanistic information could be obtained in aqueous lecithin–DDT suspensions because the proton NMR peaks for lecithin and H_2O overlapped with peaks from DDT protons, which were very broad.[27] As was discussed at the beginning of this chapter, ^{19}F NMR experiments using fluorinated analogs of DDT would overcome the peak overlap problem encountered in proton NMR. No such experiments appear to have been done, although similar ^{19}F NMR experiments in pharmacology have demonstrated the potential value of this approach. For example, ^{19}F NMR has shown that 5-fluoro-L-tryptophan and L-tryptophan compete with one another for two binding sites on HSA.[14] One of the sites has a very strong affinity for fluorotryptophan and produces an NMR signal downfield of the peak for the free species; fluorotryptophan sorbed on this type of site is in slow exchange with dissolved fluorotryptophan. On the second type of site, the sorbed and aqueous fluorotryptophan are in rapid exchange; a single broad peak is observed for the sorbed and dissolved species.[14] Although ^{19}F NMR has not been used widely in environmental chemistry, it may prove to be a valuable tool for studying interactions between environmental contaminants and natural sorbents, particularly for slowly exchanging (strongly sorbed) contaminants.

Molecular Interactions Between Fluoroaromatic Solutes and Humic Acid

Background

Proton NMR experiments described above[18–20] show that atrazine is both a hydrogen-bond donor and a hydrogen-bond acceptor. Infrared spectroscopic data[28–30] suggest that atrazine and other nitrogen-containing herbicides may hydrogen-bond to humic acid. ^{19}F NMR may provide additional information about hydrogen bonding or dipole–dipole interactions between humic acid and specific functional groups of herbicides because ^{19}F chemical shifts of *para*-substituted fluoroaromatic solutes are extremely sensitive to polar and hydrogen-bonding interactions involving *para* substituents such as $-NH_2$, $-OH$, $-COOH$, $-NO_2$, etc.[31,32] The anticipated effect of polar and hydrogen-bonding interactions on the electron distribution in fluorobenzene and three *para*-substituted fluoroaromatic solutes is shown schematically in Figure 3.5.

To verify that polar and hydrogen-bonding interactions affect ^{19}F chemical shifts as predicted in Figure 3.5 and to compare results obtained with two different internal chemical-shift standards, ^{19}F chemical shifts (δ_{sample}) of 0.5 mM fluorobenzene, 4-fluorophenol, 4-fluoroaniline, and 4-fluoronitrobenzene were measured relative to both internal *p*-DFB and internal C_6F_6 in cyclohexane, acetonitrile, and water. Cyclohexane is nonpolar; acetonitrile is polar and a weak hydrogen-bond acceptor; water is polar and acts both as a hydrogen-bond donor and a hydrogen-bond acceptor. Values of $\Delta\delta_{medium}$ were

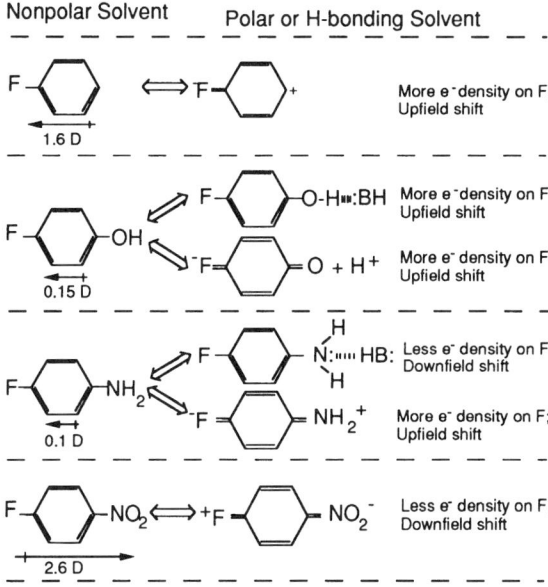

Figure 3.5 Anticipated effects of dipolar interactions and hydrogen bonding on electron distribution and ^{19}F chemical shifts of four fluoroaromatic solutes.

calculated with equation (1) using cyclohexane as the reference solvent. The effects of acetonitrile and water on $\Delta\delta_{medium}$ (Figure 3.6) were consistent with the expected trends shown in Figure 3.5, regardless of whether p-DFB (Figure 3.6(a)) or C_6F_6 (Figure 3.6(b)) was used as the internal reference compound. In acetonitrile, p-DFB and C_6F_6 gave identical results, but in water, values of $\Delta\delta_{medium}$ obtained with internal p-DFB were about 0.5 ppm upfield (less positive) than when C_6F_6 was the internal reference compound. The discrepancy between C_6F_6 and p-DFB for aqueous samples was predicted in Table 3.1 and the accompanying text (pp. 56–57).

Polar and hydrogen-bonding interactions had a greater effect on the ^{19}F chemical shifts of 4-fluoroaniline (a hydrogen-bond acceptor) and 4-fluoronitrobenzene (strongly polar) than on the chemical shifts of 4-fluorophenol and fluorobenzene (Figures 3.6(a) and (b)). Thus, 4-fluoroaniline and 4-fluoronitrobenzene were chosen as probe molecules to use in preliminary investigations to determine whether ^{19}F NMR can provide mechanistic information about interactions between humic acid and organic compounds. A series of ^{19}F NMR experiments in humate–organic solvent slurries, described below, were conducted to test the following mechanistic hypotheses:

Hypothesis 1: 4-Fluoronitrobenzene is sorbed by polar regions of humic acid.

Effect on δ: Sorption causes downfield shifts.

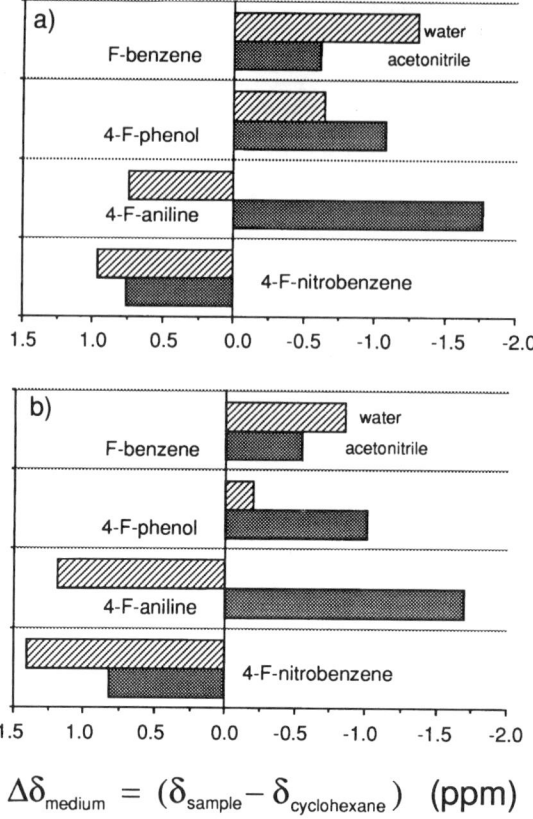

$$\Delta\delta_{medium} = (\delta_{sample} - \delta_{cyclohexane}) \; (ppm)$$

Figure 3.6 Effects of hydrogen bonding (water) and polar (water and acetonitrile) interactions on ^{19}F NMR $\Delta\delta_{medium}$ values of fluoroaromatic solutes. (a) δ_{sample} and $\delta_{cyclohexane}$ measured relative to internal 1,4-difluorobenzene; (b) δ_{sample} and $\delta_{cyclohexane}$ measured relative to internal hexafluorobenzene. F-, fluoro.

Hypothesis 2: 4-Fluoroaniline is sorbed by polar, non-hydrogen-bonding regions of humic acid.
Effect on δ: Sorption causes upfield shifts.
Hypothesis 3: The NH_2 group of 4-fluoroaniline forms hydrogen-bonds with humic acid.
Effect on δ: Sorption causes downfield shifts.

Methods

Aldrich humic acid (sodium salt) was purified by suspending the solid material in 0.01 M NaCl and immediately adding HCl to adjust the pH to 7. The suspensions were centrifuged repeatedly (27 000 *g*) to remove insoluble material (mainly quartz and clay minerals). The humate was purified further by batch reaction with Na-saturated Chelex resin at pH 6.5. The purified

Na-humate was dialyzed against 0.001 M CsCl and freeze-dried. Cesium-saturated humate was used to minimize the affinity of the humate for water; Cs^+ has a low hydration energy. Aldrich humate was used, even though it is a poor analog of soil humate,[33] because the purpose of these NMR experiments was to determine whether ^{19}F NMR can give information about humate–solute interaction mechanisms, not to characterize the reactivity of soil humic acid.

To test the hypotheses listed above, a 70 mg/mL slurry of purified Cs-humate was prepared in cyclohexane by weighing 50 mg of purified Cs-humate into a 5 mm NMR tube, then adding 0.7 mL of a cyclohexane-d_{12} solution that contained 0.5 mM hexafluorobenzene (internal reference)and 2.5 mM each of 4-fluoronitrobenzene, 4-fluoroaniline, fluorobenzene, and 1-fluoronaphthalene. Preliminary experiments showed that chemical shifts measured in a multisolute mixture were the same as when measured individually. To determine whether 4-fluoroaniline forms hydrogen bonds with humate (hypothesis 3), a 300 mg/mL Cs-humate slurry was prepared in acetonitrile with 1 mM hexafluorobenzene and 4-fluoroaniline. The other solutes were omitted because acetonitrile is a polar solvent and cannot be used to study dipolar humate–solute interactions. No aqueous humate spectra were acquired because polar and hydrogen-bonding interactions between these solutes and H_2O would make it impossible to test the hypotheses listed above and to determine whether the solutes interact with polar or hydrogen-bonding regions of humic acid.

All spectra were obtained at 470.268 MHz on a Varian VXR-500 NMR spectrometer with a 5 μs pulse (40°) and 0.8 s delay. All spectra were referenced to internal hexafluorobenzene. Samples were spun at 20 Hz, which caused most of the humate to settle to the bottom of the NMR tube because Cs-humate is not soluble in organic solvents. This produced a sharp phase boundary between a nearly clear suspension at the top of the sample and a humate-rich region (> 60% v/v humate) at the bottom of the tube. The NMR tube was positioned in the spectrometer to obtain spectra with different proportions of the clear suspension and the humate-rich region in the receiver coil (see insets in Figures 3.7 and 3.8). Approximately 400 and 660 transients were collected for the cyclohexane– and acetonitrile–humate slurries, respectively.

Results

Interpretation of solution-state NMR results for humate solutions and suspensions depends upon whether or not humate-bound fluoroaromatics are sufficiently mobile to produce a detectable NMR signal using a high resolution solution-state NMR spectrometer. Sorption of a small molecule by a macromolecule causes fast T_2 relaxation and broad NMR peaks for the sorbed species. For extremely immobile species, the NMR signal decays so rapidly that the sorbed or bound species cannot be detected. In such cases, the chemical shift and peak width reflect only the unbound, solution-phase

a)

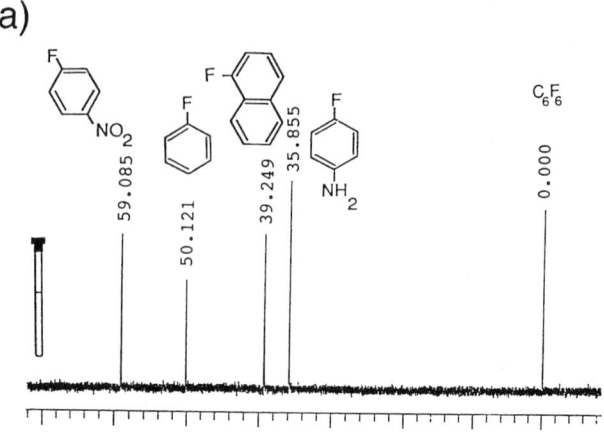

b)

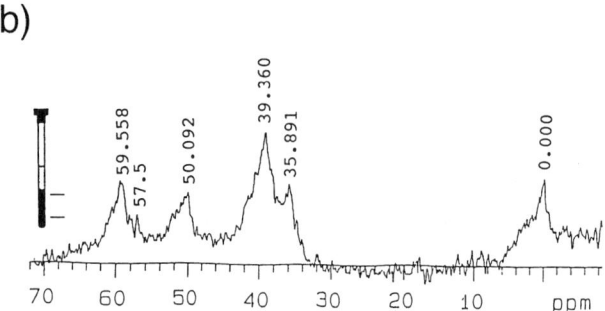

Figure 3.7 ¹⁹F NMR spectra of fluoroaromatic solutes in cyclohexane and humate–cyclohexane slurries. Peak positions are measured relative to internal C_6F_6 at 0 ppm. Humate concentration and the position of the sample in the receiver coil are shown schematically in the insets. (a) Pure cyclohexane or (b) the bottom part of the humate–cyclohexane mixture (predominantly humate) was in the receiver coil.

species, and the peak area would be decreased in proportion to the amount of sorption.

Fluorine NMR peaks for fluoroaromatic compounds in a cyclohexane–humate slurry are much broader than for the same solutes in pure cyclohexane (Figures 3.7(a) and (b)). Similarly, peaks for fluoroaniline and hexafluorobenzene in the humate-rich region of the acetonitrile–humate slurry are extremely broad (Figure 3.8(b)). If the solutes are sorbed by humate and if sorbed species are mobile enough to produce a peak, then line broadening could be caused by (1) reduced mobility of humate-bound species; (2) chemical exchange between free and bound species; (3) heterogeneity of sorption sites in humate; (4) interaction with paramagnetic sites. If humate-bound species are too immobile to be detected, then broad peaks may be caused

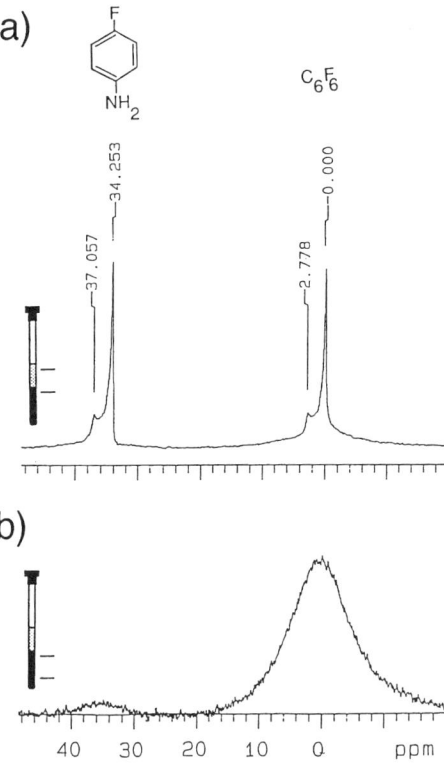

Figure 3.8 ^{19}F NMR spectra of 4-fluoroaniline and fluorobenzene in humate–acetonitrile slurries. Peak positions are measured relative to internal C_6F_6. Humate concentration and the position of the sample in the receiver coil are shown schematically in the insets. (a) 25% dilute suspension and 75% concentrated humate slurry; (b) 100% concentrated humate slurry.

by the physical and chemical heterogeneity of the slurries and by variation in solution-phase magnetic susceptibilities in proximity to humate particles. The present preliminary experiments do not distinguish between these possible sources of line broadening.

In the cyclohexane–humate slurry there is no evidence for dipole–dipole or hydrogen-bonding interactions between humate and fluorobenzene, fluoroaniline, or fluoronaphthalene; the chemical shifts in the cyclohexane–humate slurry were the same as in pure cyclohexane (Figures 3.7(a) and (b)). Either the sorbed-phase species are too immobile to contribute to the observed peaks, or only a small fraction of each of these compounds is sorbed by humate from cyclohexane. In contrast, the fluoronitrobenzene peak shifted 0.5 ppm downfield of its position in pure cyclohexane, though the breadth and asymmetry of the peak reduce the accuracy with which its position can be determined. The 0.5 ppm downfield shift, if real, supports the hypothesis that a fraction of 4-fluoronitrobenzene associates with polar regions of humate, at least in cyclohexane–humate slurries. A new peak appeared at about 57.5 ppm, possibly due to fluoronitrosobenzene, though further research is needed to verify this hypothesis. Chloronitrosobenzene is an intermediate that forms when chloronitrobenzene is reduced to chloroaniline in the presence of organic matter;[34] a similar reaction is likely with fluoronitrobenzene.

When the acetonitrile–humate sample was positioned so that mainly clear suspension was in the receiver coil (Figure 3.8(a)), both fluoroaniline and hexafluorobenzene produced two peaks separated by 2.8 ppm. That the separation was identical for both nonpolar hexafluorobenzene and polar, hydrogen-bond-accepting fluoroaniline suggests that the two peaks result from two molecular environments that differ only in bulk susceptibility; the two peaks are not caused by slow exchange between sorbed and free species. The large upfield peak for each compound is most probably caused by molecules in the relatively clear suspension, since mainly clear suspension was inside the receiver coil. The downfield peak, then, would be due to the humate-rich slurry. The fluoroaniline peaks are 34.26 ± 0.03 ppm downfield of the C_6F_6 peak for the same environment. In pure acetonitrile, fluoroaniline was 34.16 ppm downfield of C_6F_6 (data not shown). If a large fraction of 4-fluoroaniline had formed hydrogen bonds with hydrogen-bond donor sites on humate, the peak in the humate-rich slurry should have been much more than 0.1 ppm downfield of that in pure acetonitrile (cf. Figure 3.6). Thus, the 4-fluoroaniline NMR peak in acetonitrile–humate slurries is dominated by acetonitrile-solvated fluoroaniline, not by fluoroaniline that is hydrogen-bonded to humate.

In the spectrum from the humate-rich region at the bottom of the NMR tube, only a single broad peak was present for each compound (Figure 3.8(b)), and peak positions could not be determined accurately. Comparison of Figure 3.8(a) with Figure 3.8(b) reveals that the fluoroaniline peak area decreased much more with increasing humate concentration than did the hexafluorobenzene peak area. If humate-sorbed species are too immobile to contribute to the observed NMR peaks, then the greater decrease in fluoroaniline peak area is consistent with greater sorption and immobilization of fluoroaniline due to humate–fluoroaniline hydrogen bonding, and with greater mobility of hexafluorobenzene due to either less sorption of hexafluorobenzene or to C_6F_6 partitioning into liquid-like hydrophobic regions of organic matter, where the mobility may be sufficient for sorbed hexafluorobenzene to contribute to the NMR peak. Additional experiments are needed to determine conclusively whether humate-bound species can be observed in solution-state NMR experiments such as those described here. One approach would be to quantify the amount of each compound that is sorbed, then to use a solute in a coaxial capillary tube as a peak-area standard in order to determine whether the amount of sorption is proportional to the decrease in peak area.

Conclusions

Solution-phase NMR is very useful for determining interaction mechanisms, including hydrogen-bonding, dipole–dipole, and π–π interactions. Proton and fluorine NMR also can provide information about the chemical nature of sorption sites and about sorbate–sorbent interactions. Fluorine NMR has the advantage of giving simple spectra in the presence of complex, heterogeneous sorbents, whereas ^{1}H NMR peaks from protons in a complex macromolecular sorbent frequently overlap with and obscure the signal from sorbed molecules. If humate-bound species are too immobile for standard solution-phase NMR experiments, then magic angle spinning could be used to obtain information about sorbed species.

^{19}F NMR could be used in heterogeneous natural systems to identify degradation products,[34–36] since changes in molecular structure produce greater changes in chemical shifts than do intermolecular interactions. The ^{19}F NMR spectrum of a humate–cyclohexane slurry (Figure 3.7(b)) indicates that 4-fluoronitrobenzene may be reduced to fluoronitrosobenzene in the presence of high humate concentrations, although further research is needed to confirm this reaction.

Acknowledgments The NMR data presented in Table 3.1 and the section "Molecular Interactions Between Fluoroaromatic Solutes and Humic Acid" were obtained on instrumentation that was purchased in part with funds from NIH grant #1-S10-RR04750, NSF grant #CHE-88-00770, and NSF grant #92-13241. The research was supported in part by a grant from Chevron Oil Field Research under agreement LH-2655, and in part by a Michigan State University All-University Research Initiation Grant.

References

1. Chiou, C. T., in *Reactions and Movement of Organic Chemicals in Soils*, Sawhney, B. L., and Brown, K. (eds), Soil Science Society of America, Madison, WI, 1989, pp. 1–29.
2. Green, R. E., and Karickhoff, S. W., in *Pesticides in the Soil Environment: Processes, Impacts, and Modeling,* Cheng, H. H. (ed), Soil Science Society of America, Madison, WI, 1990, pp. 79–101.
3. Chiou, C. T., Peters, L. J., and Freed, V. H., *Science*, 206, 831, 1979.
4. Chiou, C. T., Porter, P. E., and Schmedding, D. W., *Environ. Sci. Technol.* 17, 227, 1983.
5. Hassett, J. J., Means, J. C., and Banwart, W. L., US Environmental Protection Agency, EPA-600/3-80-041, US Government Printing Office, Washington DC, 1980.
6. Karickhoff, S. E., *J. Hydraul. Eng.* 110, 707, 1984.
7. Rao, P. S. C., and Davidson, J. M., in *Environmental Impact of Nonpoint Source Pollution*, Overcash, M. R., and Davidson, J. M. (eds), Ann Arbor Scientific, Ann Arbor, MI, 1980, pp. 23–67.

8. Welhouse, F. J., Barak, P., and Bleam, W. F., *J. Phys. Chem.* 97, 11583, 1993.
9. Laszlo, P., *Progr. NMR Spectrosc.* 3, 231, 1967.
10. Emsley, J. W., and Phillips, L., *Progr. NMR Spectrosc.* 7, 1, 1971.
11. Emsley, J. W., and Phillips, L., *Mol. Phys.* 11, 437, 1966.
12. Foster, R., and Fyfe, C. A., *Chem. Commun.* 642, 1965.
13. Witanowski, M., Stefaniak, L., and Webb, G. A., *Annu. Rep. NMR Spectrosc.* 18, 1, 1986.
14. Gerig, J. T., and Klinkenborg, J. C., *J. Am. Chem. Soc.* 102, 4267, 1980.
15. Packer, K. J., *Progr. NMR Spectrosc.* 3, 87, 1967.
16. Pfeifer, H., *Phys. Rep. (Phys. Lett. C)* 7, 293, 1976.
17. Wasylishen, R. E., in *NMR Spectroscopy Techniques*, Vol. 5, Dybowski, C., and Lichter, R. L. (eds), Marcel Dekker, New York, NY, 1987, p. 45.
18. Welhouse, G. J., and Bleam, W. F., *Eviron. Sci. Technol.* 26, 959, 1992.
19. Welhouse, G. J., and Bleam, W. F., *Environ. Sci. Technol.* 27, 494, 1993.
20. Welhouse, G. J., and Bleam, W. F., *Environ. Sci. Technol.* 27, 500, 1993.
21. Abraham, M., Grellier, P., Prior, D., Duce, P., Morris, J. and Taylor, P., *J. Chem. Soc., Perkin Trans.* 2 699, 1989.
22. Abraham, M., Grellier, P., Prior, D., Morris, J., and Taylor, P., *J. Chem. Soc., Perkin Trans.* 2 521, 1990.
23. Thelen, K. D., Jackson, E. P., and Penner, D., *Proc. North Central Weed Sci. Soc.* 47, 108, 1992.
24. Wilson, N. K., *J. Am. Chem. Soc.* 94, 2431, 1972.
25. McKinney, J. D., Wilson, N. K., and Keith, L. H., in *Mass Spectrometry and NMR Spectroscopy in Pesticide Chemistry*, Vol. 4, Haque, R., and Biros, F. J. (eds), Plenum Press, New York, NY, 1974, pp. 139–160.
26. Ross, R. T., and Biros, F. J., in *Mass Spectrometry and NMR Spectroscopy in Pesticide Chemistry*, Vol. 4, Haque, R., and Biros, F. J. (eds), Plenum Press, New York, NY, 1974, pp. 263–272.
27. Haque, R., and Tinsley, I. J., in *Mass Spectrometry and NMR Spectroscopy in Pesticide Chemistry*, Vol. 4, Haque, R., and Biros, F. J. (eds), Plenum Press, New York, NY, 1974, pp. 239–261.
28. Sullivan, J. D., and Felbeck, G. T., *Soil Sci.* 106, 42, 1968.
29. Senesi, N., and Testini, C., *Soil Sci.*, 130, 314, 1980.
30. Senesi, N., Testini, C., and Miano, T. M., *Org. Geochem.* 11, 25, 1987.
31. Taft, R. W., Price, E., Fox, I. R., Lewis, I. C., Andersen, K. K., and Davis, G. T., *J. Am. Chem. Soc.* 85, 3146, 1963.
32. Taft, R. W., Gurka, D., Joris, L., Schleyer, P. V. R., and Rakshys, J. W., *J. Am. Chem. Soc.* 91, 4801, 1969.
33. Malcolm, R. L., and MacCarthy, P., *Environ. Sci. Technol.* 20, 904, 1986.
34. Dunnivant, F. M., and Schwarzenbach, R. P., *Environ. Sci. Technol.* 26, 2133, 1992.
35. Thorn, K. A., Arterburn, J. B., and Mikita, M. A., *Environ. Sci. Technol.* 26, 107, 1992.
36. Hatcher, P. G., Bortiatynski, J. M., Minard, R. D., Dec, J., and Bollag, J.-M., *Environ. Sci. Technol.* 27, 2098, 1993.

4.

A ^{19}F and ^{2}H NMR Spectroscopic Investigation of the Interaction Between Nonionic Organic Contaminants and Dissolved Humic Material

BRUCE E. HERBERT & PAUL M. BERTSCH

Interaction between nonionic organic contaminants (NOC) and natural organic matter strongly influences the fate and transport of NOC in the environment. Microscopic descriptions of NOC–organic matter interaction have been developed based on macroscopic observations of NOC sorption to organic matter and organic solute transport under varying conditions. These models include the partitioning concept describing NOC sorption to organic matter and the concept of intra-organic matter diffusion used to account for nonequilibrium organic solute transport;[1,2] however, little microscopic information exists to validate them. NMR may be a powerful method to gain information and insight concerning NOC–organic matter interaction. Chemical shifts, linewidths, and the magnitude of spin–spin couplings exhibited in the NMR spectra of a given nucleus are dependent on the characteristics of the surrounding media and therefore can be used to study the NOC–organic matter interaction.[3]

NMR characterization of the chemical interaction between NOC and organic matter can potentially provide information on important aspects of the sorption mechanism. This information may be useful to explain the influence of organic matter characteristics on NOC sorption[4,5] and the mechanisms controlling nonequilibrium sorption of NOC to organic matter and mineral phases,[4-6] and to evaluate different conceptual models of natural humic material, including the polymer concept,[7,8] where humics are considered to be flexible linear polyelectrolytes, and the micelle concept,[9] which considers humics to be aggregates of simple organic compounds, oligomers, and humic molecules ultimately forming micellar-type structures.

NMR has been used to study the interactions of small organic solutes with macromolecules and organized organic assemblies. Fluorine-19 NMR has been particularly useful to study these interactions because this magnetically active nucleus is 100% abundant, has a high sensitivity (83% of ^{1}H) and large chemical shift range, and exhibits no background signal.[10] Several NMR studies have followed the interactions between fluorinated anesthetics and phospholipids, sodium dodecyl sulfate (SDS) micelles, and intact rabbit brain tissue.[11–14] Generally, these studies have indicated both that sorption sites are less polar than aqueous solutions and that there is decreased mobility of the sorbed solute. Fluorine-19 NMR has also been used to characterize the interactions between fluorine-containing ligands, such as fluorotryptophan, and human serum albumin (HSA).[15,16] The natures of both specific and nonspecific interactions between the ligands and HSA have been inferred from these studies.

The interaction between a small molecule and a macromolecule is often reflected in NMR relaxation times of nuclei associated with the small molecule. For example, the relaxation rate of deuterium, whose relaxation is dominated by the quadrupolar relaxation mechanism, is quite sensitive to solute–macromolecule interactions.[17] The chemical shifts of deuterium and the proton are nearly identical, simplifying interpretation of deuterium NMR spectra. The interactions of two deuterated pyridine dinucleotides and lactate dehydrogenase were studied using changes in ^{2}H relaxation rates.[18] It was shown that ^{2}H NMR was much more sensitive than ^{1}H or ^{13}C NMR to these interactions. The binding of deuterated sugars to a lectin has also been followed with this technique.[19]

The objectives of this study were to: (1) apply NMR techniques to the study of the interactions of NOC and natural dissolved humic acid; (2) to evaluate the nature of NOC–humic acid interactions; and (3) to characterize the humic acid microenvironments involved in NOC sorption. Substituted fluorobenzenes, differing in their functional group content, were used as probes to study the NOC–organic matter interaction. Because natural organic matter is molecularly complex, this study also followed changes in the NMR spectra of fluorinated or deuterated organic solutes in model systems of increasing complexity. The model systems used to study the interactions of the labeled solutes included solvents of differing polarity, SDS micelle solutions, and solutions of the bile salt, taurocholic acid. Calibrating the response of NMR spectra in simpler model systems provides a framework from which to draw conclusions about NOC interactions with natural organic matter.

Experimental Methods

Materials

p-Fluoronitrobenzene (98%, Sigma Chemical, St. Louis, MO), p-fluorotoluene (98%, Sigma), p-fluorophenol (98%, Aldrich Chemical, Milwaukee, WI), hexafluorobenzene (Sigma), fluorobenzene-d_5 (98 atom% D, Aldrich), sodium dodecyl sulfate (99%, Sigma), taurocholic acid–Na$^+$ (98%, Sigma), and humic acid (lot no. O1816HH, Aldrich) were used as received. Organic solvents were HPLC-grade.

Natural organic matter was extracted from the 0 to 10 cm depth of a Lakeland soil (Typic Quartzpsamment, Aiken County, SC) and sifted at field moist conditions to pass a 2 mm sieve. Humic acid was extracted using a modified method from Schnitzer.[20] After extraction with 0.1 M NaOH, the humic material was repeatedly fractionated with 6 M HCl and 2 M NaOH. These procedures were performed under argon. The humic acid was then dialyzed in tubing of 1000 nominal molecular weight cutoff (NMWCO) against H^+-saturated Chelex 100 resin (Biorad, Richmond, CA) until the $AgNO_3$ test for Cl^- was negative. The humic acid solutions were used immediately in the NMR experiments.

Preparation of NMR Solutions

The fluorinated and deuterated probes were dissolved in isopropanol to give approximately 0.5 M solutions. Additions of 5-50 mL of the stock solutions were made to the experimental solutions such that the final concentrations of the probe molecules were approximately 1 mM.

Aliquots of the stock solutions of the fluorobenzene probes were dissolved in 5 mL of several organic solvents: hexane, cyclohexane, acetonitrile, methylene chloride, isopropanol, ethanol, methanol, and water. SDS and taurocholic acid (TA) were dissolved in deionized water containing 17% D_2O, to prepare a 0.1 M SDS and 0.047 M TA solution, respectively. The ionic strength of the taurocholic acid solution was adjusted to 0.1M with NaCl. Humic acid solutions were prepared using both Aldrich humic acid and the extracted Lakeland humic acid. Aldrich humic acid (0.1002 g) was dissolved in 50.075 mL H_2O at pH 10. The solutions were sonified for 1 h. Aliquots were adjusted to pH 6.8 and 2.2, and ionic strength of 0.1 M, with HCl and NaCl, respectively. The contributions of the humic acid to total ionic strength were ignored. D_2O was added to give 17% D_2O and a final organic matter concentration of 1640 mg L^{-1}. The organic carbon concentration of the Aldrich humic acid solutions equaled 1090 mg L^{-1}, based on average reported percentage C for Aldrich humic acids.[21,22] Aliquots of the Lakeland humic acid solutions were adjusted to pH 6.7, 4.1, and 2.2 with NaOH and HCl. The samples adjusted to pH 2.2 were observed to precipitate. Ionic strength in the humic acid solutions was adjusted to 0.1 M with NaCl, ignoring the contribution of organics. Final concentrations were 3742 mg L^{-1} and 592 mg L^{-1} organic carbon. Organic carbon concentrations were measured with a Shimadzu TOC-500 organic carbon analyzer (Shimadzu Corp., Kyoto, Japan) using dextrose standards prepared with HPLC-grade water.

NMR Analyses

Fluorine-19 NMR spectra were recorded at 25 °C using a Bruker WP-200 FT spectrometer operating at 188.3 MHz with a 15 mm fluorine probe, and a

Bruker AM-300 FT spectrometer operating at 282.4 MHz with a 5 mm fluor-ine probe. Chemical shifts and linewidths were acquired on the WP-200 using a 6.7 μs pulse (approximately 34°) with no delay. Other instrumental para-meters for this spectrometer included an acquisition time of 0.73 s and a spectral width of 5649 Hz, using 8192 data points. Spectra were processed with an exponential multiplication corresponding to line broadening of 1.3 Hz. Instrumental parameters used for data collection on the AM-300 included a pulse of 6.5 ms with no delay, an acquisition time of 1.64 s, and a spectral width of 8472 Hz, using 16 384 data points. These spectra were processed with an exponential multiplication corresponding to line broad-ening of 0.5 Hz. Aqueous solutions analyzed with ^{19}F NMR spectroscopy were run locked on D$_2$O. Samples of organic solvents were run unlocked. In the organic solvent studies, the spectrometer was locked on a sample containing D$_2$O, then the nonaqueous sample was analyzed unlocked, after which the D$_2$O sample was reanalyzed unlocked. If the fluorine peak in the D$_2$O sample changed by greater than 0.2 ppm during the experiment, the samples were reanalyzed. Chemical shifts of the fluorobenzenes in the differ-ent solutions were referenced against external solutions of hexafluorobenzene in the same solutions. Hexafluorobenezene was chosen because it is similar in molecular shape to the compounds of interest and has a zero dipole moment. External referencing with hexafluorobenzene corrects for bulk magnetic sus-ceptibility changes.

Deuterium NMR spectra were recorded at 25 °C on a Bruker AM-500 operating at 76.8 MHz and utilizing a 5 mm broadband probe. Spectra were acquired using a 3 ms pulse (approximately 55°) with no delay. Acquisition times and spectral widths for these experiments were 1.78 s, and 2304 Hz, respectively. Spectra were processed with an exponential multiplication cor-responding to line broadening of 0.5 Hz. The chemical shifts of the ^{2}H sam-ples were referenced indirectly to tetramethylsilane via the observed shift of a neat solution of D$_2$O. The spectrometer was initially locked on D$_2$O while recording the ^{1}H spectra, then run unlocked while recording the ^{2}H spectra. An internal standard of D$_2$O was used to correct the ^{2}H NMR spectra for inhomogeneities and bulk magnetic susceptibility changes.

Results and Discussions

NMR Chemical Shifts

Chemical shifts of substituted fluorobenzene in different organic solvents were used as model systems to understand the observed shifts of these solutes in micelle and humic acid solutions. The polarity, polarizability, and hydrogen-bonding characteristics of the solvents used in these experiments, as described by solvatochromic parameters,[23,24] are given in Table 4.1. $E_t(30)$ is a polarity scale in which higher numbers indicate a more polar solvent,[25] while π^* is a dipolarity/polarizability scale in which higher numbers indicate greater ability

Table 4.1 Solvatochromic Parameters for Solvent Polarity $(E_t(30))$, Polarity/Polarizability (π^*), Hydrogen-bond Donor Strength (α), and Hydrogen-bond Acceptor Strength $(\beta)^a$

Solvent	$E_t(30)$	π^*	α	β
Cyclohexane	30.9	0.00	0.00	0.00
Hexane	31.0	−0.08	0.00	0.00
Methylene chloride	40.7	0.82	0.30	0.00
Acetonitrile	45.6	0.75	0.19	0.31
Propan-2-ol	48.4	0.48	0.76	≈ 0.95
Ethanol	51.9	0.54	0.83	0.77
Methanol	55.4	0.60	0.93	≈ 0.6
Water	63.1	1.09	1.17	≈ 0.5

^{a} From Taft et al.,[23] Kamlet et al.,[24] and March.[25]

of a solvent to stabilize charge through a dielectric effect.[23] The propensity of hydrogen-bond formation by various solvents is given by the α and β parameters. Hydrogen-bond donor and acceptor strength is given by α and β, respectively, where higher numbers indicate stronger hydrogen bonding.[24]

Fluorine-19 chemical shifts of p-fluoronitrobenzene, p-fluorophenol, p-fluorotoluene, and fluorobenzene in various solvents are given in Table 4.2. Downfield (more positive) chemical shifts of fluoronitrobenzene in solvents of higher polarity were observed. Compared to cyclohexane, chemical shifts of fluorophenol in water and methanol were downfield, while small upfield shifts were observed for acetonitrile, methylene chloride, and propan-2-ol. Generally, upfield (more negative) chemical shifts were observed for fluorotoluene and fluorobenzene in solvents of greater polarity, though methanol did not follow this pattern. The largest chemical shifts of the four probe solutes were observed for fluoronitrobenzene, probably due to the greater electron withdrawing power of the nitro group.[25]

Chemical shift is a function of the shielding constants, which describe the electron density around the nucleus.[3] The observed shielding constant (σ_{obsd}) is a sum of the shielding constants arising from a number of contributions:

$$\sigma_{obsd} = \sigma_g + \sigma_b + \sigma_a + \sigma_w + \sigma_e + \sigma_c \qquad (1)$$

where σ_g is the contribution to the total shielding of the isolated gaseous molecule and will be constant during media changes, σ_b is that due to bulk magnetic susceptibility, σ_a is the anisotropy of the susceptibility of the solvent molecules, σ_w is the contribution due to van der Waals interactions, σ_e is the contribution due to the reaction field due to the medium, and s_c is the contribution due to specific solute–solvent interactions such as hydrogen bonds. Outside the contribution of σ_g, which is unaffected by surrounding molecules, the largest effect on fluorine chemical shifts is due to σ_w.[26] For this reason,

Table 4.2 ^{19}F NMR Chemical Shifts of Fluoronitrobenzene (FNB), Fluorophenol (FP), Fluorotoluene (FT), and Fluorobenzene (FB) in Solvents of Varying Polarity[a]

	δ (ppm)			
Solution	FNB	FP	FT	FB
Cyclohexane	59.1	50.1	48.9	49.5
Hexane	58.9	50.3	48.9	49.4
Methylene chloride	59.7	49.9	47.0	48.3
Acetonitrile	59.0	49.4	47.6	49.0
Propan-2-ol	59.8	49.7	48.5	49.1
Ethanol	60.0	50.1	48.6	n.d.[b]
Methanol	60.1	50.4	50.0	49.3
Water	60.5	50.7	48.4	48.8

[a] Shifts were referenced to external solutions of hexafluorobenzene.
[b] Not determined.

downfield chemical shifts of ^{19}F in solvents of increasing polarizability are generally observed.[26]

When comparing ^{19}F NMR chemical shifts in two solvents or two environments, the choice of reference is more important than in ^{1}H NMR, because some of the terms in equation (1) are larger than in the corresponding equation describing the shielding of a proton. External referencing can result in large observed chemical shifts. With an appropriate internal reference, the differences between shielding constants of two solutions can be minimized, thereby allowing the causes of chemical shifts differences between two samples to be inferred.[26]

Using an internal reference of the same shape as the solute of interest corrects for $\Delta\sigma_b$, $\Delta\sigma_a$, $\Delta\sigma_w$.[26] The shielding constant σ_e for the reference molecule is minimized if the reference compound has minimal dipole and quadrupolar moment, thereby allowing an interpretation of the $\Delta\sigma_e$ term. Given an internal reference that meets these criteria, the observed chemical shift for a nucleus in different environments is due to differences in $\Delta\sigma_e$, $\Delta\sigma_c$, and possibly $\Delta\sigma_w$. This allows the interpretation of chemical shifts differences based on knowledge of the physical chemistry of the different media. Both fluorobenzene and hexafluorobenzene have been used as internal standards for aromatic fluorine compounds.[26,27]

Several theoretical models have been proposed to explain solvent influences on the chemical shifts of nuclei corrected with suitable internal references. Emsley and Phillips[26] proposed that chemical shift differences for a nucleus in different solvents can be explained by reaction field theory, which describes the effect of solvent characteristics on shielding constants in terms of bulk solvent properties such as dielectric constant. Taft et al.[27] considered chemical shift differences observed for a number of substituted fluorobenzenes to be the result of the solvent's effect on the relative contributions of different dipolar resonance forms to the ground electronic state of the sub-stituted fluorobenzenes. Because the intramolecular electronic distributions for the various resonance forms are different, the electronic densities shielding the fluorine nucleus in the resonance forms of the solutes are also different. In the case when chemical shifts are dominated by $\Delta\sigma_e$, Giam and Lyle[28] explained solvent effects on chemical shifts as being the result of differences in the reaction fields of solvents of varying polarity. The solvents' reaction fields align with the dipole of the solutes and influence σ_e. Solvents of greater polarity have greater reaction fields and therefore a larger effect on σ_e. Whether the reaction fields increase or decrease electron density around the fluorine nucleus depends on the position of the nucleus in relation to the direction of the solute dipole moment.

Chemical shift is also dependent on the formation of hydrogen bonds between the solute and solvent through the σ_c term. In this study *para*-sub-stituted nitro-, phenol-, methyl-, and hydrogen-fluorobenzenes were used to study the effects of the functional groups on the fluorine chemical shift. Because the functional groups were in the *para* position relative to the fluor-ine nuclei, electron density around the fluorine nuclei is affected by hydrogen-bond formation between the nonfluoro functional groups and the solvent molecules. The nitro group is a strong hydrogen-bond acceptor and can with-draw electron density from the fluorine upon hydrogen-bond formation. This would cause a downfield chemical shift in hydrogen-bond-donating solvents. The phenol group can be both a hydrogen-bond acceptor or donator; there-fore hydrogen bond formation can cause both upfield and downfield shifts in different solvents.

The chemical shifts of *p*-fluoronitrobenzene were generally correlated with solvent polarity ($E_t(30)$) and hydrogen-bond donor strength (α) (Table 4.3). This effect could be due to the stabilization of dipolar resonance forms in the more polar solvents or the formation of hydrogen bonds, both of which would have decreased shielding around the fluorine nucleus. The observed chemical shift of fluoronitrobenzene in acetonitrile was lower than predicted based on solvent polarity, but was reasonable if chemical shift was primarily dependent on the hydrogen-bond donor strength of the solvents. The down-field chemical shifts of *p*-fluorophenol in water and methanol relative to cyclohexane are consistent with the formation of hydrogen bonds between the phenol group and the solvent, where the solvent is acting as a hydrogen-bond donor. The upfield chemical shifts of fluorophenol in acetonitrile, methylene chloride, and propan-2-ol are consistent with the stabilization of the fluorophenol resonance forms with increased charge separation in the solvents of higher polarity, which would increase electron density around the fluorine nucleus and cause an upfield shift. The small, upfield chemical shifts of *p*-fluorotoluene and fluorobenzene correlate with the polarity/polar-izability of the solvents, probably due to stabilization of the resonance forms of greater charge separation leading to increased electron density around the fluorine nucleus and upfield chemical shifts in solvents of higher polarity.

Table 4.3 ^{19}F NMR Chemical Shifts (δ) and Linewidths (ν) at Half-height of Fluor-otoluene (FT), Fluorophenol (FP), and Fluoronitrobenzene (FNB) in Micelle and Humic Acid (HA) Solutions

Solution	Concn.[a] (mg L^{-1})	pH	FT		FP		FNB	
			δ (ppm)	$\nu_{1/2}$ (Hz)	δ (ppm)	$\nu_{1/2}$ (Hz)	δ (ppm)	$\nu_{1/2}$ (Hz)
H$_2$O	0	6.8	48.37	3.4	50.65	4.1	60.53	n.d.
SDS	28850	6.8	48.26	3.4	50.38	n.d.	60.60	n.d.
TA	26885	6.8	48.82	n.d.	50.40	n.d.	60.46	n.d.
Aldrich HA	1094	6.8	48.35	31.2	50.63	14.9	60.53	23.9
Aldrich HA	1094	2.1	48.36	55.6	50.63	17.7	60.50	42.9
Lakeland HA	3742	6.7	48.36	33.9	50.60	21.7	60.51	33.3
Lakeland HA	3742	2.1	48.32	9.5	50.64	6.8	60.50	15.3
Lakeland HA	592	6.7	48.41	31.8	50.66	n.d.	60.55	n.d.
Lakeland HA	592	4.1	48.40	45.4	50.66	n.d.	60.55	n.d.

[a] Concentration of the HA solutions is mg L^{-1} organic carbon.
[b] Not determined.

Chemical shifts can also be caused by experimental error such as magnetic inhomogeneities, the presence of paramagnetics, and small temperature shifts. Since the range of chemical shifts observed for the fluorobenzene compounds in different solvents varied between 0 and 1.9 ppm, the possibility that experimental error was the dominant factor causing the observed chemical shifts in this study must be considered. Several factors make us believe that the chemical shifts observed in this study are valid: for example, we obtained good day-to-day reproducibility, the results were similar to published chemical shifts of fluorobenzenes in similar systems,[23,24,26] and our observations corresponded with established hypotheses of solvent influence on chemical shift.

The observed chemical shifts of p-fluoronitrobenzene, p-fluorophenol, and p-fluorotoluene in aqueous, micelle, and humic acid solutions are presented in Table 4.3 and Figures 4.1 and 4.2. In the following section, chemical shifts in the micelle and humic acid solutions are referenced to the shift observed for the fluorinated solutes in aqueous solutions. The chemical shifts of fluoronitrobenzene in SDS were shifted downfield relative to the solute in aqueous solution. A relatively large change in the chemical shift of fluoronitrobenzene at the critical micelle concentration (CMC) of SDS was observed, indicating the importance of the formation of the micelles (Figure 4.1). Since downfield chemical shifts were inferred to be effects of increased hydrogen bonding in the various model solvents, fluoronitrobenzene is probably solubilized in the palisade region of the micelle, where there are higher concentrations of water and charged functional groups available for hydrogen-bond formation than in the interior of the micelle.[31]

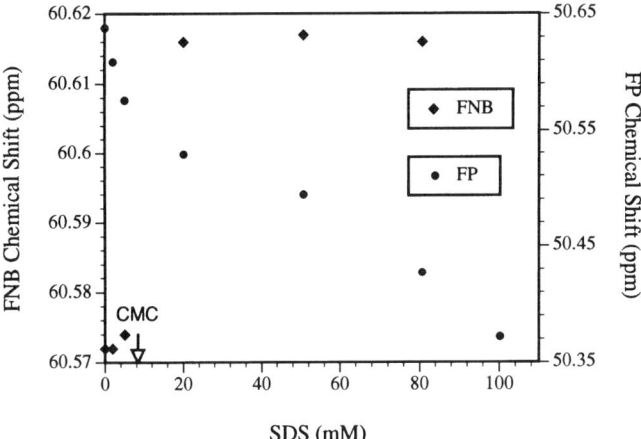

Figure 4.1   Chemical shift of *p*-fluoronitrobenzene (FNB) and *p*-fluorophenol (FP) in sodium dodecyl sulfate (SDS) referenced to an internal solution of hexafluorobenzene. The critical micelle concentration (CMC) is indicated by the arrow.

Small upfield shifts were observed for fluorophenol in SDS and fluoronitrobenzene and fluorophenol in taurocholic acid (Table 4.3). Chemical shifts were a function of surfactant concentration, with the largest changes in chemical shift observed around the CMCs of SDS and taurocholic acid (Figures 4.1, 4.2). Upfield shifts of the solutes, indicative of increased electron shielding around the fluorine, are consistent with partitioning of the solutes into the

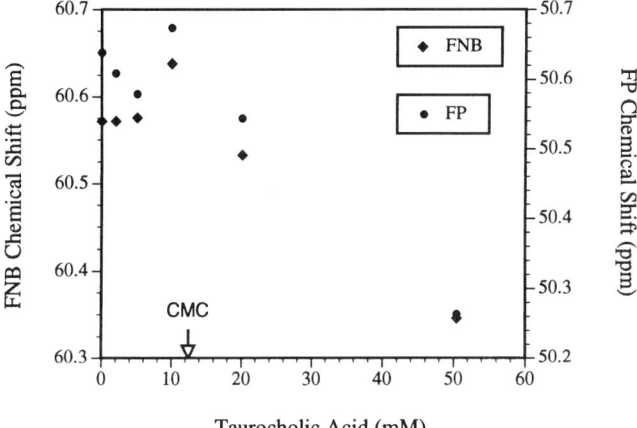

Figure 4.2   Chemical shift of *p*-fluoronitrobenzene (FNB) and *p*-fluorophenol (FP) in taurocholic acid referenced to an internal solution of hexafluorobenzene. The approximate critical micelle concentration (CMC) is indicated by the arrow.

nonpolar interiors of the micelles and a decrease in hydrogen-bond formation between the solutes and water.[30]

The chemical shifts of the fluorinated solutes in humic acid solutions were much smaller than observed for the micelle solutions (Table 4.3). When shifts were present, their directions were consistent with the corresponding shifts observed in micelle solutions. Variation of humic acid concentration or pH had a small effect on chemical shift with no apparent CMC effect, as observed in the surfactant solutions. Because the pH 2.2 Lakeland humic acid solutions precipitated, removing the organic aggregates from the receiver coil, these solutions do not reflect the characteristics of the humic material. The small observed chemical shifts indicate that sorption microenvironments are nearly as polar as bulk water and that hydrogen bonding is still significant compared to the microenvironments in the micelle aggregates involved in fluorobenzene sorption.

Like the observed chemical shifts of the fluorobenzenes in different solvents, the chemical shifts of the fluorobenzene solutes in surfactant and humic acid solutions were generally small and the influence of experimental parameters such as those listed above can not be discounted. The use of internal standards, the purity of the surfactants, and the reproducibility of the chemical shifts support the conclusion that the chemical shifts of the fluorobenzenes in the surfactant solutions were a result of real changes in the polarity or hydrogen bonding potential of different microenvironments. On the other hand, the small chemical shifts of the fluorobenzene compounds in humic solutions must be interpreted with extreme caution, given the known existence of paramagnetics in humic materials.

Chemical shift differences between the free and sorbed state of solutes in macromolecule and micelle solutions have been used to follow the sorption process in biochemical studies. Fluorine-19 NMR studies on the solubilization of fluorinated anesthetics in phospholipid membranes and SDS micelles showed single upfield fluorine shifts of less than 1 ppm, using external referencing, upon solubilization of the anesthetics indicating nonspecific interactions between the fluorinated anesthetics and the phospholipid membranes.[11,12] More specific interactions were indicated by spectra of methoxyfluran solubilized in phospholipids, which exhibited separate peaks for the solubilized and free anesthetic.[11] Fluorine-19 NMR studies have also been used to infer specific interactions between fluorinated ligands and proteins. Gerig and Klinkenborg[15] and Jenkins and Lauffer[16] studied the interactions between 5-fluorotryptophan and human serum albumin. Large chemical shift differences of 1 to 9 ppm were observed between the free and bound fluorinated ligand. The peaks representing the bound ligands were downfield of the free ligand peaks in these studies. Comparison of these reported chemical shifts and those observed in this study indicates that the interaction between the fluorobenzenes and humic acid or surfactants is generally weaker than those observed in most of the biochemical systems.

NMR Relaxation Studies

Table 4.3 also reports the ^{19}F peak widths at half-height ($\nu_{1/2}$) for fluoroni-trobenzene, fluorophenol, and fluorotoluene in aqueous, micelle, and humic acid solutions. Peak widths at half-height for the fluorinated solutes increased from 3 and 4 Hz in aqueous solutions to between 15 and 50 Hz in humic acid solutions. The largest increase in $\nu_{1/2}$ was observed for fluorotoluene, the solute of lowest polarity. The pH of the humic acid solutions had a significant effect on $\nu_{1/2}$. A decrease in pH in the Aldrich and Lakeland humic acid solutions increased $\nu_{1/2}$, except for the Lakeland sample at pH 2.1. Extensive precipitation in this solution removed most of the humic acid from the receiver coil, resulting in small observed $\nu_{1/2}$ due to the overwhelming concentration of soluble species. The concentration of organic carbon had a smaller effect on $\nu_{1/2}$ than pH. Increases in linewidth have been observed in other systems: the solubilization of fluorinated anesthetics in phospholipids resulted in an increase in $\nu_{1/2}$ of 18 Hz,[11] while the specific binding of fluorotrypto-phan to HSA resulted in an increase in linewidth of 55 Hz.[15]

Line broadening can be the result of two processes. The relaxation rate of the nuclei can change, or the nuclei can undergo exchange between two magnetic environments during an intermediate time scale.[3,15] The exchange of a fluorinated anesthetic between the interior of SDS micelles and the bulk phase was fast on the NMR time scale.[12] On the other hand, the exchange of fluoro-tryptophan between the free and bound state in a protein solution was slow on the NMR time scale.[15] Because the sorption of the fluorobenzenes to the micelles and humic acids used in this study most closely resembles the sorption of fluorinated anesthetics to SDS, exchanges of the fluorobenzenes between the sorbed and free states are probably fast on the NMR time scale.[31] This indicates that line broadening is probably the result of changes in the relaxation rates of the fluorine nucleus in the micelle and humic acid solutions. As will be shown below, line broadening can be related to the correlation times of the nucleus, which describe the molecular motion of the nucleus. The linewidths of the fluorobenzene solutes reported in Table 4.3 indicate that these solutes undergo a restriction of their molecular mobility after sorption to humic acid. Quantitative descriptions of the change in correlation times were performed using ^{2}H NMR.

This interpretation of the cause for line broadening assumes that other environmental factors besides molecular motion are not affecting linewidths. As in the case of chemical shifts, factors such as the presence of paramag-netics can cause line broadening. Compared to aqueous systems, the small changes in line broadening in surfactant solutions indicate that these other factors are not so important in these systems. On the other hand, the presence of paramagnetics in the humic solutions and their effect on line broadening must be considered. Our assumption that line broadening is not caused by paramagnetics is based on the lack of changes in the D_2O linewidths between the humic and aqueous solutions. D_2O linewidths would also be increased if paramagnetics were present. The chemical shift data presented earlier support the hypothesis that fluorobenzenes are not being separated from water on the microscopic scale by sorption to the humic macromolecule and therefore would experience similar exposure to paramagnetics.

Deuterium is well suited for NMR relaxation studies of organic solutes sorbed to macromolecules because ^2H relaxation is dominated by the quadrupolar relaxation mechanism.[17] Assuming that line broadening is a function of change in rotational motion, then the rotational correlation times, τ_c, can be calculated for the free and sorbed state of fluorobenzene in micelle and humic acid solutions. The $\nu_{1/2}$ values of fluorobenzene-d_5 in aqueous, micelle, and humic acid solutions are given in Table 4.4. The ^2H spectra of fluorobenzene-d_5 in the different solutions are given in Figures 4.3 and 4.4, with the peaks used to measure $\nu_{1/2}$ indicated by arrows. Linewidths of ^2H NMR spectra of fluorobenzene-d_5 increased from approximately 2 Hz in aqueous solution to 12 to 18 Hz in humic acid solutions, depending on pH (Table 4.4). The $\nu_{1/2}$ of fluorobenzene-d_5 in the micelle solutions was much smaller than the $\nu_{1/2}$ observed for the humic acid solutions. Similarly to the observed line broadening of the ^{19}F spectra, a decrease in pH of the humic acid solutions increased $\nu_{1/2}$ of the fluorobenzene-d_5 spectra.

The relaxation of ^2H is dominated by the quadrupolar relaxation mechanism and exhibits a single exponential decay even outside the extreme narrowing condition.[17,19,32] The spin–spin relaxation time, T_2, of fluorobenzene-d_5 in aqueous solution can be calculated from the linewidth by assuming the extreme narrowing condition, which should be applicable in aqueous solution:

$$T_2 = \frac{1}{\pi \nu_{1/2}} \tag{2}$$

$$\frac{1}{T_2} = \frac{3}{40} \frac{2I+1}{I^2(2I-1)} \left(\frac{e^2 Qq}{h}\right)^2 \tau_c \tag{3}$$

where T_2 is the spin–spin relaxation time, $\nu_{1/2}$ is the linewidth at half-height, I is the nuclear quantum spin number, $(e^2 Qq/h)$ is the quadrupolar coupling constant, and τ_c is the correlation time, as defined previously. The calculated correlation time for fluorobenzene-d_5 in aqueous solution (Table 4.4), based on equations (2) and (3), was 1.4×10^{-11} s, which is reasonable for a freely rotating solute.[3]

In systems where the ^2H nucleus is rapidly exchanging between two environments of different correlation times, the relationship between the observed change in linewidth, $\Delta\nu_{1/2}$, and the correlation times in the two environments can be described with the following equations:[18]

$$\Delta\nu_{1/2} = \nu_{sorbed} - \nu_{free} \tag{4}$$

Table 4.4 ^{2}H NMR Linewidths and Calculated Correlation Times of Fluorobenzene-d_5 in Micelle and Humic Acid (HA) Solutions

Solution	Concn.[a]	pH	D$_2$O (Hz)	FB (Hz)	τ_c (10^{-11} s)[b]
H$_2$O	0	6.8	1.76	2.25	1.4
SDS	0.1 M	6.8	1.58	2.32	1.6
Taurocholic acid	0.47 M	6.8	1.62	3.83	2.9
Lakeland HA	592 mg L^{-1}	6.7	1.90	12.28	299
Lakeland HA	592 mg L^{-1}	4.1	2.00	18.43	482

[a] Concentration of the humic acid solutions is mg L^{-1} organic carbon.
[b] Correlation times were calculated assuming $\chi = 0.75$ and $\chi = 0.021$ for the micelle and humic solutions, respectively.

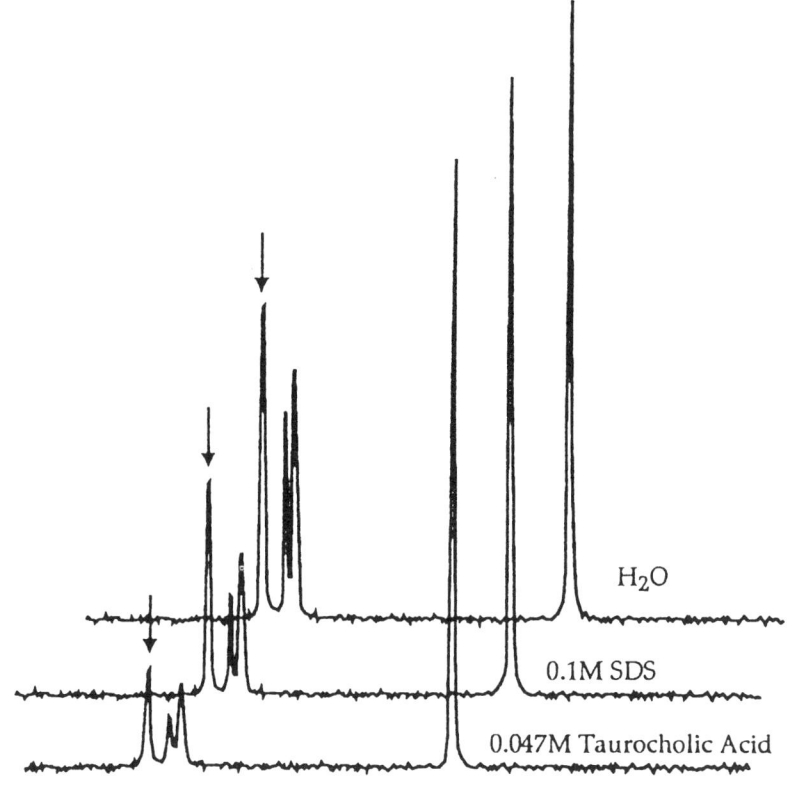

Figure 4.3 ^{2}H NMR of fluorobenzene-d_5 in micelle solutions with D$_2$O as reference. Linewidths at half-height for the fluorobenzene peaks indicated by arrows are 2.25 Hz, 2.5 Hz, and 4.0 Hz for the H$_2$O, SDS, and taurocholic acid solutions, respectively.

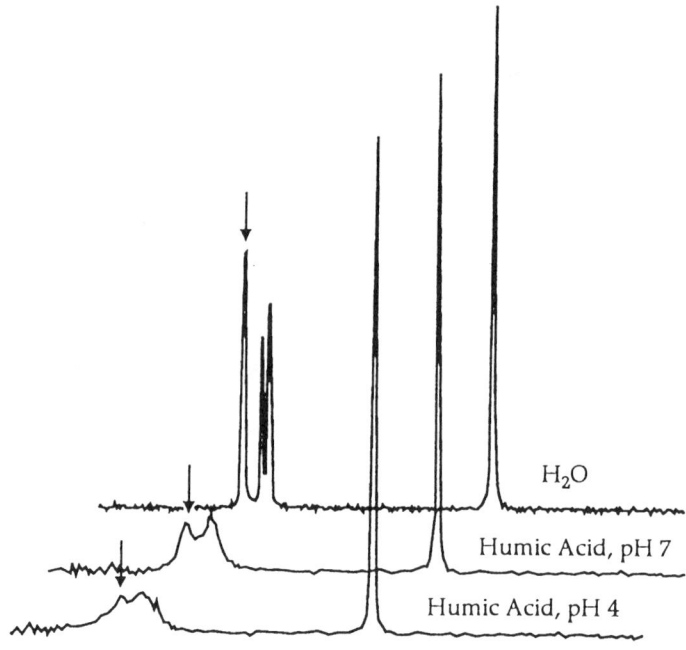

Figure 4.4 ^{2}H NMR of fluorobenzene-d_5 in Lakeland humic acid solutions with D_2O as reference. Linewidths at half-height for the fluorobenzene peaks indicated by arrows are 2.25 Hz, 12.1 Hz, and 18.2 Hz for the H_2O, humic acid pH 7, and humic acid pH 4 solutions, respectively.

$$\Delta\nu_{1/2} = \frac{3}{\pi 8}\left(\frac{e^2 Qq}{h}\right)^2 \left[\tau_f(1-\chi) + \frac{\chi}{20}\left(6\tau_s + \frac{10\tau_s}{1+\omega^2\tau_s^2} + \frac{4\tau_s}{1+4\omega^2\tau_s^2)}\right)\right] \quad (5)$$

where ω is the Larmor frequency, χ is the mole fraction of sorbed solute, and τ_f and τ_s are the free and sorbed correlation times, respectively. The mole fraction of solubilized fluorobenzene-d_5 in 0.1 M SDS ($\chi = 0.75$) was assumed to be similar to the value found by Stilbs[33] for benzene solubilized by 0.1 M SDS. The mole fraction of fluorobenzene-d_5 sorbed to dissolved humic acid was calculated using the following equation:

$$\chi = \frac{K_{oc}C_{DOC}}{1 + K_{oc}C_{DOC}} \quad (6)$$

where K_{oc} is the fluorobenzene-d_5 partition coefficient normalized to constant organic carbon concentration. K_{oc} was calculated based on the fluorobenzene log K_{ow} value of 2.27 and the relationship between log K_{ow} and log K_{om} developed by Chiou,[34] i.e., log $K_{om} = 0.904$ log $K_{ow} - 0.779$. K_{om} is the partition coefficient normalized to constant organic matter concentration. The K_{oc} was calculated from K_{om} using the assumption that Lakeland humic acid has an organic carbon content of 52%. The calculated rotational correlation times for the sorbed fluorobenzene-d_5 in humic acid and surfactant solutions are given in Table 4.4. The values of τ_c in Table 4.4 for the humic acid and surfactant solutions correspond to τ_s in equation (5).

The correlation times τ_s of fluorobenzene-d_5 in humic acid solutions, assuming $\chi = 0.021$, were approximately 3×10^{-9} s and 5×10^{-9} s for the pH 6.7 and pH 4.1 humic acid solutions, respectively (Table 4.4), while the fluorobenzene-d_5 correlation times in micelle solutions were not much larger than those calculated for the solute in aqueous solution. The correlation times of fluorobenzene-d_5 sorbed to humic acid are on the same order of magnitude as those observed for larger polymers of approximately 10^5 MW,[35] and for sugar sorbed to lectin,[19] and are an order of magnitude smaller than the observed correlation times for ligands sorbed to proteins.[18]

Given the assumptions used to calculate the mole fraction of fluorobenzene-d_5 sorbed to humic acid, the relaxation data based on the observed $\nu_{1/2}$ showed that the motion of the sorbed fluorobenzene-d_5 was restricted to a rate expected for the humic acid aggregates themselves, and was much slower than observed for the solute solubilized by SDS or taurocholic acid micelles. The influence of pH on the calculated correlation times demonstrates the effect of increasing aggregation of humic acid on the mobility of sorbed fluorobenzene-d_5.[7] Conceptually, small organic solutes, such as fluorobenzene, could exist within a three-dimensional cage-like environment. As the cage contracts due to pH changes, the mobility of the organic solute is decreased.

Summary

Following chemical shifts and line broadening, ^{19}F and ^{2}H NMR were used to characterize the interaction between nonionic organic compounds (NOC) and natural organic matter. The ^{19}F and ^{2}H NMR spectra of the substituted fluorobenzenes fluorobenzene-d_5, p-fluoronitrobenzene, p-fluorophenol, and p-fluorotoluene were collected in organic solvents of varying polarity, in micelle solutions, and in solutions of natural organic matter. The organic solvents and micelle solutions represented model systems of increasing complexity and were useful in providing a framework for interpreting NMR spectra of organic solutes sorbed to the molecularly complex natural organic matter. Chemical shifts of fluoronitrobenzene and fluorophenol in organic solvents showed the importance of hydrogen-bond formation between the solutes and the solvents. The chemical shifts of the substituted fluorobenzenes in surfactant solutions indicated that the solutes partitioned into the lower-polarity environment of the interior of the micelles and decreased their hydrogen bonding with water, except for fluoronitrobenzene in SDS, which appeared to sorb to the palisade region of the micelle where hydrogen bonding with the charged functional groups of the SDS molecule is possible. The small chemical shifts of the fluorobenzenes in humic acid solutions indicated that the molecular environment of the solutes was fairly similar, in terms of polarity and hydrogen-bond formation potential, to the aqueous solution. Hydrogen bonds could be formed between the fluorobenzene solutes and water molecules within the three-dimensional macromolecular framework or between the fluorobenzene solutes and charged functional groups on the humic macromolecules themselves.

Line broadening, of both ^{19}F and ^{2}H NMR spectra, in solutions of micelles and humic acids was observed. Observed ^{2}H line broadening was used to calculate correlation times of the sorbed fluorobenzene-d_5; these times represent the molecular motion of the solute. It was found that sorbed fluorobenzene-d_5 had correlation times (10^{-9} s) on the same order of magnitude as observed for large-molecular-weight polymers, suggesting that the solute was strongly sorbed to the humic acid molecules. A decrease in pH was found to restrict further the motion of the sorbed solute through an increase in humic acid aggregation. It was hypothesized that the sorbed organic molecule existed within a three-dimensional organic matter cage, which decreased in size with decreasing pH. This resulted in a decrease in observed solute mobility. The differences between chemical shifts and correlation times of the fluorobenzene probes solubilized by micelles and sorbed by humic acid suggest that the sorption process in these two systems may be inherently different, contrary to the hypotheses of Wershaw.[9] Humic materials are inherently more complex, with macromolecules of much higher molecular weight, than the surfactant micelles.

Acknowledgments The authors thank Drs. Gary Mills and Doug Hunter of the Savannah River Ecology Laboratory, Dr. Ron Garber of the University of South Carolina—Columbia, and Dr. Sharon Anderson of Michigan State University for their valuable discussions and suggestions. This research was partially funded by contract DE-AC09-76SR00819 between the University of Georgia and the US Department of Energy. The NMR spectrometers of the Analytical Center of the University of South Carolina were made available through funding from NSF grants CHE-8904942 and CHE-8411172 and NIH grant 1-S10-RR02425-1.

References

1. Curtis, G. P., Reinhard, M., and Roberts, P. V., in *Geomchemical Processes at Mineral Surfaces*, David, J. A., and Hayes, K. F. (eds), ACS Symp. Ser. No. 323, American Chemical Society, Washington DC, 1986, pp. 191–216.
2. Brusseau, M. L., Jessup, R. E., and Rao, P. S. C., *Environ. Sci. Technol.* 25, 134–142, 1991.

3. Becker, E. D., *High Resolution NMR. Theory and Chemical Applications*, 2nd edn, Academic Press, Orlando, FL, 1980.
4. Gauthier, T. D., Shane, E. C., and Guerin, W. F., *Environ. Sci. Technol.* 21, 243–248, 1987.
5. Garbarini, D. R., and Lion, L. W., *Environ. Sci. Technol.* 20, 1263–1269, 1986.
6. Ball, W. P., and Roberts, P. V., *Environ. Sci. Technol.* 25, 1223–1237, 1991.
7. Ghosh, K., and Schnitzer, M., *Soil Sci.* 129, 266–276, 1980.
8. Freeman, D. H., and Chung, L. S., *Science*, 214, 790–792, 1981.
9. Wershaw, R. L., *J. Contam. Hydrol.* 1, 29–45, 1986.
10. Jenkins, B. G., *Life Sci.* 48, 1227–1240, 1991.
11. Koehler, K. A., Jain, M. K., Stone, E. E., Fossel, E. T., and Koehler, L. S., *Biochim. Biophys. Acta* 510, 177–185, 1978.
12. Yoshida, T., Takahashi, K., Kamaya, H., and Ueda, I., *J. Coll. Int. Sci.* 124, 177–185, 1988.
13. Wyrwicz, A. M., Pszenny, M. H., Schofield, J. C., Tillman, P. C., Gordon, R. E., and Martin, P. A., *Science* 222, 428–430, 1983.
14. Evers, A. S., Berkowitz, V. A., and d'Avignon, D. A., *Nature (London)* 328, 157–160, 1987.
15. Gerig, J. T., and Klinkenborg, J. C., *J. Am. Chem. Soc.* 102, 4267–4268, 1980.
16. Jenkins, B. G., and Lauffer, R. B., *Mol. Pharmacol.* 37, 111–118, 1990.
17. Smith, I. P., in *NMR of Newly Accessible Nuclei. Chemically and Biochemically Important Elements*, Vol. 2, Laszlo, P. (ed.), Academic Press, New York, NY, 1983, pp. 1–26.
18. Zens, A. P., Fogle, P. T., Bryson, T. A., Dunlap, R. B., Fisher, R. R., and Ellis, P. D., *J. Am. Chem. Soc.* 98, 3760–3764, 1976.
19. Neurohr, K. J., Lacelle, N., Mantsch, H. H., and Smith, I. C. P., *Biophys. J.* 32, 931–938, 1980.
20. Schnitzer, M., in *Methods of Soil Analysis*, Part 2, Page, A. L., Miller, R. H., and Keeney, D. R. (eds), Agronomy Monograph No. 9, ASA and SSSA, Madison, WI, 1982, pp. 581–594.
21. Chiou, C. T., Kile, D. E., Brinton, T. L., Malcolm, R. L., Leenheer, J. A., and MacCarthy, P., *Environ. Sci. Technol.* 21, 1231–1234, 1987.
22. Malcolm, R. L., and MacCarthy, P., *Environ. Sci. Technol.* 20, 904–911, 1986.
23. Taft, R. W., Abboud, J. L. M., Kamlet, M. J., and Abraham, M. H., *J. Solut. Chem.* 14, 153–175, 1985.
24. Kamlet, M. J., Abboud, J. L. M., Abraham, M. H., and Taft, R. W., *J. Org. Chem.* 48, 2877–2887, 1983.
25. March, J., *Advanced Organic Chemistry, Reactions Mechanisms, and Structure*, 3rd edn, John Wiley, New York, NY, 1985, pp. 17, 360–362.
26. Emsley, J. W., and Phillips, L., *Prog. NMR Spectrosc.* 7, 1–62, 1971.
27. Taft, R. W., Price, E., Fox, I. R., Lewis, I. C., Andersen, K. K., and Davis, G. T., *J. Am. Chem. Soc.* 85, 3146–3156, 1963.
28. Giam, C. S., and Lyle, J. L., *J. Am. Chem. Soc.* 95, 3235–3239, 1973.
29. Reichardt, C., *Solvents and Solvent Effects in Organic Chemistry*, 2nd edn, VCH, Weinheim, Germany, 1988, p. 325.
30. Meyerhoffer, S. M., and McGown, L. B., *Anal. Chem.* 63, 2082–2086, 1991.
31. Wennerstrom, H., and Lindman, B., *Phys. Rep.* 52, 1–86, 1979.
32. Halle, B., and Wennerstrom, H., *J. Magn. Reson.* 44, 89–100, 1981.
33. Stilbs, P., *J. Coll. Int. Sci.* 87, 385–393, 1982.

34. Chiou, C. T., in *Reactions and Movement of Organic Chemicals in Soils*, Sawhney, B. L., and Brown, K. (eds), ASA Special Publication No. 22, ASA, CSSA, and SSSA, Madison, WI, 1989, pp. 1–30.
35. Andersson, T., Drakenberg, T., Forsen, S., Thulin, E., and Sward, M., *J. Am. Chem. Soc.* 104, 576–580, 1982.

5.

Adsorption Isotherms and ^{13}C Solid-state NMR Study of Hazardous Organic Compounds Sorbed on Coal Fly Ash

DANIEL A. NETZEL, FRANCIS P. MIKNIS, DOROTHY C. LANE,

JOSEPH F. ROVANI, JAMES D. COX, & JEFFERY A. CLARK

Fly ash is a by-product from the combustion of coal. The 1985 annual US production was estimated to be about 1×10^8 metric tons. The utilization of fly ash during the 1980s remained stable at about 25% per year. Because of its pozzolanic properties, nearly 50% of the utilized fly ash is consumed in the production of cement and concrete. The vast quantity of fly ash that is not being used and its availability throughout the country and worldwide have motivated research for new uses in commerce and industry.

Little is known of the organic adsorbent properties of fly ash. However, if they are found to be favorable, the potential commercial applications of the adsorptive characteristics of fly ash could include its use as an adsorbent sandwich for organics in combination with landfill or other dump-site liners, in traps for organics in waste waters, in filters for organics in process air streams, and as a stabilizer for organic wastes in drums. Variables that may affect the adsorbability of the fly ash towards organics in water include temperature; solution pH; and interactions between solute molecules and fly ash, and between solvent molecules and fly ash. Thus, there is an essential need to characterize each coal fly ash type to enable potential correlation between coal fly ash structural properties and the effectiveness of the adsorption characteristics of coal fly ash for immobilizing organic hazardous waste compounds.

The composition and properties of pulverized fly ash depend on the type of coal burned and the nature of the combustion process. Thus, fly ashes from different origins may have significantly different sorption properties towards organic compounds of environmental interest. Eastern and western coal fly ashes differ significantly in their physical and chemical properties. The major minerals found in coal fly ash are α-quartz (SiO_2), mullite ($3Al_2O_3 \cdot 2SiO_2$), hematite (Fe_2O_3), magnetite (Fe_3O_4), lime (CaO), and gypsum

$(CaSO_4 \cdot 2H_2O)$.[1] Little is known of the coordination state and distribution of siliceous and aluminous material in coal fly ashes. Most siliceous and aluminous materials in fly ash are amorphous and thus are not detected or quantified by X-ray techniques. X-ray diffraction has been used to characterize crystalline minerals, but is less useful for the study of minerals that are amorphous. Even with crystalline materials, difficulties are encountered in ascertaining the general characteristics of fly ash minerals using X-ray diffraction.[1]

Very few (if any) methods exist to differentiate between fly ashes with regard to the adsorption of organic hazardous waste materials. It is possible, however, to characterize solid materials chemically by nuclear magnetic resonance (NMR) techniques. Solid-state NMR techniques are complementary to X-ray diffraction techniques and, in addition, are ideally suited for both amorphous and crystalline materials. The NMR techniques include multi-nuclear (^1H, ^{13}C, ^{27}Al, and ^{29}Si) single-pulse excitation with magic angle spinning, and cross polarization with magic angle spinning, (SPE/MAS and CP/MAS, respectively), and hydrogen spin–lattice relaxation spectroscopy. These techniques have been used to study coal pyrolysis residue and coal minerals under combustion-related conditions. However, NMR techniques have not been extensively applied to the study of coal fly ash.

There is a considerable amount of information in the literature concerning adsorption of organic compounds on activated charcoal,[2] zeolites,[3,4] silicates,[5] alumina,[6] and clays.[7] Even though solid waste materials that arise from combustion processes, such as coal fly ash and spent oil shale, are composed principally of alkaline-earth and alkali metal oxides, silicates, and aluminosilicates,[8–10] relatively few studies have been conducted using spent oil shale[11–15] and coal fly ash as adsorbents. Bernardin[16] reported that activated carbon made from lignite, wood, or coal having high ash contents has an adsorptive capacity equal to or greater than those materials of lower ash contents. In a review on fossil fuel combustion wastes, Rai et al.[9] suggest that combustion wastes may have a high affinity for sorbing various types of organic compounds.

Little work is known on the use of coal fly ash to adsorb or stabilize organic waste materials. Recently, Lane[17] performed adsorption studies which show the affinity of certain coal fly ash samples for organic compounds. Sharma et al.[18] discussed in detail the use of coal fly ash as a soil amendment. Their review dealt with adsorption and leaching of inorganic trace elements and the effects of fly ash on soil microorganisms, plant growth, and surface and groundwater.

Adsorption isotherms have been measured for many types of organic compounds using activated carbon as the adsorbent.[2] For coal fly ash which contains numerous minerals, the effects of these minerals on the sorption of organic waste materials are not adequately known or understood. Adsorption isotherms of organic compounds on coal fly ash have not been reported in the literature.

NMR techniques have been used extensively to characterize the adsorption of organic molecules onto various types of surfaces. The application of these techniques involves measuring the changes in nuclear relaxation times and chemical shift values when a molecule is adsorbed onto a surface. NMR has been used to follow changes in the nuclear properties of either the adsorbent or the adsorbate. Changes in the NMR parameters for hydrogen, carbon, and nitrogen are most often measured for adsorbates, and changes in the silicon and aluminum NMR parameters are measured for adsorbents.

Derouane[19] used ^{1}H NMR to study the adsorption of benzene, ethanol, ethylene glycol, and 2,2-dimethylbutane on alumina. Borovkov and Kazansky[20] investigated methanol, ethanol, and propan-2-ol adsorbed on silica gels, Aerosil, alumina, and aluminosilicates. The specific adsorption of these alcohols in the case of alumina was attributed to the interaction with Lewis acid sites. In several papers, studies of the ^{13}C NMR spectra of acetone adsorbed on silica gel have been reported.[21-23] Significant shifts to lower magnetic fields with respect to liquid acetone were observed for the $C=O$ group of acetone molecules adsorbed on silica gel surfaces. Bernstein et al.,[22] using both ^{13}C and ^{15}N NMR, studied the interaction of pyridine molecules with OH groups on a partially dehydroxylated silica gel surface. The bonds formed between the proton of the hydroxyl group and the lone-pair electrons of the nitrogen atom caused a strong ^{15}N NMR chemical shift to a higher field. Only small changes were observed for the ^{13}C NMR chemical shifts, indicating that no significant interaction occurs between the π-electrons of pyridine and the surface sites.

Few, if any, applications of NMR techniques to the study of the adsorption of hazardous organic compounds on silicates, aluminosilicates, and alumina have been reported, and even fewer techniques on coal fly ash have been described.

The objectives of this study were threefold: (1) to characterize the silicate and aluminosilicates in anhydrous coal fly ashes by elemental analyses, X-ray mineral analysis, and solid-state ^{29}Si NMR techniques; (2) to study the adsorption properties for selected organic compound types by determining the Freundlich adsorption isotherms; and (3) to study the sorption characteristics of fly ashes for organic compounds of environmental interest using solid-state ^{13}C NMR techniques.

Experimental

Materials

One coal fly ash was produced at the Dave Johnston Power Plant located near Glenrock, Wyoming, which uses a conventional pulverized coal process. The feed coal was mined near the plant which is located in the Powder River Basin. The other coal fly ash was produced at the Laramie River Power Plant located near Wheatland, Wyoming, which uses a conventional fired unit that includes a calcium-based sorbent for removal of sulfur dioxide from exhaust gases. The feed coal also is mined in the Powder River Basin of Wyoming.

The aromatic hydrocarbons and various chlorobenzenes were purchased as a single solution containing 200 µL each in methanol (Supelco Purgeable Aromatics Mixture 602-M). Pyridine, naphthalene, and tetrachloroethane were purchased as certified neat compounds from different chemical manufacturers. Pentachlorophenol was purchased as a certified stock solution in methanol from Supelco, Inc. Technical and analytical grades of pentachlorophenol were also purchased from J.T. Baker Chemical Co.

Adsorption Isotherm Procedure

Solid–Gas Equilibrium

All the adsorption experiments were performed on the Dave Johnston coal fly ash. Samples were prepared by weighing varying amounts of fly ash (0.1–10 g) directly into 20 mL headspace vials. Varying amounts of the standard mixture containing methanol as the solvent (1.0–50 µL) were pipetted directly onto the fly ash and the vials were immediately sealed.

Five standards were also prepared by pipetting varying amounts (1.0–50 µL) of the standard solution into the headspace vials, and immediately sealing them. Both samples and standards were placed in a box that was rotated end-over-end at ambient temperature for 72 h. The vials were withdrawn and placed in the autosampler for 2 h at 60 °C before the injection procedure was begun. Sample concentrations in 1 mL of headspace injected into the gas chromatograph (GC) were determined by an external standards calculation based on the responses of the five-point initial calibration.

Headspace samples were taken using a Hewlett-Packard Model 19395A Headspace Sampler. For the injection process, sample vials were pressurized for 10 s and a 1.0 mL sample was withdrawn and transferred by helium carrier gas at 15 mL/min to a Hewlett-Packard Model 5890 gas chromatograph. The GC separation was performed on a DB-624 capillary column, 0.53 mm × 30 m, 3.0 µm film thickness. The following oven temperature program was used: initial temperature, 50 °C; initial time, 1.0 min; temperature program rate, 4 °C/min (7.2 °F/min); final temperature 100 °C; final time, 1.5 min. The packed injection port was fitted with a liner to accommodate the GC column, which was held at 150 °C. A photo-ionization detector (HNU Model PI-52 operated at 150 °C with a 10.2 eV lamp) was used.

Solid-Liquid Equilibrium

Pyridine is the only one of the four compounds specifically evaluated that is miscible with water; the other three exhibit minimal solubility in water. The preparation of aqueous solutions of these three compounds required the addition of methanol as a cosolvent. Methanol was limited to 1% of the total volume of the solution to minimize its effect on the adsorption results.

To determine the concentration ranges that were used, individual standard solutions were prepared and analyzed. UV spectra for each compound were obtained using a wavelength scan from 190 to 450 nm (Shimadzu Model UV-265 spectrophotometer). For each compound, the wavelength at maximum absorbance was chosen as the quantitation wavelength, and an appropriate concentration range to be studied was determined (see Table 5.1). Ten concentrations bracketed by the lower and upper concentrations, and one blank, were studied for each compound. For each compound, two sets of solutions were prepared. The first was designated as the standard or control solution set and consisted of aqueous solutions of the organic compound. The second was designated the sample solution set and consisted of the aqueous solutions in contact with the fly ash.

Standard solutions were prepared in 50 mL screw-cap vials. Pyridine standards were prepared directly in water at a final volume of 20.0 mL. The other organics were also prepared at a final volume of 20.0 mL, using 200 µL of standard prepared in methanol added to 19.8 mL of water. Sample solutions were prepared as above and 4.0 g of fly ash was added to each solution.

Standard and sample vials were capped and rotated end-over-end on a mechanical rotator for 24 h to assure complete mixing. The vials were then allowed to sit undisturbed for 4 h so that the fly ash could settle to the bottom. The supernatant solution was withdrawn by pipette for analysis. The spectrophotometer was calibrated using the blank solution and ten standard solutions. Each compound exhibited linear response through these 11 points, so an average response factor was calculated and used for determining analyte concentration in the samples.

NMR Instrumentation and Conditions

^{13}C and ^{29}Si solid-state NMR measurements were made using a Chemagnetics CMX 100/200 solids NMR spectrometer. ^{13}C spectra were obtained at a ^{13}C frequency of 25 MHz using the techniques of single-pulse excitation with magic angle spinning (SPE/MAS) and cross polarization with magic angle spinning (CP/MAS) and ^{1}H decoupling. A pulse width of 6.2 µs was used for all experiments. The contact time for a 12.5 mm-diameter zirconia bullet spinner was varied, as well as the pulse delay and the number of scans. Spinning rates were between 3.5 and 3.6 kHz. The ^{13}C spectra were externally referenced to liquid tetramethylsilane (TMS) based on the solid-state spectrum of hexamethylbenzene (HMB) as the secondary reference and assigning 17.21 ppm to the shift of the aliphatic carbons of HMB relative to liquid TMS.[24]

^{29}Si SPE/MAS and CP/MAS NMR spectra with ^{1}H decoupling were obtained at a ^{29}Si frequency of 39.6 MHz. A 7.5 mm-diameter zirconia bullet spinner was used. Spinning rates were between 3.5 and 3.8 kHz. A 90° pulse

Table 5.1 UV Wavelength and Concentration Range Used for Pyridine, Pentachlorophenol, Naphthalene, and Tetrachloroethane

Compound	Wavelength at Maximum Absorbance, λ (nm)	Concentration Range (μg/mL)
Pyridine	255	0.250–100
Pentachlorophenol	302	1.00–400
Naphthalene	275	0.250–100
Tetrachloroethane	202	5.00–2000

width of 4.0 μs was used for all experiments. The pulse delays were 5 and 10 s
and the number of acquisitions varied from 700 to 16 000. For the CP/MAS
technique, the contact time was 3 ms. All ^{29}Si spectra were externally referenced to liquid TMS based on the solid-state spectrum of sodium 3-(tri-
methylsilyl)propane-1-sulfonate (DSS) as the secondary reference and
assigning 1.46 ppm to the shift of the silicon of DSS relative to liquid TMS.[25]

Results and Discussion

Coal Fly Ash Characterization

Elemental, Mineral, and Oxide Composition of Coal Fly Ash

The elemental and mineral composition of coal fly ash depends upon the
parent coal and the operating conditions of the power plant. Nearly all naturally occurring elements can be found in coal, depending on the sensitivity of
the method of analysis. Table 5.2 lists the concentrations of some of the more
abundant elements found in the coal fly ash from the Dave Johnston and
Laramie River power plants and Table 5.3 lists the mineral compositions
from X-ray diffraction analysis. Of interest to this NMR study are the concentrations of the elements carbon, calcium, silicon, aluminum, and iron.
Carbon, silicon, and aluminum are more abundant in the coal fly ash from
the Dave Johnston power plant than from the Laramie River power plant
coal fly ash (Table 5.2). However, the Laramie River coal fly ash contains
more calcium and iron than the Dave Johnston coal fly ash. Hydrogen was
not found in either of the coal fly ash samples, indicating that the coal fly ash
samples contained very little water and/or organic hydrocarbons. Table 5.4
lists the oxide compositions of the two coal fly ashes.

^{29}Si NMR Investigation of Coal Fly Ashes

The ease by which the solid-state ^{29}Si and ^{27}Al NMR spectra can be obtained
make this technique ideally suited for studying structures of many types of
silicate and aluminosilicate minerals.[26] The use of ^{29}Si and/or ^{27}Al NMR to

Table 5.2 Elemental Composition of Coal Fly Ash[a]

Element	Elemental Composition	
	Dave Johnston Coal Fly Ash	Laramie River Coal Fly Ash
Aluminum	12.3%	8.89%
Antimony	< 19.6	< 19.1
Barium	0.166%	0.606%
Beryllium	< 0.198	< 0.195
Bismuth	< 194	< 200
Boron	404	732
Cadmium	< 1.96	< 1.91
Calcium	12.4%	19.7%
Carbon	1.3%	0.3%
Chromium	98.3	53.0
Cobalt	24.9	34.6
Copper	128	175
Iron	2.51%	3.29%
Lead[b]	103	151
Lithium	35.7	28.9
Magnesium	1.97%	3.65%
Manganese[b]	288	373
Mercury	0.755	0.282
Molybednum[b]	6.04	3.23
Nickel	42.2	48.8
Phosphorus	0.155%	0.282%
Potassium	< 0.849%	< 0.955%
Selenium	10.3	23.3
Silicon	23.3%	14.6%
Silver	1.75	2.04
Sodium	0.134%	0.778%
Strontium	0.123%	0.387%
Sulfur	0.3%	0.9%
Thorium[b]	25.9	33.3
Vanadium	222	163
Zinc	123	91.6

[a] All values in mg/kg unless otherwise indicated. Source of data unpublished WRI Report to DOE.
[b] These elements showed poor NBS SRM recoveries: lead 65%, manganese 64%, molybdenum 67%, and thorium 127%.

study coal fly ash has not been reported. However, several papers discuss the use of ^{29}Si and ^{27}Al to study similar materials.[27–32]

The Q^n and $Q^n(mAl)$ notations are used to represent the anion structural units of silicates and aluminosilicates in both solution and solid states.[26] In this notation, Q represents the silicon atom bonded tetrahedrally to four oxygen atoms and the superscript n refers to the number of other SiO_4^{4-} groups attached to the SiO_4 tetrahedron being investigated. In the case of aluminosilicates, m represents the number of aluminum atoms bonded via the oxygen bridge to the silicon atom of the SiO_4 group under study. Schematic

Table 5.3 Mineral and Chemical Composition of Coal Fly Ashes

Mineral	Chemical Composition	Mineral Composition[a]	
		Laramie River Coal Fly Ash	Dave Johnston Coal Fly Ash
Quartz	SiO_2	X	X
Periclase	MgO	X	X
Anhydrite	$CaSO_4$	X	
Merwinite	$Ca_3Mg(SIO_4)_2$	X	
Ferrite spinel	$(Mg,Fe)(Fe,Al)_2O_4$	X	
Lime	CaO	X	X
Melilite	$(Ca,Na)_2(Mg,Al,Fe)(Si,Al)_2O_7$	X	
Brownmillerite	$Ca_4Al_2Fe_2O_{10}$	X	
Mullite	$Al_6Si_2O_{13}$	X	X
Hematite	Fe_2O_3	X	X
Sodalite	$Ca_2(Ca, Na)_6(Al, Si)_{12}O_{24}$	X	
	Ca_3SiO_5		
	Ca_2SiO_4		
	$Ca_3Al_2O_6$		

[a] The minerals present are indicated by an X.

representations and some examples of the Q^n and $Q^n(mAl)$ notation for silicates and aluminosilicates are shown in Table 5.5. Figure 5.1 gives the estimated ^{29}Si NMR chemical shift ranges for the various polymerized silicates and aluminosilicates.

The ^{29}Si NMR spectra of the Dave Johnston and Laramie River anhydrous coal fly ashes are shown in Figure 5.2(a) and (b), respectively, for the

Table 5.4 Oxide Composition of Coal Fly Ash

Oxide	Oxide Composition[a] (wt%)	
	Dave Johnston Coal Fly Ash	Laramie River Coal Fly Ash
CaO	13.6	23.7
SiO_2	38.8	26.8
Al_2O_3	36.3	28.9
MgO	2.6	5.2
Fe_2O_3	5.6	8.1
Na_2O	0.2	1.2
SrO	0.1	0.4
P_2O_5	0.6	1.1
SO_3	0.6	1.9
K_2O	1.6	2.0
BaO	0.2	0.6

[a] Calculated from elemental data in Table 5.1 and normalized to 100%.

Table 5.5 Coordination State Notation and Structure of Silicates and Aluminosilicates

Notation	Structure	Name
Q^0	O⁻ ⁻OSiO⁻ O⁻	Monomer (nesosilicates)
Q^1	O⁻ ⁻OSiOSi O⁻	End group (sorosilicates)
Q^2	O⁻ SiOSiOSi O⁻	Middle group (enosilicates)
Q^3	O⁻ SiOSiOSi O Si	Branching group (phyllosilicates)
Q^4	Si O SiOSiOSi O Si	Crosslinking group (tectosilicates)
$Q^4(0Al)$	Si O SiOSiOSi O Si	
$Q^4(1Al)$	Al O SiOSiOSi O Si	
$Q^4(2Al)$	Al O SiOSiOSi O Al	
$Q^4(3Al)$	Al O AlOSiOSi O Al	
$Q^4(4Al)$	Al O AlOSiOAl O Al	

total sweep width of the experiment (300 to −450 ppm). The resonance between 0 and −150 ppm is the ^{29}Si signal. The other broad resonances (150, 30, −195, and −300 ppm) are due to spinning sidebands of the center-band. These sidebands are the result of susceptibility and other tensorial

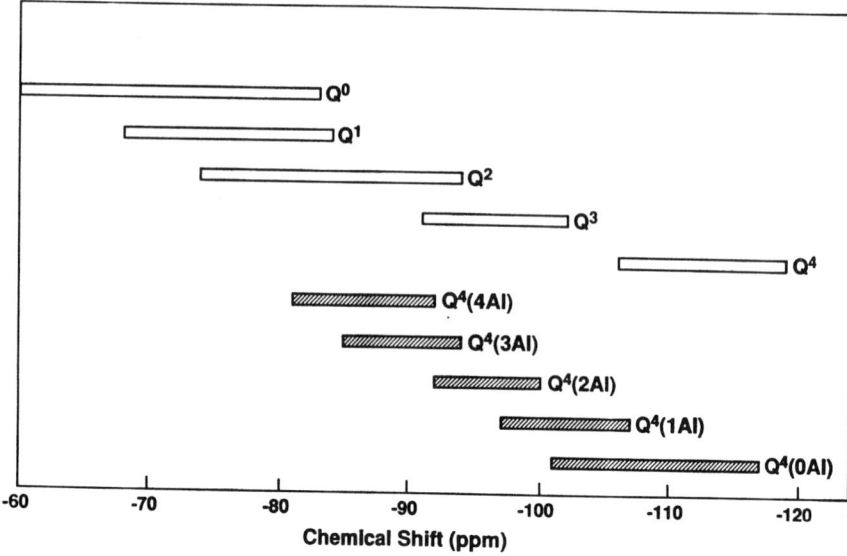

Figure 5.1 Approximate ^{29}Si chemical shift ranges for various coordination states of silicates and aluminosilicates.

effects due to paramagnetism present in the sample as impurities or constituent parts of the structure. Oldfield et al.[33] observed similar strong sidebands in some naturally occurring aluminosilicates containing paramagnetic impurities, and suggested that the sidebands may be due to the presence of large magnetic susceptibility broadening and not to chemical shift anisotropy. Watanabe et al.[34] and Grimmer et al.[35] reported dependence of the ^{29}Si linewidth in silicate minerals on the relative concentration of Fe^{3+} and Fe^{2+} ions, respectively.

The spinning sideband intensities relative to the centerband intensity are greater for the Laramie River coal fly ash than for the Dave Johnston coal fly ash. This is because of the higher Fe to Si ratio (0.226:1 versus 0.107:1) for the Laramie River fly ash. The high silicon content of the Dave Johnston coal fly ash (23.3 wt%) relative to the Laramie River (14.6 wt%) coal fly ash is reflected in the signal to noise ratio for both spectra. To obtain the ^{29}Si spectrum of the Dave Johnston coal fly ash 6000 scans were used, compared to 12 000 scans for the ^{29}Si spectrum of the Laramie River coal fly ash.

Aside from the spinning sideband intensities and the noise ratios, there are other noticeable differences in the ^{29}Si NMR spectra of the two coal fly ashes that probably are a result of their mineral composition (Figure 5.2(a) and (b); Table 5.3). However, because of their complex mineral composition and high iron content, the ^{29}Si NMR spectra of the coal fly ashes are a broad envelope of resonances due to chemical shift dispersion and paramagnetic

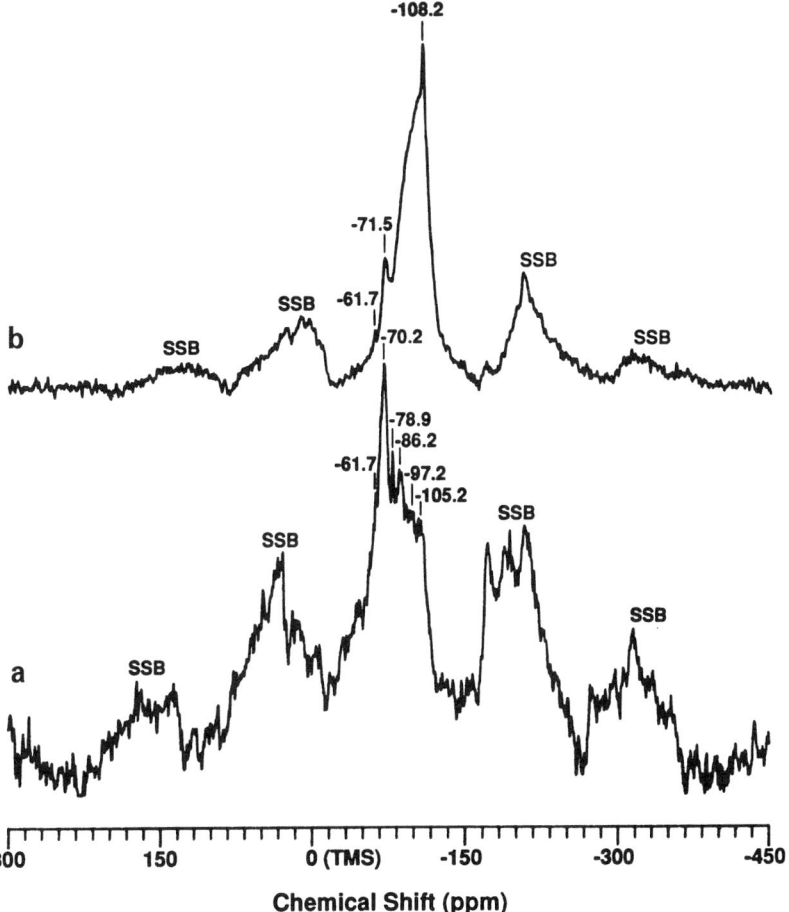

Figure 5.2 ^{29}Si NMR SPE/MAS spectra of the anhydrous coal fly ashes from: (a) the Dave Johnston Power Plant, Pd = 10 s, Acq = 64809; (b) the Laramie River Power Plant, Pd = 5 s, Acq = 12 240.

broadening. Therefore, it is not possible to assign any one resonance to a given mineral, except for resonances that appear at either edge of the chemical shift range. The poorly defined downfield signal at −61.7 ppm which appears in the spectra can be assigned to either of the minerals chondrodite [$(Mg_5(SiO_4)(OH,F)_2)$], forsterite (Mg_2SiO_4), or olivine [$(Mg,Fe)_2SiO_4$]. These appear to be the only minerals that have resonance signals in this region of the spectrum.[36]

The major differences found by NMR in the two anhydrous coal fly ash materials are the concentration of the SiO_4^{4-} anions (Q^0) at −70.2 and −71.5 ppm for the Laramie River and Dave Johnston coal fly ashes, respectively, and the concentration of SiO_2 (quartz, Q^4) at −105.2 and −108.3 ppm, respectively. The Laramie River fly ash has a much higher concentration of monomeric silicate anion relative to the concentration of the fully polymerized quartz, whereas with the Dave Johnston fly ash, the concentration of the monomeric SiO_4^{4-} anion is much less than the concentration of quartz.

Because of the broad, featureless resonances in the region of -75 to -105 ppm, no other spectral assignments can be made for the anhydrous Dave Johnston coal fly ash. However, in this spectral region of the Laramie River coal fly ash, some well-defined resonances and broad resonances are noted. The resonances at -78.9 and -86.2 ppm are assigned the coordination units of Q^1 (Si end group) and Q^2 (Si middle group), and $Q^4(4Al)$ (Si cross-linking groups with 4Al), respectively. The broad resonances centered at -97.2 and -105.2 ppm are assigned the coordination units of Q^3 (Si branching group) and $Q^4(1Al)$ (Si crosslinking group with 1Al), and Q^4 (Si crosslinking group) and $Q^4(0Al)$ (Si crosslinking group with no aluminum), respectively.

To extract additional information from the ^{29}Si spectra of coal fly ash, a mathematical smoothing function was used to reduce the spectral noise level and to define more precisely the major silicate and/or aluminosilicate components. With the major components defined, a curve-fitting routine was used to determine the distribution and quantity of silicates and aluminosilicate species within the coal fly ash sample. Figure 5.3(a) and (b) shows the expanded and deconvoluted spectra of the anhydrous coal fly ashes. Table 5.6 lists the relative percentages of anion types and the silicate coordination state assignment for each coal fly ash. The three resonances below -60 ppm (Figure 5.3(b)) are silicates and aluminosilicates which have not been reported by any other technique. Although the curve-fitting techniques determining the silicate species within a fly ash sample are only semiquantitative, they do provide relative concentration information that cannot be obtained by any other method.

Adsorption Isotherms of Hazardous Organic Compounds Sorbed on Coal Fly Ash

Several equilibrium models have been developed to describe adsorption data. The most common adsorption isotherm equations used for describing adsorption data are those derived by Langmuir[37] and by Freundlich.[38]

The logarithmic expression for the Freundlich adsorption isotherm is given in equation (1).

$$\log (x/m) = \log K + (1/n)(\log C_e) \tag{1}$$

where x = amount of adsorbate adsorbed, m = weight of adsorbent, C_e = adsorbate equilibrium concentration, and K and $1/n$ are constants characteristic of the system. The equation is widely used to describe adsorption in aqueous systems. A plot of $\log q$, where $q = x/m$, versus $\log C_e$ gives an intercept of $\log K$ and a slope of $1/n$. The value of K can be taken as a relative

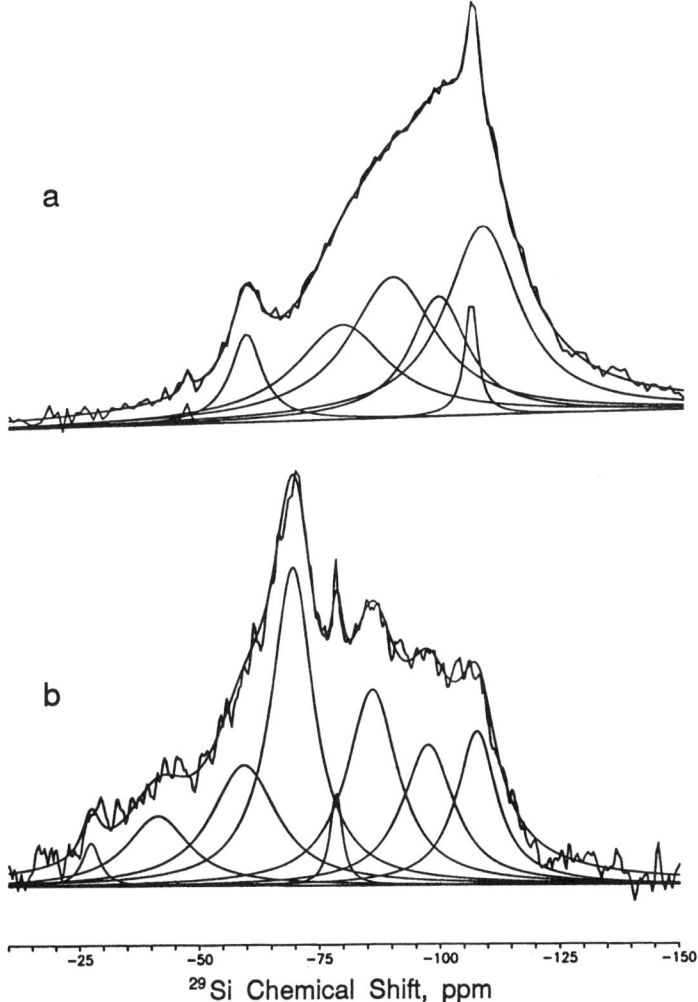

Figure 5.3   Deconvoluted ^{29}Si NMR spectra of coal fly ashes from: (a) Dave Johnston Power Plant; (b) Laramie River Power Plant.

indicator of the adsorption capacity, while $1/n$ is indicative of the energy or intensity of the reaction.[39] A value of $1/n$ of ~ 1 indicates high adsorptive capacity at high equilibrium concentrations. For $1/n \ll 1$, the adsorptive capacity is only slightly reduced at the lower equilibrium concentrations.

Solid–Gas Equilibrium

The Freundlich adsorption isotherms were obtained for three aromatic hydrocarbons and several chlorobenzene compounds using the headspace

Table 5.6 Relative Percentages of Silicate Anions in Dave Johnston and Laramie River Power Plant Coal Fly Ashes

	Dave Johnston Coal Fly Ash		Laramie River Coal Fly Ash	
Deconoluted Peak No. Left to Right[a]	Silicate Coordination State	Relative Area (%)	Silicate Coordination State	Relative Area (%)
1	—	0.3	—	1.6
2	Q^0, Q^1	6.1	—	8.3
3	Q^2	23.8	—	15.2
4	Q^3	28.5	Q^0	26.8
5	Q^3	10.6	Q^1, Q^2	2.3
6	Q^4	4.2	Q^2	19.5
7	Q^4	26.5	Q^3	14.0
8			Q^4	12.4

[a] See Figure 5.3 and text.

analysis technique (Figures 5.4 and 5.5). In these experiments a dilute solution of all of the compounds in methanol was applied directly to the Dave Johnston coal fly ash in a vial and allowed to equilibrate for 72 h at ambient temperature. Because the amount of solution added to the fly ash was totally adsorbed, only a solid–gas equilibrium existed. The headspace above the fly ash contained the vapor phase of the organic compounds in equilibrium with the adsorbed compounds on the fly ash. Table 5.7 lists the Freundlich adsorption isotherm constants (K and $1/n$) for the compounds adsorbed on the Dave Johnston coal fly ash and the same compounds adsorbed on activated carbon. The Freundlich adsorption isotherm constants for the organic compounds on activated carbon were obtained from the literature.[2] These constants represent a solid–liquid adsorption equilibrium. However, to a first approximation they can be compared to the solid–gas adsorption equilibrium because both equilibrium processes depend on monolayer coverage.

The comparison of the adsorption capacity (q) of coal fly ash to that of activated carbon for the hydroaromatics and chlorobenzenes at an equilibrium concentration of 10 mg/L is given in Table 5.8. The adsorption capacity was calculated using the respective Freundlich constants and the Freundlich adsorption isotherm equation. As one would expect, activated carbon adsorbs the compounds to a much greater degree than does coal fly ash; in fact, it adsorbs the hydroaromatics and chlorobenzenes with a capacity more than 3000-fold greater than that of fly ash at an assumed equilibrium concentration of 10 mg/L. Although 1,3-dichlorobenzene appears to be adsorbed on coal fly ash to a much greater extent than the other compounds, the higher value represents essentially two data points over a very narrow concentration range and thus the confidence level is low even though the regression coefficient is high (Table 5.7).

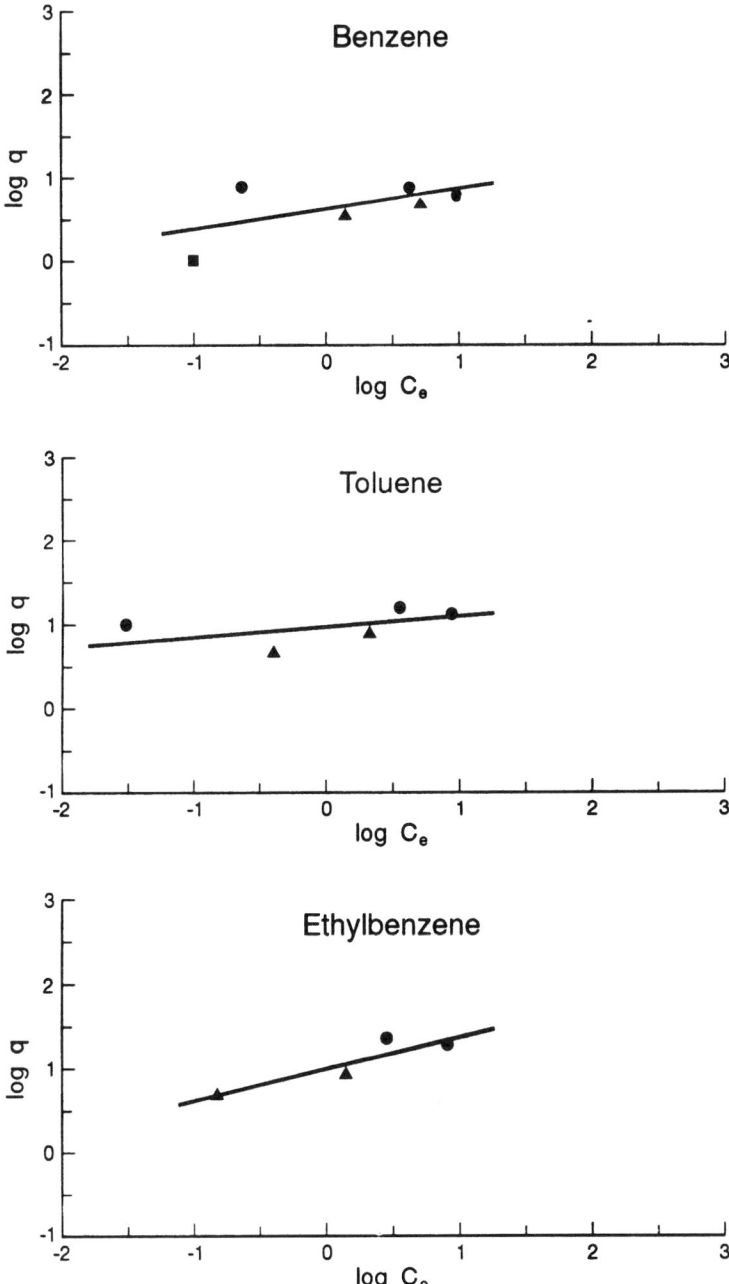

Figure 5.4 Freundlich adsorption isotherms for benzene, toluene, and ethylbenzene adsorbed on Dave Johnston coal fly ash (adsorption capacity, q, in mg/kg; equilibrium concentration, C_e, in mg/L). Symbols ■, ●, and ▲ represent different experimental runs.

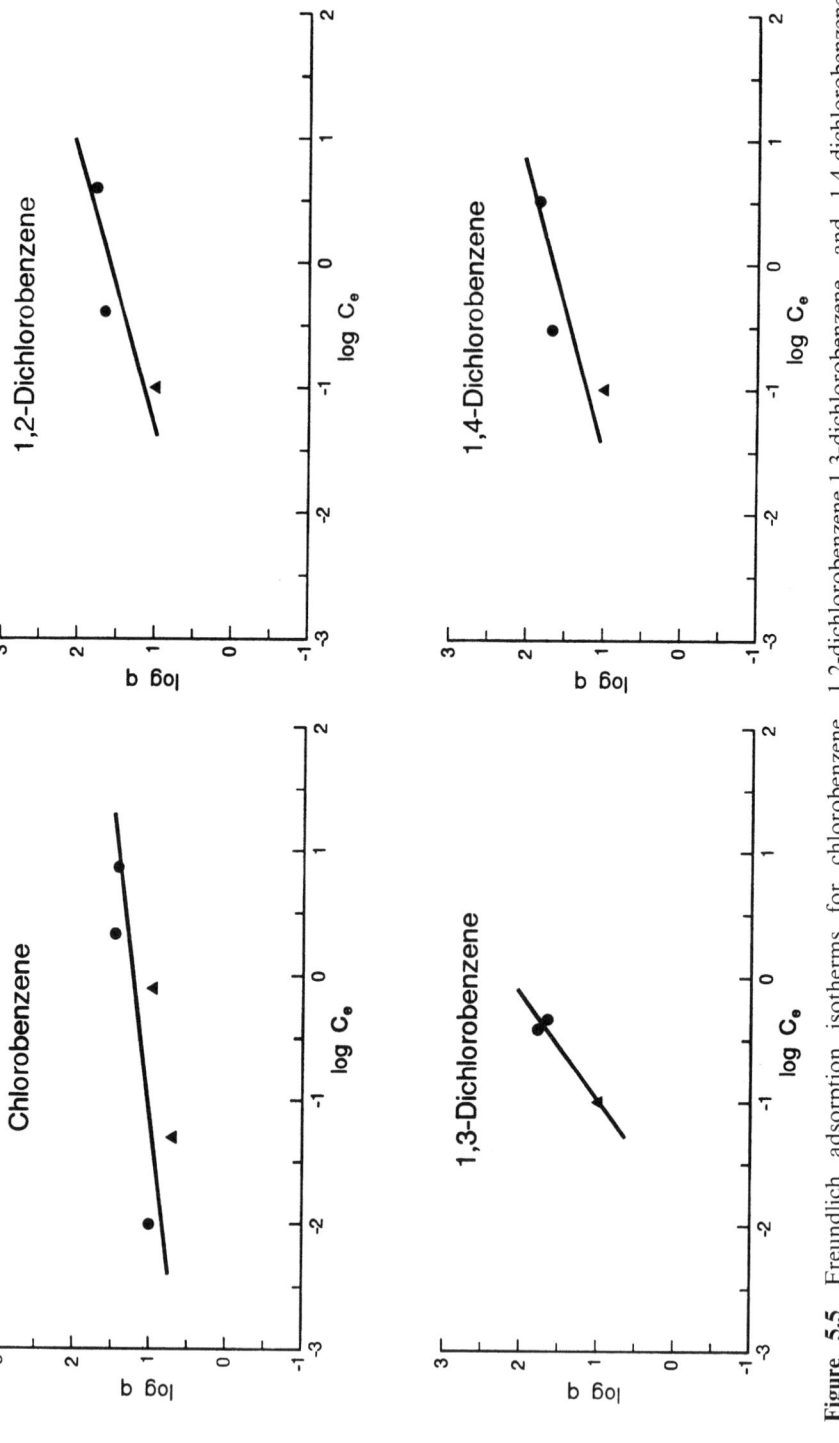

Figure 5.5 Freundlich adsorption isotherms for chlorobenzene, 1,2-dichlorobenzene, 1,3-dichlorobenzene, and 1,4-dichlorobenzene adsorbed on Dave Johnston coal fly ash (adsorption capacity, q, in mg/kg; equilibrium concentration, C_e, in mg/L). Symbols ●, and ▲ represent different experimental runs.

Table 5.7 Freundlich Adsorption Isotherm Constants for Organic Compounds on Dave Johnston Coal Fly Ash and Activated Carbon

Compound	Coal Fly Ash			Activated Carbon[a]	
	K (mg/kg)	$1/n$	r^{2b}	K (mg/kg)	$1/n$
Benzene	3.94	0.2533	0.363	1×10^3	1.6
Toluene	9.38	0.0872	0.169	16.1×10^3	0.44
Ethylbenzene	10.04	0.3842	0.803	53×10^3	0.79
Chlorobenzene	15.67	0.2055	0.565	91×10^3	0.99
1,2-Dichlorobenzene	40.25	0.4595	0.754	129×10^3	0.43
1,3-Dichlorobenzene	140.80	1.1431	0.945	118×10^3	0.45
1,4-Dichlorobenzene	46.46	0.4922	0.727	121×10^3	0.47

[a] Source: Ref. 10 in Faust and Aly.[2]
[b] Regression coefficent.

It should be noted that the comparison of the adsorption capacities for coal fly ash with those for activated carbon are based upon a solid–gas equilibrium and a solid–liquid equilibrium. Additional experiments are needed to confirm that the solid–gas adsorption isotherms can indeed be directly compared to the solid–liquid adsorption isotherms.

Solid–Liquid Equilibrium

Freundlich adsorption isotherms that were obtained for pyridine, pentachlorophenol, naphthalene, and 1,1,2,2-tetrachloroethane in aqueous solution in the presence of known amounts of Dave Johnston coal fly ash are given in Figure 5.6. In this Figure the straight lines were drawn through the points that best represent the adsorption process. The data are a composite of several experimental runs (represented by different symbols) and the data points

Table 5.8 Adsorption Capacities of Dave Johnston Coal Fly Ash and Activated Carbon at an Equilibrium Concentration of 10 mg/L for Organic Compounds

Compound	Adsorption Capacity, q (mg/kg)	
	Coal Fly Ash[a]	Activated Carbon[b]
Benzene	7	40×10^3
Toluene	11	72×10^3
Ethylbenzene	24	327×10^3
Chlorobenzene	25	889×10^3
1,2-Dichlorobenzene	116	347×10^3
1,3-Dichlorobenzene	1957	332×10^3
1,4-Dichlorobenzene	144	357×10^3

[a] Solid–gas equilibrium.
[b] Solid–liquid equilibrium.

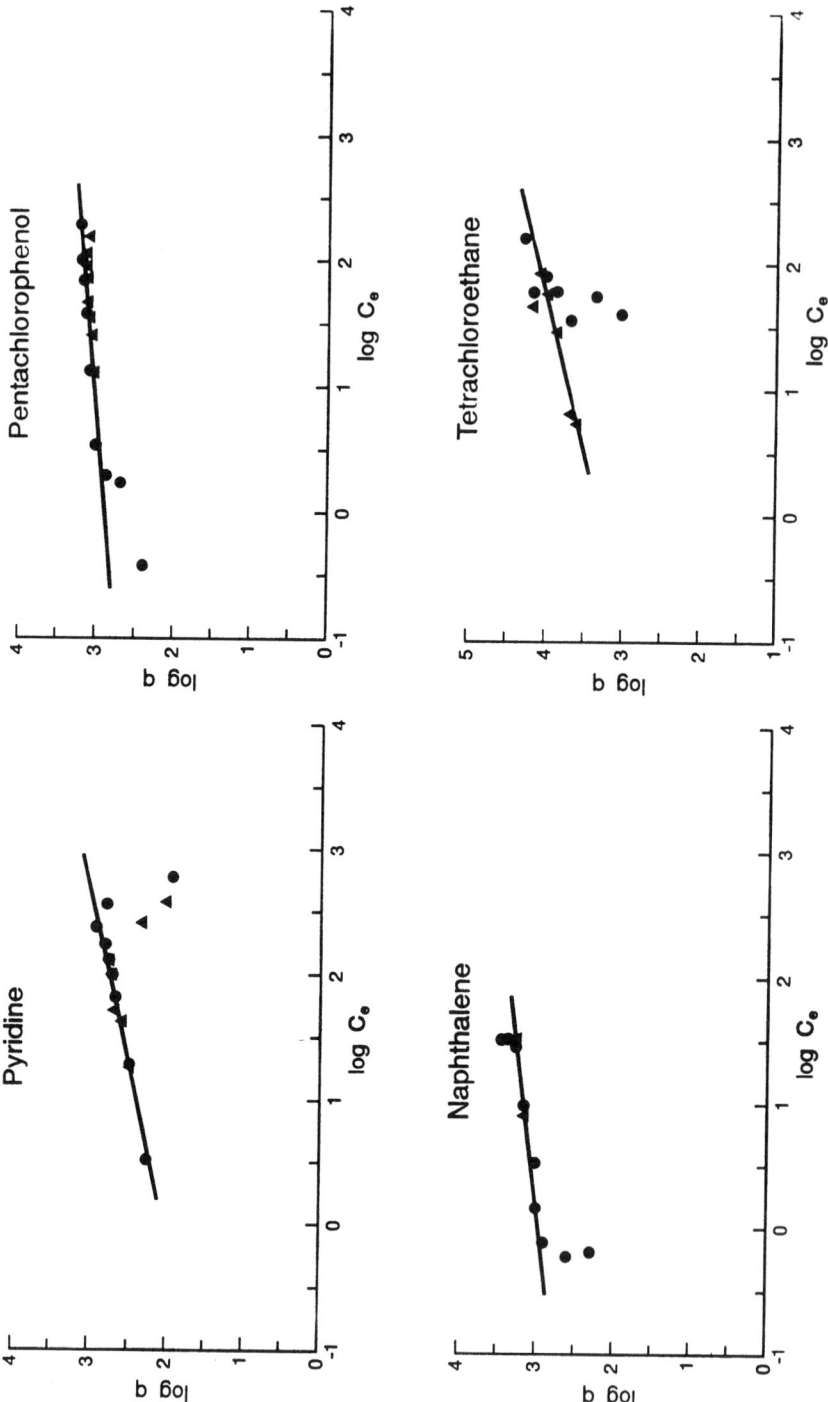

Figure 5.6 Freundlich adsorption isotherms for pyridine, pentachlorophenol, naphthalene, and tetrachloroethane adsorbed on Dave Johnston coal fly ash (adsorption capacity, q, in mg/kg; equilibrium concentration, C_e, in mg/L). Symbols ● and ▲ represent different experimental runs.

that appeared to be off the lines were dropped in computing the Freundlich adsorption isotherm constants. Table 5.9 lists the adsorption constants for the four compounds adsorbed on coal fly ash and activated carbon as reported in the literature.[2]

To compare the adsorption data, the adsorption capacities q for the coal fly ash and activated carbon were calculated for each compound using the Freundlich constants and the Freundlich equation, and an equilibrium concentration of 10 mg/L. These data are reported in Table 5.10. The adsorption capacities for both the coal fly ash and the activated carbon are based upon a solid–liquid equilibrium in which only a monolayer adsorption is assumed possible because of the rapid motion of molecules in the liquid phase. This prevents a multilayer buildup of the adsorbed molecules at the surface of the fly ash and activated carbon.

As shown in Table 5.10, the Dave Johnston coal fly ash appears to adsorb pyridine and tetrachloroethane nearly as well as activated carbon (within a factor of 5 to 10). The coal fly ash adsorption capacity is less by a factor of about 300, relative to activated carbon, for pentachlorophenol and naphthalene. Thus, it appears that the Dave Johnston coal fly ash can be a good adsorber for some types of compounds.

The relationship of the adsorption properties of the Dave Johnston coal fly ash to its chemical composition cannot be made with any certainty yet. Similar adsorption data were not obtained for the Laramie River coal fly ash, which is known to have significantly different chemical properties.

^{13}C Solid-state NMR Study of Hazardous Organic Compounds

Observing the ^{13}C NMR signal of a compound in the solid state can be more difficult than observing the ^{13}C NMR signal of the same compound in the liquid state. Major factors contributing to the difficulties are: (1) sample phase (amorphous versus crystalline); (2) long hydrogen and carbon spin–lattice relaxation times (T_1^H and T_1^C); (3) short spin–lattice relaxation times in the rotating frame ($T_{1\rho}$); (4) long polarization-transfer time (T_{CH}); and (5) Hartmann–Hahn matching conditions.[40,41]

Phenol and Polychlorophenols

Figure 5.7(a) through (j) shows the SPE/MAS and CP/MAS solid-state ^{13}C NMR spectra of phenol and polychlorophenols using ∼2.5 g of sample in a jumbo probe. The SPE/MAS spectrum of solid phenol (Figure 5.7(a)) was obtained using a pulse delay of 20 s and 200 scans. The resonance signal at ∼155 ppm is that of carbon with OH attached. The resonance signals at ∼130, ∼121, and ∼115 ppm are due to carbons at the *meta*, *para*, and *ortho* positions, respectively, within the aromatic ring. The ^{13}C CP/MAS spectrum of phenol is shown in Figure 5.7(b). The spectrum shows only two resonance signals: the phenolic carbon at ∼155 ppm and the *para* carbon

Table 5.9 Freundlich Adsorption Isotherm Constants for Pyridine, Pentachlorophenol, Naphthalene, and Tetrachloroethane on Dave Johnston Coal Fly Ash
and Activated Carbon

	Coal Fly Ash			Activated Carbon	
Compound	K (mg/kg)	$1/n$	r^{2a}	K (mg/kg)	$1/n$
Pyridine	120	0.33	0.979	1220	0.2^b
Pentachlorophenol	780	0.12	0.811	150×10^3	0.42^c
Naphthalene	896	0.22	0.982	132×10^3	0.42^c
Tetrachloroethane	2003	0.40	0.659	10.6×10^3	0.37^c

[a] Regression coefficient.
[b] Source: Ref. 13 in Faust and Aly.[2]
[c] Source: Ref. 10 in Faust and Aly.[2]

at ~121 ppm (contact time 500 μs and pulse delay time 20 s). Varying the
contact time from 50 μs to 8 ms and the delay time from 1 to 60 s did not
result in observing the resonance signals for either the *ortho* or *meta* carbons.
The observed spectrum suggests that the phenol molecule is rotating rapidly
in the solid state about the C_{2v} symmetry axis. Carbon and hydrogen atoms in
the *ipso* and *para* positions remain relatively static, and thus effective cross
polarization occurs between the hydrogen and carbon atoms, resulting in the
two signals observed. Low-temperature studies will be required to slow the
rotational motion of the phenol molecule so that the *ortho* and *meta* carbon
resonances can be observed by CP/MAS.

The ^{13}C SPE/MAS spectra of 4-chlorophenol, 2,6-dichlorophenol and
2,4,6-trichlorophenol shown in Figure 5.7(c), (e), and (g), respectively, were
obtained using a pulse delay of 120 s and 60 scans. The spectrum of pentachlorophenol (Figure 5.7(i)) was obtained with a pulse delay of 20 s and 200
scans. It is readily apparent that as the number of chlorine atoms increases,
the spectral resolution decreases. In fact, for pentachlorophenol, the phenolic
carbon is no longer evident in the spectrum. The long pulse delay times (20–

Table 5.10 Adsorption Capacities of Dave Johnston Coal Fly Ash and Activated
Carbon at an Equilibrium Concentration of 10 mg/L for Pyridine, Pentachlorophenol, Naphthalene, and Tetrachloroethane

	Adsorption Capacity, q (mg/kg)	
Compound	Coal Fly Ash	Activated Carbon
Pyridine	0.25×10^3	1.934×10^3
Pentachlorophenol	1.04×10^3	395×10^3
Naphthalene	1.48×10^3	347×10^3
Tetrachloroethane	5.05×10^3	24.8×10^3

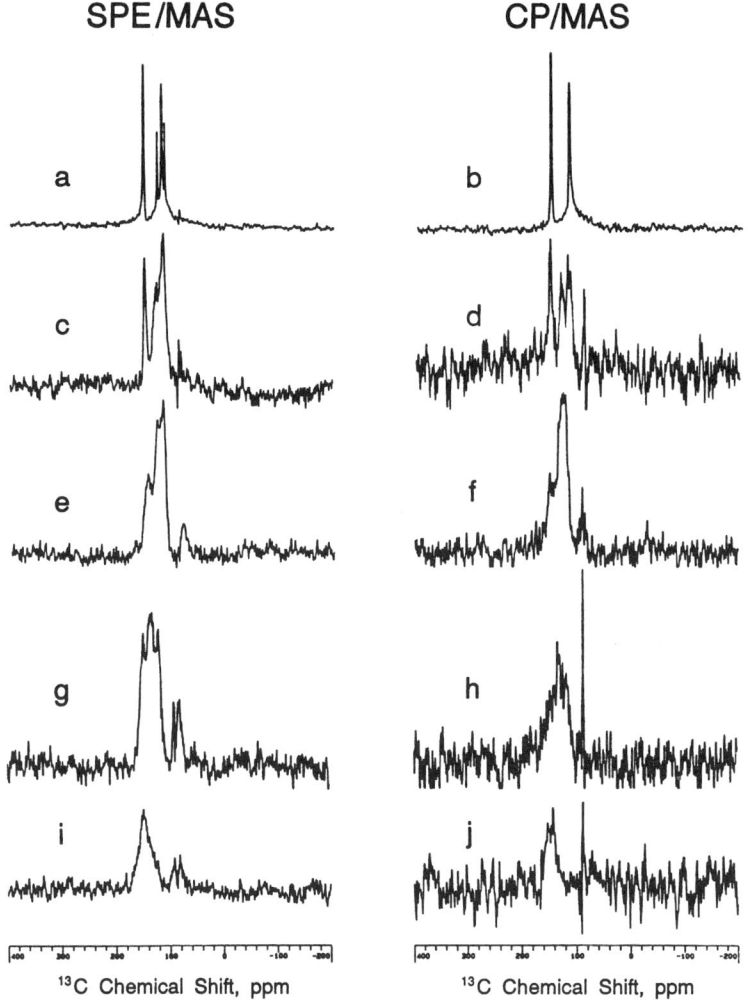

Figure 5.7 ^{13}C solid-state NMR spectra: (a,b) phenol; (c,d) 4-chlorophenol; (e,f) 2,6-dichlorophenol; (g,h) 2,4,6-trichlorophenol; (i,j) pentachlorophenol.

120 s) used to obtain the ^{13}C SPE/MAS spectra of these chlorinated aromatic compounds indicate that the relaxation times for the carbons are extremely long. That is, inter- and intramolecular distances between carbons and nearest hydrogens are large, decreasing the effective dipolar coupling of the nuclei. The decrease in coupling results in an increase in the relaxation time and a decrease in the nuclear Overhauser effect, producing a spectrum with low signal to noise ratio. For quantitative results the delay should be greater than 300 s (5 min) between pulses.

The ^{13}C CP/MAS spectra of the polychlorophenols (Figure 5.7(d), (f), (h), and (j)) were obtained using a contact time of 500 μs, a pulse delay of 120 s and approximately 500 scans. As the number of chlorine atoms increases, the signal to noise ratio and the resolution decrease. With an increase in the number of chlorine atoms, there is a corresponding decrease in the number of hydrogen atoms, and cross polarization becomes ineffective. It is apparent that any adsorption studies using solid-state ^{13}C NMR to observe the adsorbed chlorinated phenols directly should be carried out in the SPE/MAS mode.

The purity of the starting material can affect the spectral resolution observed even if the material is highly chlorinated. In the case of pentachlorophenol, the spectral resolution for technical-grade pentachlorophenol is better than for the pure material. Figure 5.8(a) through (d) show the SPE/MAS and CP/MAS ^{13}C NMR spectra of technical and pure pentachlorophenol. The four carbon resonances (148, 131, 125, and 119 ppm) are clearly resolved in the SPE/MAS spectrum of the technical-grade pentachlorophenol (Figure 5.8(a)) and discernible in the CP/MAS spectrum (Figure 5.8(b)). High-resolution ^{1}H NMR spectra of the technical and pure grades of pentachlorophenol showed little or no difference in composition. Attempts to obtain ^{13}C spectra of pure pentachlorophenol doped with CrAcAc were unsuccessful. Obviously, additional experiments are needed to ascertain the reason for the enhanced resolution in the ^{13}C NMR spectrum of technical-grade pentachlorophenol.

Sorption of Organic Compounds

Silica gel is a siliceous adsorbing material having two ^{29}Si NMR resonance signals at -101.3 and -110.5 ppm.[42] These signals correspond to the Q^3 and Q^4 coordination states of the silicate anions. The Q^3 and Q^4 coordination states also account for 70% of the silicate anions in the Dave Johnston coal fly ash (see Table 5.6). Thus, silica gel was used as a model adsorbent to investigate the use of ^{13}C NMR to measure the adsorption capacity and bonding interactions of organic materials with the adsorbent.

Figure 5.9(a) and (b) shows the ^{13}C SPE/MAS and CP/MAS NMR spectra of 20 mg of pure pentachlorophenol sorbed on 1 g of silica gel. The pentachlorophenol was intimately mixed with silica gel using methylene chloride, which was then evaporated. The only resonance signals that can readily be observed are due to sorbed methylene chloride ($\sim$50 ppm). The resonance peak at $\sim$170 ppm in Figure 5.9(b) may the signal from the *ipso* carbon.

The coal fly ashes were exposed to pyridine vapor for 11 days. The ^{13}C CP/MAS NMR spectra of the ashes were obtained using a contact time of 5 ms, a pulse delay time of 1 s and 65 000 scans (Figure 5.10(a) and (b)). The spectra are the difference spectra obtained by subtracting the probe background signals from the composite spectra. The pyridine does not appear to be sorbed on the Laramie River coal fly ash (Figure 5.10(a)) while the Dave Johnston fly ash does sorb pyridine, as determined by adsorption isotherm experiments, but the resonance signals are broadened over the range of

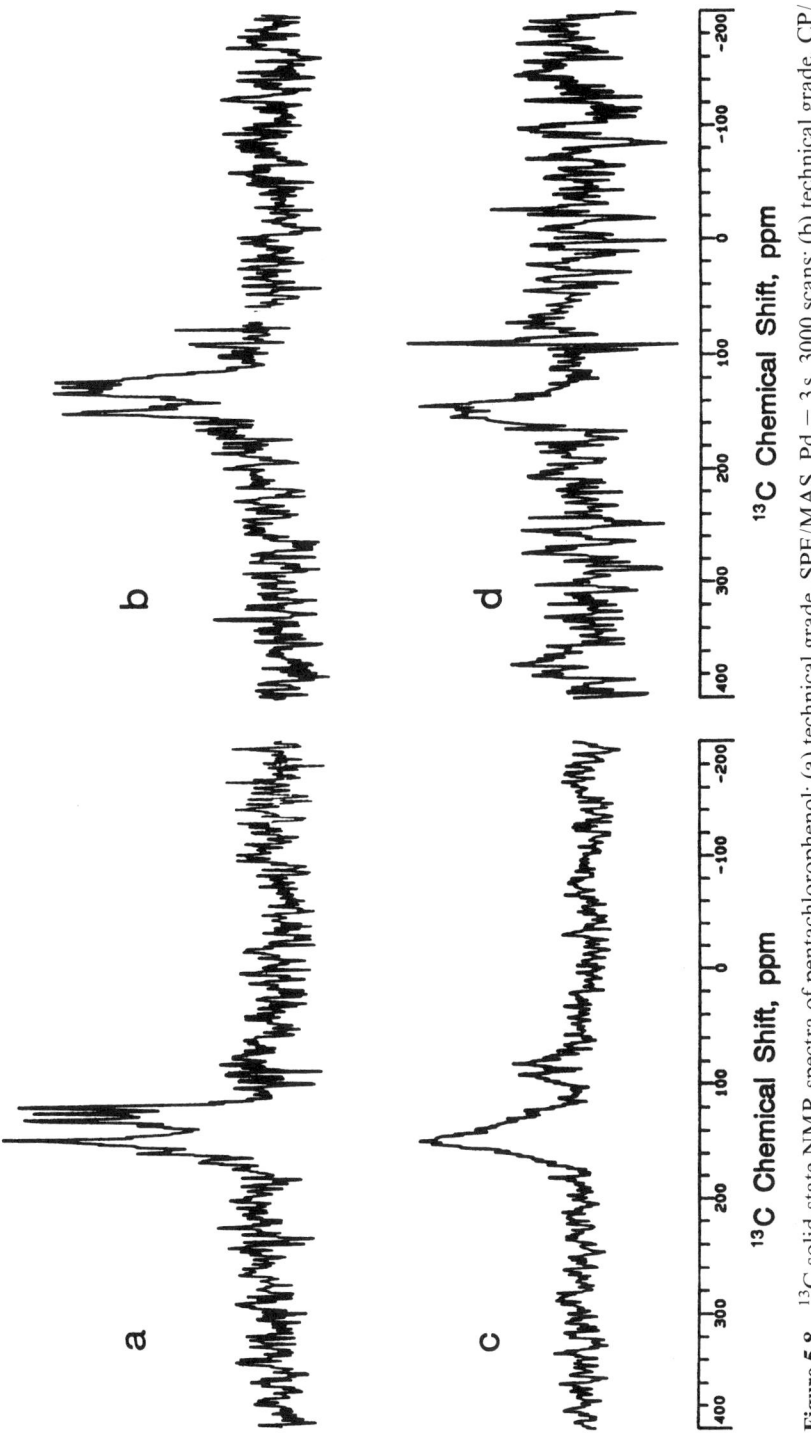

¹³C Chemical Shift, ppm

Figure 5.8 ¹³C solid-state NMR spectra of pentachlorophenol: (a) technical grade, SPE/MAS, Pd = 3 s, 3000 scans; (b) technical grade, CP/ MAS, ct = 3 ms, Pd = 1 s, 3000 scans; (c) reagent grade, SPE/MAS, Pd = 20 s, 200 scans; (d) reagent grade, CP/MAS, ct = 500 µs, Pd = 120 s, 480 scans.

113

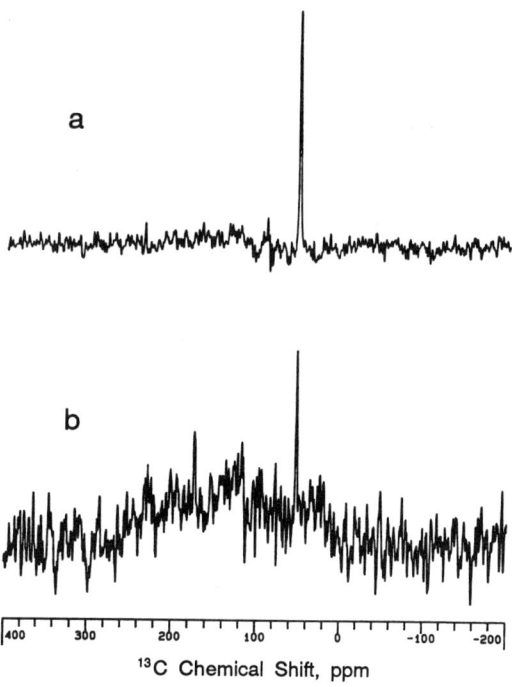

Figure 5.9 ^{13}C solid-state NMR spectra of pentachlorophenol: (a) CH_2Cl_2 solution, on silica gel, SPE/MAS, Pd $= 30$ s, 1800 scans; (b) CH_2Cl_2 solution on silica gel, CP/MAS, ct $= 3$ ms, Pd $= 1$ s, 10 800 scans.

120–160 ppm (Figure 5.10(b)). In contrast, pyridine is readily adsorbed on silica gel, as shown in Figure 5.10(c). The silica gel was exposed to pyridine vapor for only a few hours and the spectrum was obtained in less than 3 h.

Attempts were made to observe pentachlorophenol sorbed on Dave Johnston fly ash using SPE/MAS and CP/MAS ^{13}C NMR. A broad signal at ~130 ppm (barely visible above the noise level) was observed for pentachlorophenol. The *ipso* carbon, the carbon of most interest, was not observed in the chemical shift region from 150 to 170 ppm.

In a related experiment, technical-grade pentachlorophenol in methylene chloride was added to a Tennessee Valley Authority (TVA) coal fly ash to give a concentration of 20 mg of pentachlorophenol per gram of fly ash. Figure 5.11(a) shows the ^{13}C SPE/MAS spectrum of technical-grade pentachlorophenol, coal fly ash with sorbed pentachlorophenol (Figure 5.11(b)), and the mixture leached with water (Figure 5.11(c)). In Figure 5.11(b), the *ipso* carbon of pentachlorophenol sorbed on the fly ash had shifted to 169 ppm from 148 ppm for the pentachlorophenol alone. This shift is quite significant and suggests that there is a strong bonding interaction of pentachlorophenol with the TVA coal fly ash.

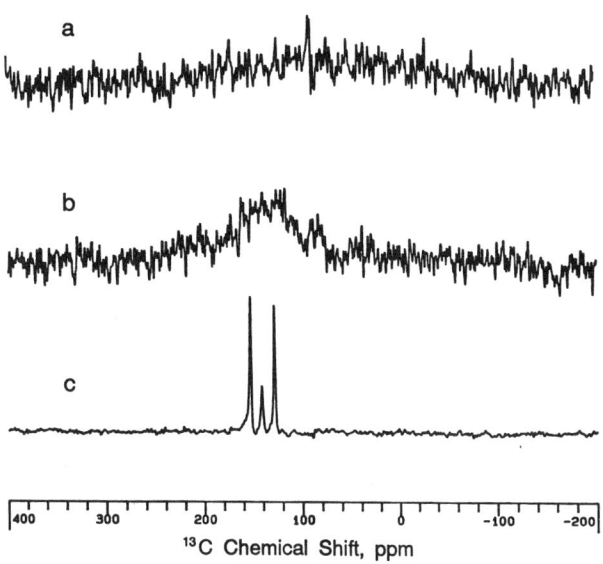

Figure 5.10 ^{13}C solid-state NMR spectra of pyridine adsorbed on: (a) Laramie River coal fly ash, CP/MAS, ct = 5 ms, Pd = 1 s, 65 000 scans; (b) Dave Johnston coal fly ash, CP/MAS, ct = 5 ms, Pd = 1s, 65 000 scans; (c) silica gel, SPE/MAS, Pd = 10 s, 900 scans.

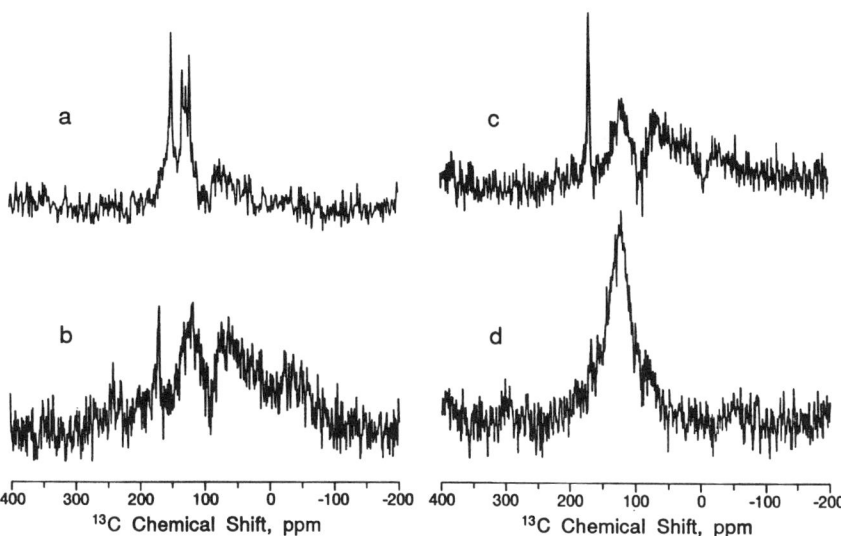

Figure 5.11 ^{13}C solid-state NMR spectra of pentachlorophenol: (a) technical grade, SPE/MAS, Pd = 1 s, 43 000 scans; (b) adsorbed on TVA coal fly ash, SPE/MAS, Pd = 1 s, 72 000 scans; (c) after leaching fly ash with water, SPE/MAS, Pd = 1 s, 72 000 scans; (d) after leaching fly ash with water, CP/MAS, ct = 5 ms, Pd = 1 s, 65 000 scans.

115

Figure 5.11(c) shows the ^{13}C SPE/MAS spectrum of the pentachlorophenol adsorbed on the fly ash after leaching with water. An enhancement of the carbon with the OH attached is observed at ~170 ppm. The CP/MAS spectrum of the same sample (Figure 5.11(d)) shows an enhanced aromatic carbon intensity but no *ipso* carbon resonance. These spectra suggest that water molecules may be near carbon sites, providing rapid relaxation of the carbon spin state and a source of hydrogens for cross polarization of carbon.

Conclusions

Dave Johnston and Laramie River Power Plant coal fly ashes were characterized by elemental, X-ray, and ^{29}Si NMR analyses. These fly ashes differ significantly in their chemical and mineral composition. Although this study on adsorption isotherms and ^{13}C NMR spectra of adsorbed hazardous organic compounds was based on Dave Johnston coal fly ash, one experiment with the adsorption of pyridine on both types of ashes suggests that the Dave Johnston fly ash, which contains a higher proportion of polysilicate anions, is a better adsorbent than the Laramie River fly ash. Many of the experiments carried out using the Dave Johnston fly ash need to be repeated using the Laramie River fly ash. If this were to be done, it would be possible to correlate the chemical, mineral, and structural properties of the ashes with their adsorption capacity for hazardous organic compounds. This was outside the scope of the current project.

Adsorption isotherm studies of hydroaromatics and chlorobenzenes have shown that the Dave Johnston coal fly ash is less effective as an adsorbent than activated carbon by factors ranging from 3000 to 35 000. This conclusion was based upon comparisons of the solid–gas adsorption isotherm equilibrium for coal fly ash and the solid–liquid adsorption isotherm equilibrium for activated carbon. Additional investigation is needed to ascertain the degree of validity for comparing the adsorption isotherm data for two different phase equilibrium conditions, even though the adsorption process in both cases depends on the monolayer coverage model of the adsorbent.

Solid–liquid phase adsorption isotherm studies were conducted using pyridine, pentachlorophenol, naphthalene, and 1,1,2,2-tetrachloroethane in aqueous solution sorbed on Dave Johnston coal fly ash. The results were compared to the adsorption capacity of activated carbon. The adsorption capacity of the four compounds on the fly ash was found to be less than that reported for activated carbon. The adsorption capacity for pyridine and tetrachloroethane on fly ash and activated carbon was found to differ by less than a factor of 10. However, for pentachlorophenol and naphthalene, the factor was found to be between 200 and 400. Clearly, the Dave Johnston coal fly ash is a relatively good adsorber of organic hazardous compounds of the type studied, but it is probably less effective than activated carbon.

Solid-state ^{13}C NMR studies were conducted in an effort to determine directly the sites where adsorbed organic compounds were taken up on coal fly ash, and to measure the strength of the bonding interaction between coal fly ash and the organic compound adsorbed. It is possible to observe the ^{13}C NMR resonances of the adsorbed compounds on fly ash with the use of (1) a large-volume probe (~2.5 g of sample) and (2) a long pulse delay time (120 s or longer) to compensate for long relaxation times. The most successful experiments were those using single pulse excitation. Although such experiments take a considerable amount of time to perform, the resulting data should be quantitative. ^{13}C cross-polarization experiments are most successful for those compounds having hydrogen atoms. The experiments are less time-consuming. However, the optimum contact time must be determined for each compound.

Because the fly ash contains considerable amounts of paramagnetic impurities, the observed resonances are broadened and the chemical shift of the carbon resonance is ill-defined. This will preclude any effort to determine bonding strength for some adsorbed species. It has been observed, however, that the ^{13}C chemical shift of the phenolic carbon in pentachlorophenol, when adsorbed on coal fly ash, was downfield relative to its position in the "free" solid state. The magnitude of this shift should be directly related to the bonding strength. Additional solid-state NMR experiments are needed to evaluate more quantitatively the technique as a means to study directly the adsorption properties of organic hazardous compounds.

Acknowlegements We express appreciation to the US Department of Energy for funding of this work under Cooperative Agreement No. DE-FC21-86MC11076 and the Instrumentation Grant No. DE-FG05-89ER75506.

Disclaimer Mention of specific brand names or models of equipment is for information only and does not imply endorsement of any particular brand.

References

1. White, S. C., and Case, E. D., Characterization of fly ash from coal-fired power plants. *J. Mat. Sci.* 25(12), 5215–5219, 1990.
2. Faust, S. D., and Aly, O. M., *Adsorption Process for Water Treatment*, Butterworth, Boston, MA, 1987.
3. Meiler, W., and Wutscheck, T., *Isotopenpraxis* 25, 41–47, 1989.
4. Deininger, D., Thesis, Leipizig, 1981.
5. Parfitt, G. D., and Rochester, C. H., *Adsorption from Solution at the Solid/Liquid Interface*, Academic Press, New York, NY, 1983.
6. Dawson, W. H., Kaiser, S. W., Ellis, P. D., and Inners, R. R., *J. Phys. Chem.* 86, 867, 1982.
7. Essington, M. E., Bowen, J. M., Wills, R. A., and Hart, B. K., *Adsorption of Aniline and Toluidines on Montmorillonite: Implications for the Disposal of Shale Oil Production Wastes*, WRI-92-R019, Western Research Institute, Laramie, Wyoming, 1992.

8. Essington, M. E., and Spackman, L. K., *Inorganic Geochemical Investigation of Spent Oil Shales*, DOE Report DOE/MC/11076-2701, NTIS, Springfield, Virginia, 1988.

9. Rai, D., Ainsworth, C. C., Eary, L. E., Mattigod, S. V., and Jackson, D. R., EPRI EA-5276, Research Report Center, Palo Alto, CA, 1987.

10. Netzel, D. A., *Multinuclear NMR Approach to Coal Fly Ash Characterization*. WRI report to DOE, WRI-92-R015, Western Research Institute, Laramie, Wyoming, 1992.

11. Fox, J. P., Jackson, D. E., and Sakaji, R. H., *13th Oil Shale Symposium Proceedings, Golden CO*, Colorado School of Mines, Golden, CO, 1980, pp. 311–320.

12. George, M., and Jackson, L., *Leach Potential of Codisposed Spent Oil Shale and Retort Water Using Two Extraction Methods*, DOE Report DOE/FE/60177-1882, NTIS, Springfield, Virginia, 1985.

13. Boardman, G. D., Godrej, A. N., Cawher, D. M., and Lu, Y. W., DOE Report DOE/LC/10964-2037, NTIS, Springfield, Virginia, 1985.

14. Essington, M. E., and Hart, B. K., *Adsorption of Pyridine by Combusted Oil Shale*, DOE/MC/11076-2889, NTIS, Springfield, Virginia, 1990.

15. Sorini, S. S., and Lane, D. C., *Organic and Inorganic Hazardous Waste Stabilization Using Combusted Oil Shale*, DOE/MC/11076-3019, NTIS, Springfield, Virginia, 1991.

16. Bernardin, Jr, F. E., *Adsorption Technology*, Slejko, F. L. (ed.), Marcel Dekker, New York, NY, 1985.

17. Lane, D. C., 1992, unpublished data.

18. Sharma, S., Fulekar, M. H., and Jayalakshmi, C. P., *Crit. Rev. Environ. Control* 19(3), 251, 1989.

19. Derouane, E. G., *Bull Soc. Chim. Belg.* 78, 89, 101, 159, 1969.

20. Borovkov, V. J., and Kazansky, V. B., *Kinet. Katal.* 15, 705, 1974.

21. Gay, I. D., *J. Phys. Chem.* 78, 38, 1974.

22. Bernstein, T., Michel, D., Pfeifer, H., and Fink, P., *J. Colloid Sci.* 84, 310, 1981.

23. Borovkov, V. Y., Zaiko, A. V., Kazansky, V. B., and Hall, W. K., *J. Catal.* 75, 219, 1982.

24. Hayashi, S., and Hayamizu, K., Shift references in high-resolution solid-state NMR. *Bull. Chem. Soc. Jpn.* 62, 2429–2430, 1989.

25. Hayashi, S., and Hayamizu, K., Chemical shift standards in high-resolution solid-state NMR (1) ^{13}C, ^{29}Si, and ^{1}H nuclei. *Bull. Chem. Soc. Jpn.* 64, 685–687, 1991.

26. Englehardt, G., and Michel, D., *High Resolution Solid-State NMR of Silicates and Zeolites*. John Wiley, New York, NY, 1987.

27. Pradip, Subbarao, E. C., Kapur, P. C., Jagannathan, N. R., and Rao, C. N. R., *Mat. Res. Bull.* 22, 1055–1062, 1987.

28. Parry-Jones, G., Al-Tayyib, A. J., and Al-Mana, A. I., *Cement Concrete Res.* 18, 229–234, 1988.

29. Groves, G. W., and Rodger, S. A., *Adv. Cement Res.* 2(8), 135–150, 1989.

30. Young, J. F., *J. Am. Ceram. Soc.* 71(3), C-118–C-120, 1988.

31. Wilson, M. A., Young, B. C., and Scott, K. M., *Fuel* 65, 1584–1587, 1986.

32. Barnes, J. R., Clague, A. D. H., Clayden, N. J., Dobson, C. M., and Jones, R. B., *Fuel* 65, 437–441, 1986.

33. Oldfield, E., Kinsey, R. A., Smith, K. A., Nichole, J. A., and Kirkpartrick, R. J., *J. Magn. Reson.* 51, 325, 1983.

34. Watanabe, T., Shimizu, H., Masusu, A., and Saito, H., *Chem. Lett.* 1293, 1983.

35. Grimmer, A.-R., Lampe, F. V., Mägi, M., and Lippmaa, E., *Z. Chem.* 23, 342, 1983.
36. Goberdhan, D. G., *Solid State NMR Studies of Silicates, Minerals and Clays*, The British Library, West Yorkshire, UK, 1988.
37. Langmuir, I., *J. Am. Chem. Soc.* 37, 1139, 1915.
38. Freundlich, H., *Colloid and Capillary Chemistry*, Methuen, London, 1926.
39. Weber, Jr, W. J., *Physicochemical Processes for Water Quality Control*, Wiley–Interscience, New York, NY, 1972.
40. Axelson, D. E., *Solid-State Nuclear Magnetic Resonance of Fossil Fuel*, Multiscience, Montreal, 1985.
41. Botto, R. E., and Sanada, Y., (eds), *Magnetic Resonance of Carbonaceous Solids*, ACS Adv. Chem. Ser. No. 229, American Chemical Society, Washington DC, 1993.
42. Netzel, D. A., 1991, unpublished data.

SOLUTION AND CONDENSED PHASE CHARACTERIZATION

6.

Solution and Condensed Phase Characterization

ROGER A. MINEAR & MARK A. NANNY

Improvements in nuclear magnetic resonance (NMR) instrumentation, magnetic field strength, pulse sequences, and computer technology and software have increased the range of applications and specific elements available for study by NMR. The five chapters in this Part clearly indicate the benefits of these advances, especially regarding studies of aquatic, environmental significance. Each of the studies focuses on environmentally significant issues. For example, chlorination is widely used to disinfect drinking waters, a method that can produce undesirable disinfection by-products. This was first recognized in 1974 with the discovery of trihalomethanes in most finished drinking waters where hypochlorite was used for disinfection. Chapter 7 examines the chlorination of alanine and relates it to the chlorination reactions of acetaldehyde and ammonia, a topic of importance with respect to drinking water disinfection.

Aluminum is also widely used in drinking water treatment, and understanding its hydrolysis chemistry and complexation behavior is of great importance to aquatic chemistry. In addition, the aquatic chemistry of aluminum is important because acid rain can release soluble aluminum ions from clay into soil water, possibly damaging terrestrial plant life. Aluminum may eventually reach and accumulate in hydrological systems where it can be toxic to aquatic life. Chapters 8 and 9 focus on ^{27}Al NMR in defining aqueous aluminum speciation in a mildly acidic solution or in the presence of complexing organic compounds. Furthermore, aluminum is of environmental and geochemical significance since it is an integral component of clays, another ubiquitous constituent of natural waters (surface and ground). Interaction between clays, cations, and internal water molecules can be significant in understanding the fate and transport of chemicals through the environment. Since colloidal suspensions of clay materials frequently represent challenges

123

to water and wastewater treatment, understanding of physical and chemical processes are of tantamount importance to the environmental scientist and engineer. Chapter 10 explores cation behavior in clay matrices by using "uncommon" nuclei such as ^{7}Li, ^{23}Na, and ^{133}Cs as probes. This is unique in that many NMR studies of complexation in clay have focused primarily upon the nuclei ^{27}Al and ^{29}Si.

In Chapter 11, ^{2}H and ^{17}O NMR are used to explore the formation and characterization of colloidal clay gels in ultrafine solids that are produced during processing Athabasca oil sands. This study is an example of NMR being used to investigate a very difficult and practical problem, the formation of ultrastable suspensions of waste solids.

The remainder of this overview chapter highlights the general background and importance of each of the chapters in this Part with respect to environmental science and technology. Several of the authors of these chapters supplied material for incorporation. Each of these contributions follows below, with attribution where appropriate.

Aqueous Chlorination Reactions of Organic Compounds

As noted previously, drinking water disinfection usually employs chlorination processes. The reaction mechanisms and by-products are not well understood. Since some of the identified chlorination by-products give rise to health concerns, it is of great importance, not only from a chemical point of view, but also from a concern for public health, to understand chlorination reactions fully and to identify the reaction by-products. One reason why the identification of chlorination by-products is such a difficult task is that many of these by-products are presumably polar and hydrophilic, in addition to being present in trace amounts. Gas chromatography (GC) coupled with mass spectrometry (GC/MS) require compounds of sufficient volatility and thermal stability. Derivatization techniques can be used to create volatile derivatives of the polar compounds, but ensuring that all polar compounds are detected, identification of the original polar molecule, and problems with sample modification are all of concern.

Even though chlorine isotopes cannot currently be used for NMR studies, enhancements in NMR sensitivity have made it possible to use ^{13}C and ^{1}H studies as adjuncts to GC/MS analysis where lack of thermal stability or high polarity has obscured reaction intermediates. F. E. Scully, B. Conyers, E. Forrer and E. P. Locke demonstrate this nicely in Chapter 6 for studies on the chlorination of amino acids. Even so, it is made clear that NMR has yet to become a true trace technique. Their results led to evidence of a previously unexpected pathway in aqueous chlorination reactions. Even though they were hindered by low sensitivity of the NMR and had to use reactant concentrations that were greater than those found in typical drinking water disinfection systems, their work begins to establish a basic understanding of aqueous chlorination chemistry.

Aluminum Speciation and Chemistry in Dilute Aqueous Solutions*

Because aluminum is ubiquitous in the environment and also due to concerns regarding effects of dissolved aluminum upon the health of plants, animals, and human beings, there is great incentive to examine the speciation and chemistry of aluminum at low concentrations in aqueous solutions. Liquid-state ^{27}Al NMR has proved invaluable for NMR studies, because of the 100% abundance of the ^{27}Al isotope which possesses a 5/2 spin quantum number. This strong advantage, allied to technological evolution in NMR—for example, the increase in magnetic fields and the introduction of high resolution magic angle spinning (MAS) solid-state NMR—has aroused numerous investigations on the chemistry of aluminum in liquid and solid states. In NMR spectrometry, obtaining ^{27}Al spectra with reduction in probe background (i.e., "baseline roll" due to acoustic ringing) is now possible with NMR probes fabricated with aluminum-free and low-aluminum materials, as reported by Simeral et al.[1] and Macfall et al.[2,3] Whether or not such probes become commercially available, their use would certainly improve the quality of ^{27}Al NMR spectrometric analyses of dilute solutions for hydroxyaluminum species in order to produce more accurate data about the fate of the dissolved species upon formation of aluminum (oxy)hydroxide solid phases.

The most abundant knowledge of the chemistry of soluble aluminum using ^{27}Al NMR has been brought by Akitt and coworkers since the late 1960s (e.g., refs. 4 and 5). They demonstrated the existence of the tridecamer A_{13} found by Johansson in 1960,[6] by the 62.5 ppm resonance downfield from 0 ppm for $Al(H_2O)_6^{3+}$ which is supposedly due to the central tetrahedral AlO_4 ion in the metastable Al_{13} polyion structure. Later NMR work by Bottero and coworkers in the early 1980s confirmed this point.[7,8] The use of ^{27}Al NMR in environmental problems was initiated by Bertsch and coworkers,[9,10] by studying the speciation of aluminum at low concentrations. More recently, Faust et al.[11] related chemical shifts and linewidths of the broader spectral peak due to the hydroxy-Al monometric complexes to calculated concentrations of the monomeric species. They worked at environmental levels of dissolved aluminum in acidic solutions and made use of the instrument incorporating NMR probes giving reduced Al background as described by Macfall et al.[2,3] Signals shifted 62.5 ppm downfield from 0 ppm were not encountered in their NMR experiments (Labiosa, private communication).

*With contributions from D. V. Vivit, K. A. Thorn, and J. D. Hern and from F. Thomas, A. Masion, J. Y. Bottero, and P. Tekely.

Being able to work at concentrations at or near environmental concentrations is significant since aluminum speciation and quantification are concentration-dependent. Therefore, working at high aluminum levels and extrapolating the results to environmentally significant concentrations is tenuous.

Complexation between aluminum and organic ligands has not been studied extensively by NMR, although this technique could be used qualitatively to establish the structure of the complexes, and quantitatively to derive complexation constants. The interactions between aluminum and organic ligands during the hydrolysis process have been addressed only in the 1990s, using liquid- and solid-state ^{27}Al NMR.[12–14] These studies have shown that the competition between the hydroxyl ions and the organic ligands results in inhibition of the tridecamer formation and precipitation of organo-aluminum products containing polymerized aluminum.

Interactions of Pollutants with Clays*

To date there have been few NMR studies of interactions of small molecules with clay systems. The probable reason for the lack of reported work is that ^{29}Si and ^{27}Al NMR are used extensively in characterizing the structure of minerals and especially in studies of zeolites. In our laboratory it has been found that the NMR spectra of these two nuclei appear to contain very little information about cation adsorption. Researchers have most likely started with these two nuclei and become discouraged about the potential of NMR for studies of adsorption.

The first published report using modern NMR of ^{113}Cd was by Bank and coworkers;[15] this was followed by work with ^{133}Cs from Kirkpatrick's group.[16,17] The most recent publications are associated with Prost and coworkers[18–20] examining the cations ^{23}Na, ^{39}K, ^{113}Cd, and ^{133}Cs. Pratum used solid-state ^{13}C NMR to study the interaction of tetraalkylammonium cations with clay complexes.[21] These publications prove that such studies contain useful information about both the structure and dynamics of cations in clays. We fully expect them to presage further work as research groups in environmental science become aware of these investigations and obtain access to modern high-field NMR instrumentation.

Ultrafine Particles and Gels†

Ultrafine materials (particles < 0.3 μm) have an important role in environmental research that is as yet little understood. In soils, sediments, and sludges such fractions usually account for most of the surface area, and the surface chemistry becomes much more important than the chemistry of the

---

*Contributed by W. L. Earl in collaboration with A. Labouriau and C. T. Johnston.
†Contributed by J. A. Ripmeester, L. S. Kotlyar, B. D. Sparks, and R. Schutte.

bulk material. Clays are some of the better known ultrafine materials, with surfaces that can act as ion exchangers and interact with hydrophilic as well as hydrophobic organics. It is clear that suspensions of ultrafines play an important role in the mobilization and transport of both heavy-metal and organic pollutants.

The properties of clay–water gels are of great industrial importance and have been studied at length, although mainly by macroscopic methods such as viscometry.[22] Early NMR work focused on the study of the interaction of water molecules with surfaces,[23-25] although Woessner realized that he had to consider mesoscopic length scales in order to explain his data. The Alberta oil sands have been exploited for some 30 years or so without solving the problem of producing vast quantities of sludge (now known as "fine tails"). The oil sands tailings were studied for many years,[26,27] and since a solution to the tailings problem was not easily found, a consortium was set up to study the fundamentals of fine tails formation. Our paper describes the application of NMR methods to the study of one component of the fine tails. The observation of gel-like properties led to the isolation of a fraction (ultrafines) which, although present in relatively small quantities, seems to dominate the behavior of the tails. In the very early stages of the work we became acquainted with the work of Grandjean and Laszlo,[28,29] who found that they could study relatively dilute suspensions of swelling clays with ^{2}H NMR through the distinct H doublet signature of a bound water fraction on oriented clay platelets. Examination of suspensions of the gel-forming fractions of ultrafines showed that these too showed the characteristic signature, thus giving an immediate clue as to the nature of the ultrafines. This led us to try and develop a number of methods which would give information on a microscopic level on the structure and dynamics of gel formation in suspensions of ultrafines. On the theory side, the work of Halle and Wennerstrom[30] provides a framework for explaining most of the results. The only concept that needs to be added is that of "domains," as used in earlier work on clays by Woessner,[23-25,31] because gels and suspensions of ultrafines are locally anisotropic. In order to learn more about domain sizes, one would also have to measure diffusion constants.

The NMR results did much to help define models for the oil sands fine tails, which in turn have helped in other problems of structure and dynamics on a mesoscopic scale[32,33] and are currently of great interest. Complementary information from different techniques should, of course, be considered. Obvious extensions of the work reported here include the effect of various cations and organics on gel formation, and the effects of various gel-breaking methods on the NMR observables. Ultrafines are likely to show up in other industrial sludges and sediments. Looking ahead, there will also be a role for NMR methods in studying the role of clays in the transport of pollutants, although low concentrations may require extensive isotopic labeling to enhance sensitivity.

References

1. Simeral, L. S., Zens, T., and Finnegan, J., Al-27 NMR without probe background. *Appl. Spectrosc.* 47, 1954, 1993.
2. Macfall, J. S., Ribeiro, A. A., Cofer, G. P., Dai, K. G., Faust, B. C., and Richter, D. D., Paper delivered at the 33rd Experimental Nuclear Magnetic Resonance Conference, Asilomar Conference Center, California, 1992.
3. Macfall, J. S., Ribeiro, A. A., Cofer, G. P., Dai, Ko-H., Labiosa, W., Faust, B. C., and Richter, D. D., Design and use of background-reduced ^{27}Al NMR probes for the study of dilute samples from the environment. *Appl. Spectrosc.* 49, 156, 1995.
4. Akitt, J. W., Greenwood, N. N., Lester, G. D., Aluminum-27 nuclear magnetic resonance studies of acidic solutions of aluminum salts. *J. Chem. Soc. A* 803, 1969.
5. Akitt, J. W., Greenwood, N. N., Khandelwal, B. L., and Lester, G. D., [27]Al nuclear magnetic resonance studies of the hydrolysis and polymerization of the hexa-aquo-aluminum(III) cation. *J. Chem. Soc. Dalton Trans.* 604, 1972.
6. Johansson, G., On the crystal structures of some basic aluminum salts. *Acta Chem. Scand.* 14, 769, 1960.
7. Bottero, J. Y., Cases, J. M., Fiessinger, F., and Poirier, J. E., Studies of hydrolyzed aluminum chloride solutions. 1. Nature of aluminum species and composition of aqueous solutions. *J. Phys. Chem.* 84, 2933, 1980.
8. Bottero, J. Y., Marchal, J. P., Poirier, J. E., Cases, J. M., and Fiessinger, F., Étude par RMN de l'aluminium-27, des solutions diluées de chlorure d'aluminium partiellement hyrolysées. *Bull. Soc. Chim. Fr.* 11–12, Part 1, 439, 1982.
9. Bertsch, P. M., Thomas, G. W., and Barnhisel, R. I., Characterization of hydroxy-aluminum solutions by [27]Al nuclear magnetic resonance spectroscopy. *Soil Sci. Soc. Am. J.* 50, 825, 1986.
10. Bertsch, P. M., Layton, W. J., and Barnhisel, R. I., Speciation of hydroxy-aluminum solutions by wet chemical and aluminum-27 nuclear magnetic resonance spectroscopy. *Soil Sci. Soc. Am. J.* 50, 1449, 1986.
11. Faust, B. C., Labiosa, W. B., Dai, Ko'H., Macfall, J. S., Browne, B. B., Rebeiro, A. A., and Richter, D. D., Speciation of aqueous Al(III) complexes of hydroxide, sulfate, and oxalate at pH 2–5 and at concentrations of geochemical relevance by Al-27 NMR spectroscopy. In Abstracts with Programs, 1994 Annual Meeting, The Geological Society of America.
12. Thomas, F., Masion, A., Bottero, J. Y., Rouiller, J., Genevrier, F., and Boudot, D., Aluminum (III) speciation with acetate and oxalate. A potentiometric and ^{27}Al NMR study. *Environ. Sci. Technol.* 25, 1553, 1991.
13. Thomas, F., Masion, A., Bottero, J. Y., Rouiller, J., Montigny, F., and Genevrier, F., Aluminum(III) speciation with hydroxy-carboxylic acids. ^{27}Al NMR study. *Environ. Sci. Technol.* 27, 2511, 1993.
14. Masion, A., Thomas, F., Bottero, J. Y., Tchoubar, D., and Tekely, P., Formation of amorphous precipitates from aluminum–organic ligands solutions: macroscopic and molecular study. *J. Non-Cryst. Solids*, 171, 191, 1994.
15. Bank, S., Bank, J. F., and Ellis, P. D., Solid-state [113]Cd nuclear magnetic resonance study of exchanged montmorillonites. *J. Phys. Chem.* 93, 4847, 1989.
16. Weiss, C. A., Kirkpatrick, R. J., and Altaner, S. P., Variations in interlayer cation sites of clay minerals as studied by [133]Cs MAS nuclear magnetic resonance spectroscopy. *Am. Miner.* 75, 970, 1990.

17. Weiss, C. A., Kirkpatrick, R. J., and Altaner, S. P., The structural environments of cations adsorbed onto clays: [133]Cs variable-temperature MAS NMR spectroscopic study of hectorite. *Geochim. Cosmochim. Acta* 54, 1655, 1990.

18. Lapreche, V., Lambert, J. F., Prost, R., and Fripiat, J. J., High-resolution solid-state NMR of exchangeable cations in the interlayer surface of a swelling mica: [23]Na, [111]Cd, and [133]C vermiculites. *J. Phys. Chem.* 94, 8821, 1990.

19. Tinet, D., Faugere, A. M., and Prost, R., [113]Cd NMR chemical shift tensor analysis of cadmium-exchanged clays and clay gels. *J. Phys. Chem.* 95, 8804, 1991.

20. Lambert, J. F., Prost, R., and Smith, M. E., [39]K solid-state NMR studies of potassium tecto- and phyllosilicates: the in situ detection of hydratable K^+ in smectites. *Clays Clay Miner.* 40, 253, 1992.

21. Pratum, T. K., A solid-state, ^{13}C NMR study of tetraalkylammonium/clay complexes. *J. Phys. Chem.* 96, 4567, 1992.

22. van Olphen, H., *An Introduction to Clay Colloid Chemistry*, Interscience, New York, NY, 1963.

23 Woessner, D. E., and Snowden, B. S., NMR doublet splitting in aqueous montmorillonite gels. *J. Chem. Phys.* 50, 1516, 1969.

24. Woessner, D. E., NMR and structure in aqueous heterogeneous systems. *Mol. Phys.* 34, 899, 1977.

25. Woessner, D. E., An NMR investigation of the range of the surface effect on the rotation of water molecules. *J. Magn. Reson.* 39, 297, 1980.

26. Scott, J. D., Dusseault, M. B., and Carrier III, W. D., Behaviour of the clay/bitumen/water system from oil sands extraction plants. *Applied Clay Sci.* 1, 207, 1985.

27. Pierre, A. C., Zou, J., and Barker, C., Structure comparison of an oil sands tailings sludge with a montmorillonite gel model. *Fuel* 71, 1373, 1992.

28. Grandjean, J., and Laszlo, P., Deuterium NMR studies of water molecules restrained by their proximity to a clay surface. *Clays Clay Miner.* 37, 403, 1989.

29. Delville, A., Grandjean, J., and Laszlo, P., Order acquisition by clay platelets in a magnetic field. NMR study of the structure and microdynamics of the adsorbed clay layer. *J. Phys. Chem.* 95, 1383, 1991.

30. Halle, B., and Wennerstrom, H., Interpretation of magnetic resonance data from water nuclei in heterogeneous systems. *J. Chem. Phys.* 75, 1928, 1981.

31. Woessner, D. E., and Snowden, Jr., B. S., A pulsed NMR study of dynamics and ordering of water molecules in interfacial systems. *Ann. N.Y. Acad. Sci.* 204, 113, 1973.

32. Ziman, J. M., *Models of Disorder*, Cambridge University Press, Cambridge, 1979.

33. Lekkerkerker, H. N. W., Crystalline and liquid crystalline order in concentrated colloidal dispersions: an overview. *NATO ASI Ser. B* 211, 165, 1989.

7.

NMR Studies of the Reaction of Amino Acids with Aqueous Chlorine

FRANK E. SCULLY, JR., BARBARA CONYERS, ERIKA FORRER,

& EDWARD P. LOCKE

Over the past 20 years, gas chromatography/mass spectroscopy (GC/MS) has been widely used to identify trace organic environmental contaminants and to study the mechanisms of the formation or transformation of organic compounds either by natural or man-made processes. In the area of water and wastewater disinfection, GC/MS has been highly successful in identifying numerous volatile organic chlorination by-products, some of which may pose undesirable health risks to humans and aquatic organisms at concentrations found in some waters. However, despite a considerable amount of research in this area much of the chemistry continues to be poorly understood. Analysis of trace organics by GC/MS relies on the assumption that the compounds to be analyzed are (1) volatile and (2) thermally stable to GC temperatures as high as 300 °C. Because nuclear magnetic resonance spectroscopy (NMR) is a mild and nondestructive method of analysis, it can reveal reactions that occur in water that cannot be observed by GC/MS.

Until recently the reactions of amino acids with two or more equivalents of aqueous chlorine were believed to produce aldehydes and nitriles according to equation (1).[1-8] LeCloirec and Martin[9] have reported that the formation of nitriles in such situations may come in part from the reaction of monochloramine with aldehydes (equation (2)). Because reaction (2) may affect the distribution of products in reaction (1), it was important to determine the relationship between these two reactions. This chapter will review the applications of NMR we have used in studies of the products formed upon chlorination of amino acids.

$$\underset{\overset{|}{R-CH-COO^-}}{\overset{NH_2}{}} \xrightarrow{\geq 2HOCl} \underset{\overset{\|}{R-C-H}}{\overset{O}{}} + R-C\equiv N \tag{1}$$

$$R-\overset{\overset{\displaystyle O}{\|}}{C}-H + NH_2Cl \rightleftharpoons R-\overset{\overset{\displaystyle OH}{|}}{\underset{\underset{\displaystyle NH-Cl}{|}}{C}}-H \xrightarrow{-H_2O} R-\overset{\overset{\displaystyle N-Cl}{\|}}{C}-H \xrightarrow{-HCl} R-C\equiv N \quad (2)$$

$$\mathbf{I}$$

Methods

All manipulations of moisture-sensitive compounds, mixtures, and solutions were carried out in a glove bag maintained dry with P_2O_5 and purged with dry nitrogen. Broadband decoupled carbon-13, nitrogen-15, and proton NMR spectra were recorded with a JEOL FX-90Q spectrometer or with a Varian Unity Plus 400 MHz NMR instrument. Proton spectra were generated after collecting 16 transients (45° tip angle, 8192 points, spectral width 900 Hz or 4000 Hz, acquisition time 2 s, pulse delay 2 s). Carbon spectra were generated after collecting 500–2000 transients (45° tip angle, 8192 points, spectral width of 5000 Hz, acquisition time 0.8 sec, pulse delay 1.2 s).

The standard used for all ^{15}N NMR experiments was a solution containing 2.9 M [15]NH$_4$Cl (78 mg), 12 µL concentrated HCl, and 488 µL H$_2$O. The standard was placed in a coaxial insert in a 5 mm tube containing samples or reactions mixtures. The nitrogen NMR spectrum of the standard inserted into a tube containing D$_2$O displayed a single peak. That peak was set at 24.93 ppm.

Chlorine gas was dried by bubbling it through concentrated sulfuric acid and passing the effluent through a 30 cm column of 5 Å molecular sieves. The gas was then bubbled into a solution of 1 g sodium borate in 10 mL of D$_2$O until the concentration of active chlorine was 17 000 mg Cl$_2$/L. The pH was above 6.

Exchangeable hydrogen atoms on the amino acids were converted to deuterium atoms by isotope exchange. Alanine (0.89 g, 10 mmol) and sodium borate (0.040 g) were dissolved in 9 mL D$_2$O and the resulting solution was lyophilized. The process was repeated to ensure greater than 95% exchange in order to minimize the interference from DOH in the proton spectrum.

For NMR experiments employing ^{15}N-labeled alanine, 17 mg of H$_2$O-free sodium borate was dissolved in 0.5 mL Clorox along with 2 drops NaOD. The solution was chilled in an ice bath prior to adding 14 mg of ^{15}N-alanine to the tube, followed by 0.15 mL D$_2$O. The mixture was stirred in the ice bath for 5 min before taking the NMR spectrum.

Working solutions of alanine were prepared by dissolving the powdered mixture of buffer and alanine described above in 1 mL of D$_2$O and adjusting the pH to 10.3 with concentrated NaOD. This solution was added rapidly with stirring to 9 mL of the standardized solution of NaOCl (17 000 mg Cl$_2$/L in D$_2$O). The final concentration of alanine was 1.0 M. The pH was monitored with a calibrated glass microelectrode as the solution became clear and colorless and the pH decreased to approximately 7.5. A portion was placed in an NMR tube and the remainder was diluted 3.5-fold with H_2O and stored in the dark for later extraction. The NMR spectra of these concentrated D_2O solutions were recorded immediately. After NMR spectra of the reaction mixtures in D_2O had been recorded, the portion stored in the dark was saturated with KCl and extracted with 0.5 mL of $CDCl_3$. The spectrum of the extract was recorded immediately.

A dilute solution of aqueous chlorine was prepared from a commercial solution (Clorox) and standardized with thiosulfate. Reagent-grade acetaldehyde was distilled prior to use and stored under nitrogen. Milli-Q water was deoxygenated by bubbling N_2 through it for at least 10 min. A 1.0 M stock solution of acetaldehyde was prepared in 1000 mL of deoxygenated Milli-Q water. A stock solution of monochloramine was prepared by mixing in a syringe pump equal volumes of a 0.1 M ammonium chloride solution with an 8.0×10^{-2} M solution of aqueous chlorine. A working solution was prepared by diluting 1 mL of the stock solution to 10 mL with 0.05 M sodium phosphate or sodium acetate buffer containing 0.5 M sodium perchlorate. The reaction with acetaldehyde was initiated by mixing 3 mL of this solution with 120 μL (or a similar volume) of 1 M acetaldehyde in a 1 cm quartz UV cell (thermostatted to 20 ± 0.1 °C) and monitoring the change in absorbance at 244 nm. Kinetic studies were conducted using a Cary 219 spectrophotometer. Simultaneously the decrease in oxidant concentration was monitored by iodometric titration.

Results and Discussion

When nonpolar amino acids react with aqueous chlorine they form N-chlorinated amino acids which can react with a second equivalent of aqueous chlorine to form N,N-dichloramino acids. Stanbro et al.[10] and Stelmazsynska et al.[11] proposed that N,N-dichloramines decompose through N-chloroaldimine intermediates in order to explain the kinetics of the decomposition of N,N-dichloramino acids. NMR was the first spectroscopic technique to demonstrate unequivocally that in model solutions chlorination of the nonpolar amino acids isoleucine, valine, and phenylalanine produces unusually stable N-chloroaldimines,[12–15] as well as previously identified nitriles and aldehydes. With more than two equivalents of aqueous chlorine, N-chloroaldimines represent the major product of the chlorination of the amino acids studied thus far. These compounds were first identified by NMR after extraction into nonaqueous solvent, because early attempts at GC/MS analysis (injection temperatures ≥ 250 °C) revealed that they were thermally unstable. The proton NMR spectrum of the reaction mixture of the chlorination of alanine in D_2O (Figure 7.1(a)) reveals the characteristic resonance of the N-chloroaldiminic proton (quartet) at approximately 8 ppm. The resonance of the singlet superimposed on the methyl doublet of the N-chloroaldimine (2.0 ppm) is due to acetonitrile, another chlorination product. Despite efforts

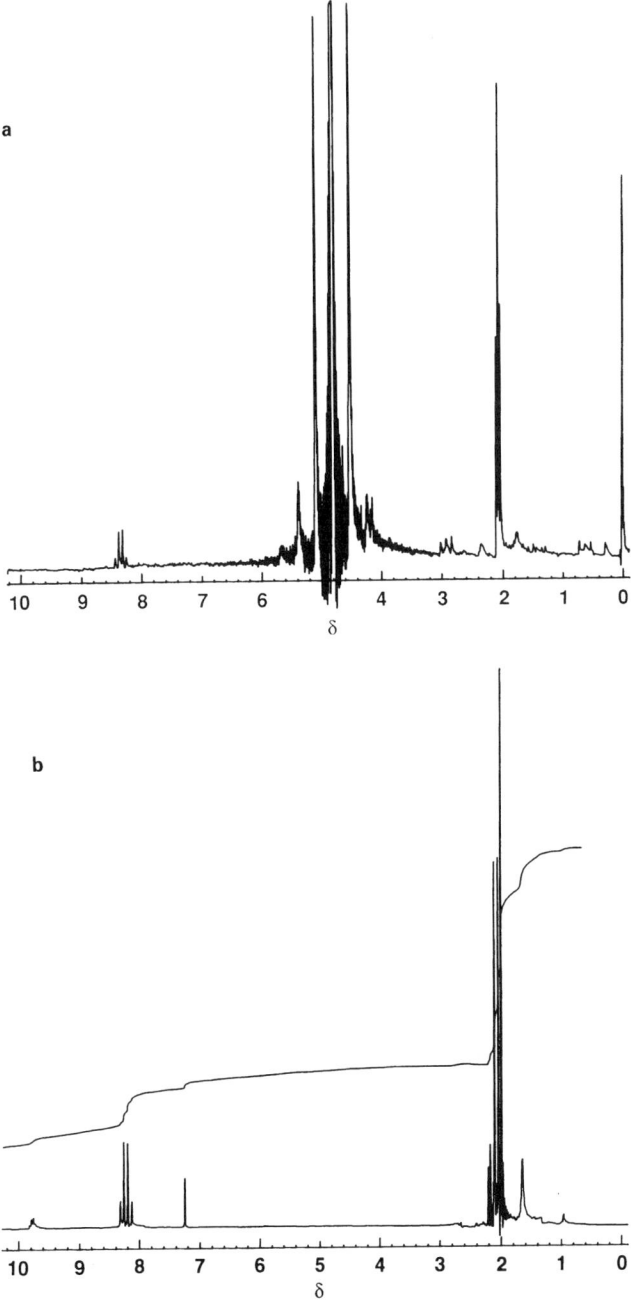

Figure 7.1 (a) ¹H NMR spectrum of a 1 M solution of alanine in D₂O containing carbonate buffer (pH 10.3) chlorinated until the pH decreased to pH 6.5. (b) The same solution extracted into CDCl₃.

133

to exclude H_2O from the reaction mixture, a large resonance at approximately 5 ppm due to DOH was routinely observed. When the reaction mixture is extracted into deuterated chloroform, a similar spectrum (Figure 7.1(b)) is observed.

The reaction of ^{15}N-labeled alanine in borate buffer at pH 7.8 with two equivalents of chlorine revealed resonances at 236 ppm, 291 ppm, and 312 ppm in the ^{15}N NMR spectrum (Figure 7.2). Additional experiments with standard solutions identified the resonance at 312 ppm to be monochloramine, and the resonance at 236 ppm to be acetonitrile. The resonance at 291 ppm is believed to be the N-chloroaldimine. When the reaction mixture was extracted with 0.7 mL $CDCl_3$, the proton spectrum of the extract revealed the characteristic chloroaldimine resonance at 8.3 ppm seen in previous spectra. Previous studies have shown that N-chloroaldimines decompose in aqueous solution to form aldehydes, possibly undergoing hydrolysis with the loss of monochloramine. The presence of monochloramine in the nitrogen-15 spectrum supports this hypothesis.

More extensive studies of the reactions of aqueous chlorine with amino acids showed that the primary reactions are understood to be as illustrated in Figure 7.3 for phenylalanine.

Because several of the amino acids are very polar or thermally labile, their chlorination products are not amenable to extraction and GC/MS analysis. Threonine reacts readily with aqueous chlorine to form a reaction mixture which, after extraction into $CDCl_3$, gives the NMR spectrum shown in Figure 7.4. The spectrum reveals the presence of an N-chloroaldimine (doublet at 8.2 ppm, part of the multiplet at 4.5 ppm, and the doublet at 1.4 ppm). The presence of acetaldehyde was suggested by a quartet at 9.8 ppm and a doublet at 2.2 ppm. Its presence was confirmed by enriching the solution with acetaldehyde and re-recording the spectrum, and also by analyzing the solution by headspace GC/mass spectroscopy. The resonance at 1.8 ppm corresponds to the OH hydrogen in the chloroaldimine. The broad resonance at 4.5 ppm has the correct chemical shift and integration for the methine hydrogen atoms in this compound. Integration of the peaks suggests that the two products, N-chlorolactaldimine and acetaldehyde, are formed in a 3:1 ratio.

Based on previous work, the formation of lactonitrile was expected. Although it is extractable into $CDCl_3$, the NMR did not support the formation of this compound. The presence of acetaldehyde suggested that under the conditions of the reaction the cyanohydrin (lactonitrile) lost HCN to form the aldehyde.

N—Chlorolactaldimine   Lactonitrile   Lactaldehyde

Figure 7.2 The ^{15}N NMR spectrum of the product mixture from the reaction of aqueous chlorine with ^{15}N-labeled alanine at pH 7.8.

Because of the lability of the N-chloroaldimine moiety and the non-volatility of compounds containing the —OH functionality, attempts to identify the products by extraction and GC/MS analysis have been unsuccessful.

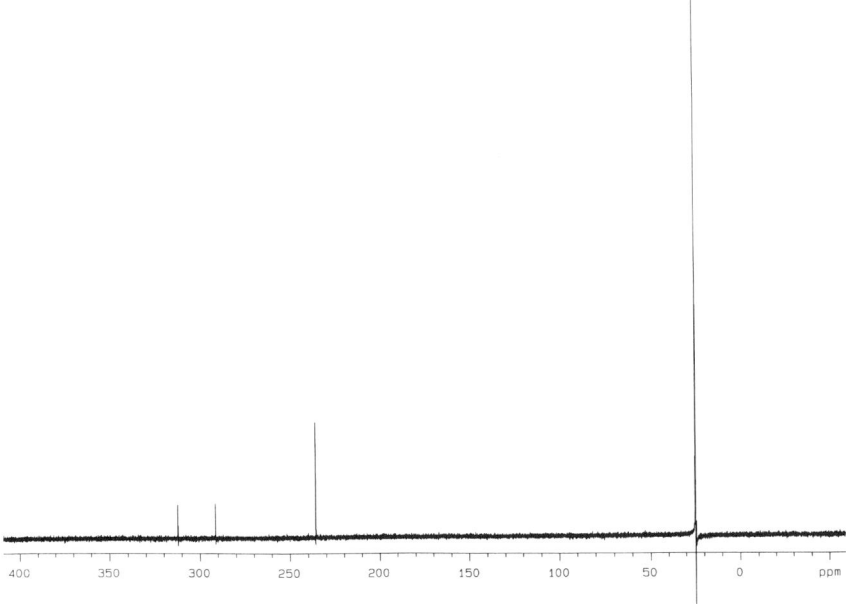

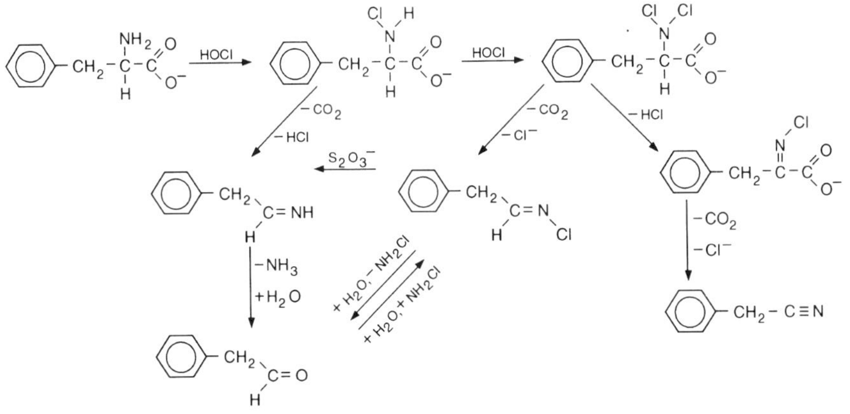

Figure 7.3 Scheme of the major reaction pathways for the chlorination of phenyl-alanine and the decomposition of the chlorination products.

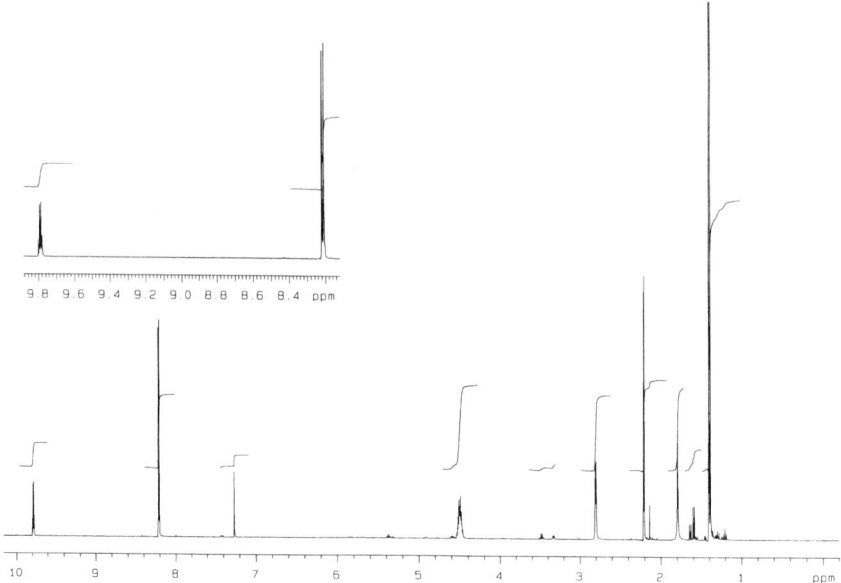

Figure 7.4   The ^{1}H NMR spectrum of a CDCl$_3$ extract of a solution of threonine in carbonate buffer (pH 10.3) chlorinated with excess aqueous chlorine and saturated with KCl.

The ^{13}C and ^{1}H NMR spectra of some of the N-chloroaldimines confirm the structural assignment by revealing the presence of *syn* and *anti* stereoisomers of the chloroaldimines.[13,14]

$$
\begin{array}{cc}
\text{Cl—N} & \text{N—Cl} \\
\| & \| \\
\text{CH}_3\text{—CH—C—H} & \text{CH}_3\text{—CH—C—H} \\
| & | \\
\text{H}_3\text{C} & \text{H}_3\text{C} \\
Syn & Anti
\end{array}
$$

Because evidence in our laboratory suggested that, at least at pH 7–10, N-chloroaldimines slowly decompose by hydrolysis to aldehydes rather than by dehydrohalogenation to nitriles,[14] we were intrigued by the proposal of LeCloirec and Martin[9] that monochloramine reacts with aldehydes to form nitriles through the intermediacy of an N-chloroaldimine (equation (2) above). They also propose that part of the acetonitrile formed on chlorination of alanine is due to this reaction. The reaction of monochloramine with aldehydes would be of great environmental significance, as an increasing number of drinking water utilities replace chlorination with ozonation (which produces aldehydes[16]) followed by chloramination for residual disinfection.

This observation would suggest that the dehydration step in equation (2) involves an equilibrium.

When 4.0×10^{-3} M monochloramine is mixed with an excess of acetaldehyde (3.8×10^{-2} M) at pH values ranging from 5 to 6.5, the monochloramine slowly reacts. Under these pseudo-first-order conditions, the observed rate of loss of monochloramine increases with decreasing pH (Figure 7.5), and is first order in H_3O^+ concentration. The presence of excess ammonia, which would affect the speciation of the chloramine (mono- vs. dichloramine), had no effect on the reaction rate. The half-life of the reaction at pH 6.5 is 31 min at 20 °C. Extrapolating back to environmentally significant concentrations, the half-life of acetaldehyde under pseudo-first-order conditions in the presence of 1×10^{-5} M monochloramine is estimated to be 82 days at pH 6.5. The half-life would be even longer at higher pH. The conclusion from these observations is that the direct reaction of monochloramine with aldehydes is too slow to be of much significance in a drinking water distribution system.

Conclusion

The major advantage of NMR is its value as a tool for elucidating the structure of complex or unusual unknown organic compounds. Furthermore, the structural information is acquired under very mild, nondestructive conditions and in solutions where the compounds are more apt to appear in the form in which they exist in the environment. The major limitation of NMR in the environmental sciences is the amount of analyte required to obtain meaningful

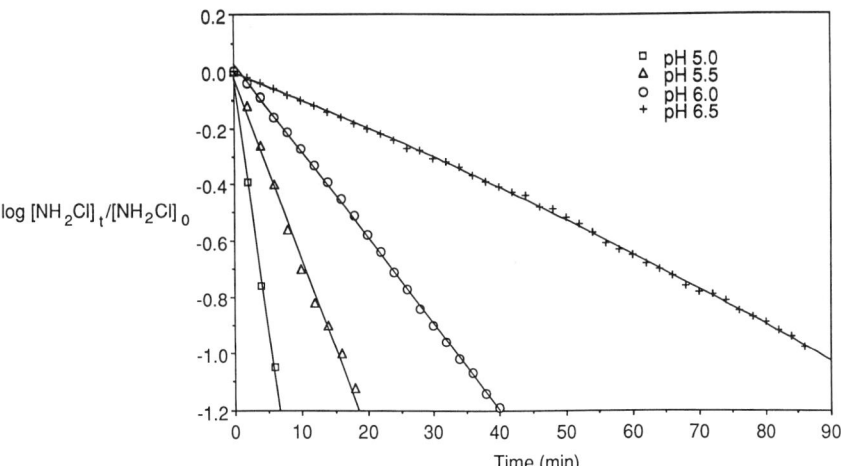

Figure 7.5 Decrease in the concentration of monochloramine (initial concentration $= 4 \times 10^{-3}$ M) in the presence of excess acetaldehyde (3.8×10^{-2} M) at different pH values.

spectra. However, for compounds of environmental significance which are formed in high yield, NMR is more likely than GC/MS to reveal their structure as they appear in a natural aqueous environment and without the risk of thermolytic degradation. In the case of the *N*-chloroaldimines the structure of many of them could not have been determined by any other current analytical method.

Acknowledgments This research was supported by the National Science Foundation Grant BCS-9002442, project manager Dr. Edward Bryan. Any opinions, findings, and conclusions or recommendations expressed in this publication are those of the authors and do not necessarily reflect the views of the National Science Foundation.

References

1. Langheld, K., Über den Abbau der α-Aminosaüren zu fetten Aldehyden mittels Natriumhypochlorit. *Chem. Ber.* 42, 392, 1909.
2. Langheld, K., Über das Verhalten von α-Aminosaüren gegen Natriumhypochlorit. *Chem. Ber.* 42, 2360, 1909.
3. Dakin, H. D., The oxidation of amino acids to cyanides. *Biochem. J.* 10, 319, 1916.
4. Dakin, H. D., Cohen, J. B., Danfresne, M., and Kenyon, J., *Proc. R. Soc. London, Ser. B* 89, 232, 1916.
5. Dakin, H., On the oxidation of amino acids and of related substances with chloramine-T. *Biochem. J.* 11, 79, 1917.
6 Wright, N. C., The action of hypochlorites on amino-acids and proteins. The effect of acidity and alkalinity. *Biochem. J.* 30, 1661, 1936.
7. Pereira, W. E., Hoyano, Y., Summons, R. E., Bacon, V. A., and Duffield, A. M., Chlorination studies II. The reaction of aqueous hypochlorous acid with α-amino acids and dipeptides. *Biochim. Biophys. Acta* 313, 170, 1973.
8. Stanbro, W. D., and Smith, W. D., Kinetics and mechanism of the decomposition of *N*-chloroalanine in aqueous solution. *Environ. Sci. Technol.* 13, 446, 1979.
9. LeCloirec, C., and Martin, G., Evolution of amino acids in water treatment plants and the effect of chlorination on amino acids. In *Water Chlorination: Chemistry, Environmental Impact, and Health Effects*, Jolley, R.L. et al. (eds), Lewis Publishers, Chelsea, MI, 1985, p. 821.
10. Stanbro, W.D., and Lenkevich, M. J., Kinetics and mechanism of the decomposition of *N,N*-dihalopeptides. *Int. J. Chem. Kinet.* 17, 401, 1985.
11. Stelmaszynska, T., and Zgliczynski, J.M., *N*-(2-Oxoacyl)amino acids and nitriles as final products of dipeptide chlorination mediated by myeloperoxidase/H_2O_2/ Cl^- system. *Eur. J. Biochem.* 92, 301, 1978.
12. Nweke, A., and Scully, Jr., F. E., A stable organic *N*-chloroaldimine and other products of the chlorination of isoleucine in model solutions and in a wastewater. *Environ. Sci. Technol.* 23, 989, 1989.
13. McCormick, E. F., Conyers, B., and Scully, Jr., F. E., N-Chloroaldimines Part II: Chlorination of valine in model solutions and in a wastewater. *Environ. Sci. Technol.* 27(2), 255, 1993.

14. Conyers, B., and Scully, Jr., F. E., N-Chloroaldimines Part III: Chlorination of phenylalanine in model solutions and in a wastewater. *Environ. Sci. Technol.* 27(2), 261, 1993.
15. Walker, E., Conyers, B., and Scully, Jr., F. E., N-Chloroaldimines Part IV: Identification in a chlorinated municipal wastewater by GC/MS. *Environ. Sci. Technol.* 27(4), 720, 1993.
16. Rice, R.G., in *Safe Drinking Water: The Impact of Chemicals on a Limited Resource*, Rice, R.G. (ed.), Lewis Publishers, Alexandria, VA, 1985, p. 123.

8.

Comparative Results of ^{27}Al NMR Spectrometric and Ferron Colorimetric Analyses of Hydroxyaluminum Hydrolysis Products in Aged, Mildly Acidic, Aqueous Systems

DAVISON V. VIVIT, KEVIN A. THORN, & JOHN D. HEM

The abundant presence of aluminum (Al) in the environment, its commercial importance, and the potential toxic effects of dissolved forms of Al on living organisms have brought about extensive studies of Al hydrolysis products that originate from natural processes and industrial activities. A recent review of past studies of hydroxyaluminum (hydroxy-Al) solutions illustrates the complexity of aluminum hydrolysis.[1] Because of the need to identify and quantify individual Al ion species, initial investigations of concentrated solutions of hydrolyzed aluminum salts using ^{27}Al nuclear magnetic resonance (NMR) spectrometry began more than 20 years ago in the United Kingdom with continuous-wave instrumentation.[2] Since then, groups in Switzerland,[3,4] France,[5,6] and the United States[7-12] as well as in the UK,[13,14] involved with geochemical and pharmacological investigations, have applied pulse NMR instruments to conduct research mostly with concentrated Al hydrolysates.

In NMR spectrometry, shapes and positions of spectral peaks indicate aluminum nuclei in different chemical environments, and the peak areas are proportional to the number of nuclei in each given symmetry. Figures 8.1(a) and (b) and 8.2(a) and (b) illustrate liquid-state ^{27}Al NMR spectra of aluminum hydrolysis products. The well-defined peak at 62.5 ppm, downfield from the 0 ppm reference position for $Al(H_2O)_6^{3+}$, has been attributed to the tetrahedrally coordinated aluminum ion within the highly symmetrical configuration of the $[AlO_4Al_{12}(OH)_{24}(H_2O)_{12}]^{7+}$ species or "Al$_{13}$" polyion[2,3,8] in aqueous solution. This geometrical arrangement, also termed the "Keggin" structure,[15] was originally deduced from X-ray diffraction studies of aluminum hydroxysulfate and hydroxyselenate salts precipitated from heated concentrated aluminum solutions after being partly neutralized

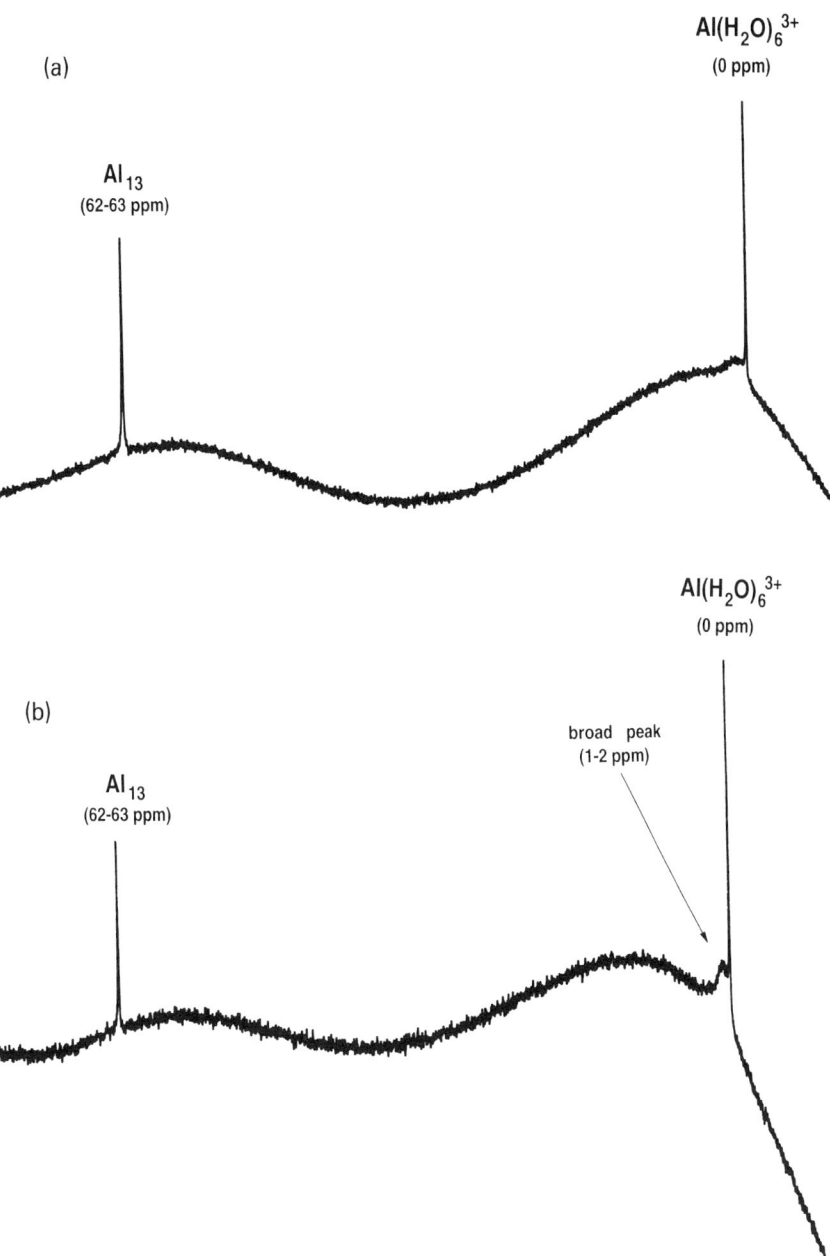

Figure 8.1 ^{27}Al NMR spectra of product solution from Experiment 33, 0.002 M Al (ClO$_4$)$_3$ titrated with 0.01 M NaOH at pH 4.90: (a) initial spectrum of solution, aged 9 days (pH 4.61); (b) spectrum of same solution, aged 51 days (pH 4.48).

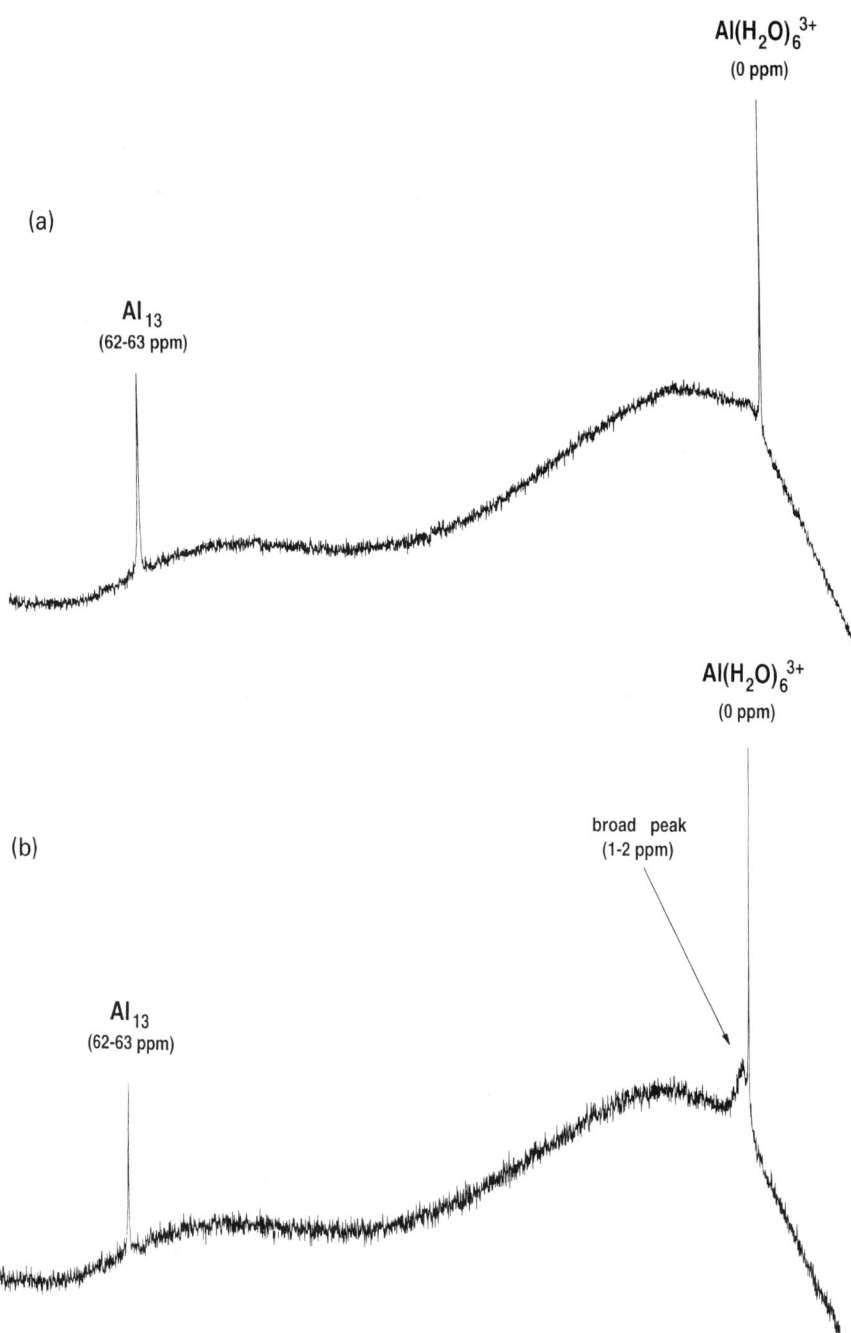

Figure 8.2 ^{27}Al NMR spectra of product solution from Experiment 34, 0.002 M Al (ClO$_4$)$_3$ titrated with 0.01 M NaOH at pH 4.90: (a) initial spectrum of solution, aged 1 day (pH 4.57); (b) spectrum of same solution, aged 54 days (pH 4.28).

with base.[16] The Al_{13} structural unit that was postulated has 12 Al ions octahedrally coordinated with OH^- ions enclosing the central tetrahedral Al ion which is bonded to four of the octahedral ions through bridging by O^{2-} ions.[16]

For a given sample specimen, broad NMR spectral peaks indicate the existence of (1) chemical-exchange processes and (2) species having less symmetry and larger electric field gradients than those which give rise to narrower peaks. Mildly acidic solutions of aluminum hydrolysis products exhibit a broad peak downfield from 0 ppm centered at $\sim$1 to $\sim$2 ppm (Figures 8.1(b) and 8.2(b)). This peak has been interpreted as being a consolidation of resonances from the $AlOH^{2+}$, $Al(OH)_2^+$, and $Al(H_2O)_6^{3+}$ monomeric equilibrium species.[17] If the pH is lowered by addition of acid to the system, monomeric Al increases in concentration, and $Al(H_2O)_6^{3+}$ becomes the predominant form. The broad peak narrows, shifts towards 0 ppm, and increases in intensity.[17] Upon aging of systems supersaturated with respect to gibbsite, the pH also decreases with further hydrolysis of freshly formed metastable hydroxy-Al polymers as double hydroxide bridge bonds are formed in the evolution of microcrystalline gibbsite.[18–20] In both cases of acidification, the Al_{13} signal disappears as the pH decreases.[7,17,20]

Use of ferron in spectrophotometric determinations of hydroxyaluminum species predates the first use of ^{27}Al NMR spectrometry.[18,22,23] In simple inorganic aqueous systems having no aluminum complex ions other than hydroxy-Al forms, total dissolved aluminum (Al_t) can be expressed as follows:

$$Al_t = Al_a + Al_b + Al_c \qquad (1)$$

Respectively, Al_a, Al_b, and Al_c are the monomeric, polymeric, and nonreactive hydroxy-Al fractions as discerned by their respective psuedo-first-order rates of reaction with ferron.[22,23] Al_a consists of the $Al(H_2O)^{3+}$, $AlOH^{2+}$, $Al(OH)_2^+$, and $Al(OH)_4^-$ species and reacts with ferron almost intantaneously (on the order of a minute or less). Al_b comprises the fraction of the Al_t, aside from Al_a, that reacts with ferron within a period 15 min to 2 h. Al_c is taken to be a fraction of the Al_t that behaves as a solid by reacting very slowly beyond 2 h or not reacting at all with ferron.

This chapter compares NMR results with companion speciation data obtained with ferron on dilute hydroxy-Al solutions ranging in total aluminum concentration from $\sim$0.4 to $\sim$1 mM Al. Hydroxy-Al polymers present in these solutions were produced in aqueous chemical systems which were not at thermodynamic equilibrium but at steady state. These laboratory systems simulate such systems that are frequently encountered in the environment where rates of formation of metastable products are counterbalanced with reactant fluxes.[19]

Experimental

Reagents

Reagent-grade chemicals, except for ferron, were used throughout the study. The ferron powder (J.T. Baker Chemical Co., Phillipsburg, NJ)* was not purified further prior to preparation of reagent solutions. Solutions were prepared with deionized, distilled water. Sodium hydroxide solutions were prepared with deionized, distilled water which had been boiled to expel carbon dioxide. Solutions were kept in polyethylene bottles unless noted otherwise.

pH-stat Titrations

$Al(ClO_4)_3$ Stock Solution, 1000 mgAl/L   On a steam bath 0.1000 g Al wire was dissolved in 2.3 mL 35% $HClO_4$ followed by cooling and diluting to 100.0 mL.

$NaClO_4$ Solution (0.0100 M)   This solution was prepared by dissolving 0.6123 g $NaClO_4$ and diluting to 500.0 mL.

Solution A (2.00 mM Al)   After 0.307 g $NaClO_4$ was added to and dissolved in 26.98 mL Al stock solution, the resulting solution was diluted to 500.0 mL with H_2O. The H^+ concentration was determined by accurately measuring the pH and correcting for the ionic strength (e.g., H^+ ion activity coefficient $\gamma = 0.91$ at 25 °C for ionic strength, $I = 0.01$).[24]

Solution B (0.01 M NaOH)   About 400 mL of CO_2-free H_2O was used to dissolve 0.6123 g $NaClO_4$. Then 400 μL 50% NaOH solution was added followed by dilution to 500.0 mL with CO_2-free H_2O. The solution was protected from atmospheric CO_2. The OH^- concentration was determined by titrating 75.0 mL of a standard potassium biphthalate (KHP) solution ($\sim$1.5 mM KHP) with the prepared solution.

Ferron Analysis

Ferron Solution, 0.10% (w/v)   After 0.50 g ferron was allowed to dissolve overnight in 500 mL H_2O with the aid of a magnetic stirrer, the resulting solution was passed through a 0.45 μm filter prior to use.

Sodium Acetate, 35% (w/v)   Using the anhydrous salt, 175 g $NaC_2H_3O_2$ was dissolved in 400 mL H_2O and diluted to 500.0 mL with H_2O. A glass reagent bottle was used for storage.

---

*Use of trade names is for identification only and does not imply endorsement by the US Geological Survey.

Hydroxylamine Hydrochloride Solution, 10% (w/v)   A 40 mL volume of concentrated HCl was used to dissolve 100 g $NH_2OH \cdot HCl$. Then 1.00 g $BeSO_4$ was added to the concentrate, followed by dissolution and dilution to 1000 mL.

Al Working Standard Solution, 5 mg/L   A 2.500 mL aliquot of the 1000 mg/L Al stock solution, described previously, was diluted to 500.0 mL with 0.1 M $HClO_4$.

^{27}Al NMR Spectrometric Analysis

Working Standard Solution, 0.1852 mM $Al(H_2O)_6^{3+}$ A 1.000 mL volume of stock solution containing 1000 mg/L Al in 5% HCl (Fisher Scientific) was diluted to 100.0 mL with distilled H_2O. Then 1.000 mL of the dilute solution was mixed with 1.000 mL D_2O, which was utilized for locking the NMR signal. This solution was used to establish the 0 ppm reference position in the NMR spectra and to quantify the spectral peaks.

Procedures

pH Stat Titrations

A 150.0 mL volume of 0.0100 M $NaClO_4$ was pipetted into a 400 mL or 1000 mL beaker or bottle made of polyethylene or polypropylene having a loosely fitting plastic cover. The size of container used depended on the length of time involved for the actual dispensing of titrants. The cell cover supported a motorized stirrer, a combined glass pH electrode (Radiometer Model GK2421C) attached to a pH meter (Radiometer Model PHM84), and a gas delivery tube for nitrogen purging (~80 mL/min). The beaker was immersed just deep enough in a 25 °C water bath to keep the contents at constant temperature. Automatic burets (Radiometer Model ABU80) were used to deliver Solutions A and B into the $NaClO_4$ solution.

Solution A was delivered with one buret at a constant rate of 0.125 mL/min in order to provide a constant supply of monomeric aluminum to the reactor. The control unit (Radiometer TTT80) attached to the buret containing Solution B, operating in pH stat mode, was started as soon as the holding pH of 4.900 was attained. During titration, automatic addition of Solution B occurred whenever a pH condition lower than the holding pH was detected.

Strip chart recorders (Radiometer REC80) connected to the autoburets provided continuous records of titrant additions at any time point. During synthesis, growth of the Al_b polymeric fraction was determined periodically by stopping the titration and withdrawing a measured volume of product solution for spectrophotometric analysis.

Before the titration was continued, a measured volume of solution was removed, if necessary, to make enough room in the reactor for titrants to be added during the next interval.

Addition of Solution B was resumed to restore the holding pH by starting the buret control unit. Addition stopped automatically at the holding pH, but the control unit remained active. The next titration period was initiated by activating the buret containing Solution A. The titration (Experiment 34) was concluded after a cumulative time period of 5200 min. In a preceding titration (Experiment 33) Solution A was delivered at a rate of 0.500 mL/min for 1200 min.

At the end of the titration, a sample aliquot was withdrawn for ferron analysis; then a 75 mL aliquot was withdrawn and was transferred to a polyethylene bottle. The remaining solution was stored in a polyethylene bottle.

The 75 mL portion of the product solution was sent to the NMR laboratory (Arvada, CO) for periodic ^{27}Al NMR analyses. Initial NMR analyses on the aged solutions were performed as soon as the instrument was available. Both bottled portions were stored at room temperature for additional measurements by both methods.

Companion NMR and ferron analyses were performed on a weekly or biweekly basis during the first month of aging and then monthly until the Al_{13} signal was observed to be just above background.

Prior to preparation of these two product solutions for NMR analysis, other Al hydrolysate solutions had been prepared for analysis with ferron only. These titrations (Experiments 24 to 32) were carried out, essentially, by the procedure described above. However, other holding pH values (4.75, 5.00, 5.30, and 5.60) were used in chloride as well as perchlorate systems. The Al concentration first used in Solution A was 0.45 mM (as Al), followed by 1.00 mM Al for Experiments 31 and 32. Appropriate amounts of NaCl or $NaClO_4$ were added to the different batches of Solution A to maintain nearly the same ionic strength conditions during production of the metastable Al hydrolysis products.

Ferron Analysis

A 25.00 mL sample aliquot containing ≤ 0.04 mg Al was pipetted into a 50 mL glass beaker. A reagent blank composed of H_2O, acetate buffer, ferron, and $NH_2OH \cdot HCl$ solutions is strongly absorbing at 370 nm ($\sim$0.3 absorbance units when compared to H_2O). Therefore, when sample aliquots contained > 0.04 mg Al, an appropriate volume of sample was diluted to 25.00 mL with H_2O. Then a 2.00 mL volume of 35% sodium acetate was added to each beaker followed by 5.00 mL of ferron solution. A stir rod was used for mixing after each addition. An interval timer was started immediately after adding 2.00 mL of the $NH_2OH \cdot HCl$ solution. After the blank and analyte solutions had been transferred to 1 cm spectrophotometer cells kept at a constant temperature of 25 °C, blank and sample absorbances (referenced to H_2O at zero absorbance) were read at 370 nm every 3 min up to 33 min; then at 40, 50, and 60 min; and then every 20 min. Collection of data was concluded after 160 min, when there was a period of very little change in the rate of increase of absorbance of the ferron–aluminum complex (usually following 80 to 120 min of data collection). Standard absorbance readings were obtained with 25 mL aliquots containing 0.01, 0.02, 0.03, and 0.04 mg Al after the Al–ferron complex had developed for $\sim$35 min. These standards were prepared by diluting 2.00, 4.00, 6.00, and 8.00 mL of the 5 mg Al/L working standard to 25.0 mL with H_2O.

The timed absorbance readings were treated in the manner described earlier,[22,23] and a computer spreadsheet program was used to calculate reaction rate data for the ferron–aluminum interaction. Figure 8.3 is a typical plot of the concentration of Al in the analyte that reacted with ferron over time.

The amount of Al_b unreacted with ferron at any time point was calculated by taking the difference between the respective concentration of reacted Al taken at that particular point and an extrapolated value lying on the line defined by the concentrations measured during the last 40 to 80 min of the reaction period (Figure 8.3). Extrapolation to zero time of the semilogarithmic plot of unreacted Al_b concentration against elapsed time (Figure 8.4), followed by volume corrections for dilutions used in analyte preparation, yielded the concentration of Al_b. Although the accuracy of determining Al_b by this procedure had not been established with a suitable standard reference material or by determination of Al_b by an alternative method, a relative standard deviation of $\sim$5% at the 20 μM level was determined with four replicate determinations, and a detection limit of $\sim$3 μM (as Al) was estimated.

The monomer (Al_a) reacted completely with ferron within 1 min,[23] and was determined as follows:

$$Al_a = Al_{a+b} - Al_b \tag{2}$$

Al_{a+b} was derived from the absorbance determined by extrapolating the plotted line used to define the unreacted Al_b concentration values to zero time (Figure 8.3).

Nonreactive colloidal material and hydroxy-Al polyions of dimensions larger than species constituting the Al_b fraction make up Al_c. These soluble macro-ions were the remaining Al species to complex with ferron, as seen with the steady slow increase in absorbance after the first hour of analysis when the reaction of Al_b with ferron was at or near completion (Figure 8.3). Al_c concentration was calculated by

$$Al_c = Al_t - Al_{a+b} \tag{3}$$

For a given sample, Al_{a+b} is the same as the value used in equation (2), and Al_t is the total Al concentration based on the Al concentration of Solution A and the volumes of Solutions A and B added to the titration vessel.

^{27}Al NMR Spectrometric Analysis

Specimens were prepared for analysis by placing about 2 mL of sample inside a 10 mm diameter NMR tube. A 5 mm tube, containing the 0.1852 mM $Al(H_2O)_6^{3+}$ working standard, was placed concentrically inside the larger

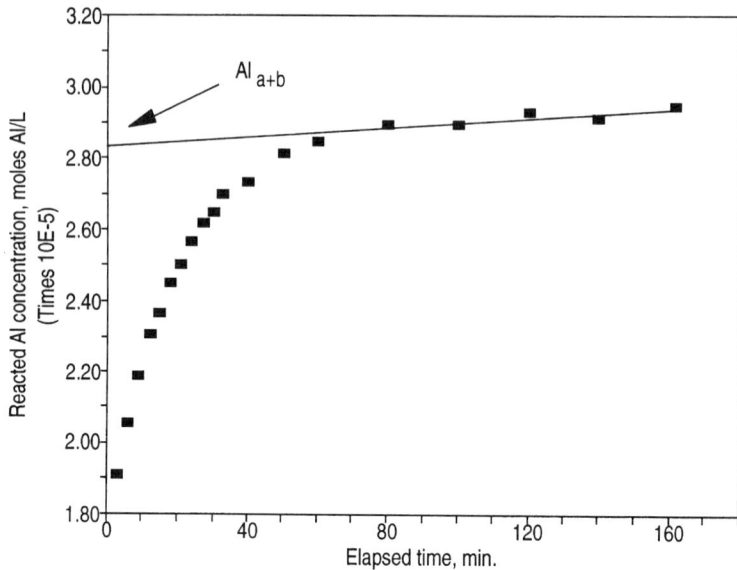

Figure 8.3 Concentration of Al reacted with ferron vs. time for the production solution from Experiment 33, aged 117 days.

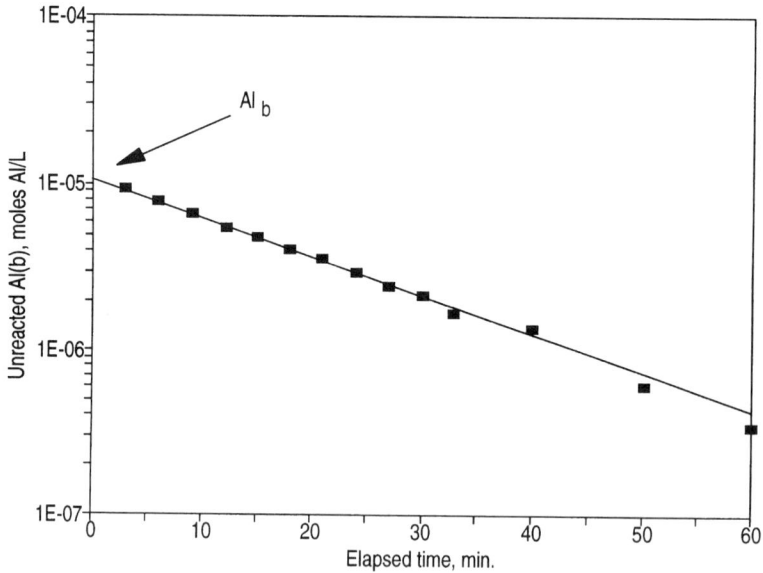

Figure 8.4 Unreacted Al_b concentration vs. time during ferron analysis of product solution from Experiment 33, aged 117 days.

tube containing the sample. This procedure was chosen because the instrument was routinely used for NMR analysis of other nuclides.

Pulse ^{27}Al NMR spectra were obtained with a Varian Model XL300 NMR spectrometer having a 10 mm probe for liquid samples. Typical operating conditions included a spectral window of 6743.1 Hz (86.28 ppm); a pulse angle of 45°; an acquisition time of 0.498 s; a pulse delay of 1.000 s; number of transients approximately 36 000; and a line broadening of 1.000 Hz. These parameters were suitable for avoiding saturation of signals. Spin–lattice relaxation times of 52 ms and 140 ms were reported, respectively, for Al_{13} and a 100 µg Al/mL (3.7 mM) standard.[17] The relaxation time for Al_{13} was found to be independent of concentration, but the relaxation time for a standard Al solution was observed to decrease upon dilution with H_2O.[17]

Normal electronic integration was impeded by distortions in the spectral baselines. These distortions were pronounced and most probably caused by acoustic ringing of the instrument probe because of the low Al concentrations in the sample specimens. Individual peaks were quantified by visual inspection (i.e., "cut-and-weigh" integration) after expanding and recording spectra on paper. The detection limit of Al_{13} by this method is in the range of ~20 to ~30 µM (as Al), but precision and accuracy could not be determined due to the lack of a stable Al_{13} standard solution having an Al concentration on the order of 1 mM or less. Further expansion of the spectral width was disregarded since it did not improve the signal to noise characteristics of the spectra.

Results and Discussion

Hydroxy-Al Monomers

Initial NMR spectra were obtained on the Al hydrolysate solutions after they had aged 1 day (Experiment 34) and 9 days (Experiment 33) (Figures 8.1(a) and 8.2(a)). The prevailing pH was ~4.6 at the time of the NMR analyses. Results of thermodynamic calculations[25] indicated that the monomeric (Al_a) fractions in these solutions were ~75% $Al(H_2O)_6^{3+}$ (Table 8.1). The broad signal (Figures 8.1(b) and (8.2(b)), attributed to the consolidation of resonances from hydroxy-Al monomers and located just downfield from 0 ppm,[17] became more distinct after 7–8 weeks. By this time, the pH was 4.5 for the solution from Experiment 33 and 4.3 for the solution produced from the slower titration of Experiment 34, with respective Al_a fractions that contained 80% and 87% $Al(H_2O)_6^{3+}$ (Table 8.1). However, quantitative NMR analyses of the monomeric fraction were problematic because of (1) differences in the peak widths between the very sharp signal of the $Al(H_2O)_6^{3+}$ standard solution and the much broader composite signal of the different monomeric species, (2) distortions of the baseline in the spectral region of the octahedral Al species caused by acoustic ringing of the NMR probe since Al metal was

Table 8.1 Calculated Percentages of Free Aluminum Ion and Hydroxy-Al Ions in the Monomeric Fraction (Al_a) at 25 °C and Ionic Strength 0.01

Species[a]	γ[b]	Percentage of Free Ion			
		pH 4.90	pH 4.60	pH 4.50	pH 4.30
Al^{3+}	0.444	56.7	75.7	80.4	87.3
$AlOH^{2+}$ [c]	0.686	29.2	19.6	16.5	11.3
$Al(OH)_2^+$ [d]	0.899	14.0	4.7	3.1	1.4
$Al(OH)_4^-$ [e]	0.899	0.0	0.0	0.0	0.0

[a] Water of hydration is not included in the formulas.
[b] Ion activity coefficient computed from data in Hem.[24]
[c] Equilibrium constant $K_{eq} = 10^{-5.01}$ for $Al^{3+} + H_2O \rightleftharpoons Al(OH)^{+2} + H^+$.[25]
[d] $K_{eq} = 10^{-10.1}$ for $Al^{3+} + 2H_2O \rightleftharpoons Al(OH)_2^+ + 2H^+$.[25]
[e] $K_{eq} = 10^{-22.7}$ for $Al^{3+} + 4H_2O \rightleftharpoons Al(OH)_4^- + 4H^+$.[25]

used in its fabrication, and (3) overlap of the sharp $Al(H_2O)_6^{3+}$ standard peak by the broader peak attributed to the hydroxy-Al monomers.

Nonetheless, monomeric Al levels were estimated by cut-and-weigh quantitation after the solutions had aged for 3 months when the 1 ppm signals became distinct. The NMR results exceeded the corresponding Al_a results obtained with ferron (Table 8.2), implying that the broad peak could be a feature which included hydroxy-Al species other than monomers, such as bridged dimers of octahedral Al ions. These species, together with monomers, constituted the dissolved hydroxy-Al fraction as equilibrium with gibbsite was approached. Previous workers[8] observed a very broad peak shifted ~3 ppm downfield from 0 ppm as evidence for the occurrence of these dimers in a more concentrated hydroxy-Al solution (0.5 M $AlCl_3$ partially neutralized with base).

Hydroxy-Al Polymers and Solids

Calculated conditional first-order rate constants (expressed as log units/s) for development of Al_b from polymerization of $AlOH^{2+}$ species[19] were −3.2 to −3.3 for the initial titration (Experiment 33) and −3.7 to −4.0 for the latter titration (Experiment 34) where the reactant fluxes were one-quarter of those in the first experiment. The latter range of reaction rate constants overlaps the range of −4.25 to −4.00 previously reported by Hem and Roberson[19] with different but nearly proportional reactant fluxes for a similar titration experiment (Experiment 12) performed at 25 °C. Solution A in Experiment 12 (0.452 mM Al, 0.338 mM H^+) was delivered at a rate of 0.125 mL/min (Al flux = 0.056 μmol/min) for 4 to 6 h per day, and the 0.01 M NaOH was dispensed at a flux of 0.115 μmol/min. The titration was allowed to proceed for several hours daily before it was stopped. Then the product solution was allowed to stand overnight at 25 °C before the titration was resumed the

Table 8.2 Hydroxy-Al Monomer Concentrations ($\times 10^{-4}$ M Al) in Aged Solutions Determined by NMR and Ferron Procedures

	Procedure		
Aging Period (days)	Ferron	NMR	pH
Experiment 33			
94	2.7	4.5	4.28
115	—	5.5	—
117	3.0	—	4.26
Experiment 34			
83	3.2	—	4.24
84	—	3.7	—
122	—	5.7	—
124	3.0	—	4.22

next day. The titration was discontinued when the cumulative time was in excess of 20 h.

Solution A in Experiment 34 (2.00 mM Al, 0.773 mM H^+) was delivered continuously at 0.125 mL/min (Al flux = 0.25 µmol/min) for more than 86 h. Daily stoppages lasted from 10 to 110 min in order for volumes of product solution to be withdrawn, pH electrode accuracy to be verified, and the steady-state condition for the product solution to be re-established (as was the case during the first day when the buret dispensing Solution B had been improperly set to manual refill mode instead of to the automatic refill setting). The flux of NaOH was 0.6 µmol/min. The initial titration (Experiment 33) had concentrations of 2.00 mM Al and 0.43 mM H^+ in Solution A which was delivered at 0.500 mL/min and reactant fluxes of 1.0 µmol Al/min and 2.3 µmol OH^-/min.

During ∼4 months of aging, the molar concentrations of Al_b and tetrahedral Al in Al_{13} (as Al) decreased steadily. For Experiment 34 the half-life of Al_b found in the product solution was 9 weeks, which was the same as the half-life for tetrahedral Al. For Experiment 33 the half-life of Al_b was also 9 weeks, but half of the starting concentration of tetrahedral Al (first measured after nine days of aging) remained 7 weeks later. If the tetrahedral Al level in the freshly prepared solution were about the same as that initially found after 9 days of aging, then the estimated half-life of tetrahedral Al would be 9 weeks. For a typical product solution in the batch experiments of Smith and Hem,[26] rates of loss of Al_b found at the outset and near the end of the aging period were smaller than the first-order rate of loss observed during most of the aging period.

Bourrie[27] and Bourrie et al.[28] suggested that Al_{13} formation under Earth-surface conditions is a geochemical "cul-de-sac" where a buildup of polymer is favored in systems having a pH of >5 and a slight degree of supersaturation with respect to gibbsite. These conditions slow the precipitation of gibbsite and prevent saturation with respect to amorphous $Al(OH)_3$. The activity of Al^{3+} is then buffered by the metastable Al_{13} fraction, which is not in true equilibrium with monomers. As gibbsite precipitation progresses, depolymerization of Al polymeric species ensues, providing additional monomers for growth of crystalline gibbsite until the Al_{13} disappears; the true equilibrium state between gibbsite and monomers would remain. Previously, Bottero et al.[6] had worked with solutions of relatively high hydroxy-Al concentrations, and they proposed that the disappearance of Al_{13} was linked to rearrangement of the central Al in the polyion structure from tetrahedral to octahedral coordination in the solid state in the development of bayerite.

Sizable fractions of polymeric Al were found in the dilute solutions prepared in pH-stat Experiments 33 and 34, which were conducted at pH 4.90. Undoubtedly, conditions of excessive degrees of supersaturation with respect to gibbsite and pH levels greater than 5 existed in the vicinity of the point of injection of base solution into the stirred contents of the reaction vessel during the titration experiments.

For Experiments 33 and 34 the ratios of the concentration of tetrahedrally coordinated Al presumed to be associated with Al_{13} to that of total Al in Al_b are generally constant during aging, as seen in Figure 8.5. However, these ratios differ from the value of 1:13 expected for Al_{13}.[2,10,13] Linear regressions of the plotted data were computed through the origin (the extreme right-hand point in Figure 8.5 was excluded from the treatment of data from Experiment 33). Respectively, ratios of 2.3:13 and 2.8:13 were found for Experiments 33 and 34. These ratios suggest that some symmetrical configuration other than Al_{13}, or partial formation of Al_{13} units, was responsible for the distinct signal 62.5 ppm downfield from the 0 ppm position of $Al(H_2O)_6^{3+}$. This unknown form seemed to have more tetrahedral Al in relation to octahedral Al than the proportion prescribed with the Al_{13} structure.

Al_{13} hydroxysulfate salts can be characterized by X-ray diffraction (XRD) analyis.[16] However, the retention of the complete Al_{13} polyion configuration upon preparation of dilute, mildly acidic solutions from these salts is questionable.

Studying more concentrated hydroxy-Al solutions (0.1 M and 1.0 M Al) at elevated temperatures (70 to 87 °C), other investigators[29,30] observed NMR signals positioned ~70 ppm downfield from 0 ppm. They ascribed these broad peaks to formation of polymers containing tetrahedral Al other than Al_{13}. These signals were not evident at temperatures less than 55 °C.

Bleaching effects by hydroxylamine on the color development of the Al–ferron complex had been observed with use of a premixed ferron reagent solution.[31] These effects could result in underestimating the absorbances due to the developing complex. In this study hydroxylamine hydrochloride was not combined with ferron until initiation of the sample analysis. Beer's Law was closely obeyed up to 0.04 mM Al (0.04 mg in 34 mL analyte) with a

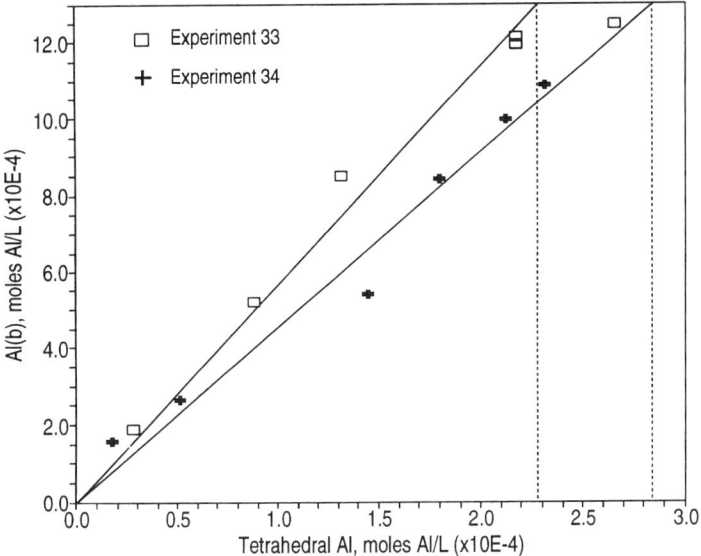

Figure 8.5   Correlation between Al_b concentration determined with ferron and tetrahedral Al concentration determined by ^{27}Al NMR spectrometry.

deviation of -1 to -2% from the absorbance value projected from the absorbance at $0.02\,\text{mM}$ Al (0.02 mg in $34\,\text{mL}$ analyte). During the entire period of analysis, the absorbance of the reagent blank was continually observed and subtracted from the absorbance of the analyte solution in order to arrive at a corrected absorbance value.

Parker and Bertsch[10] devised a procedure based on Al–ferron kinetics for quantitative determinations of Al_b that were equatable to Al_{13} determinations by NMR spectrometry. A 50:1 mole ratio of total ferron to total Al in the analyte solution and treatment of the data collected during the first $7\,\text{min}$ of the reaction of Al with ferron were required in this version of the ferron method. Table 8.3 compares the Al_b results obtained on samples from Experiment 34 by Parker and Bertsch's procedure with those by the method previously described in the Experimental section. No significant differences between the sets of results were found.

Unfiltered and filtered aliquots of previously prepared hydrolyzed aluminum perchlorate solutions[19] aged 932 to 1400 days were subjected to the ferron procedure as a preliminary study to determine the effect of filtration through a $25\,\mathring{A}$ pore size membrane on the result of the Al_b determination (Table 8.4). There were no significant differences between the results obtained with the filtered samples and those of the respective unfiltered samples. The size range of reactive hydroxy-Al polymers is reportedly 10 to 20 $\mathring{A}$.[3,26,32] ^{27}Al NMR spectrometric analysis was not yet utilized at the time of this preliminary study.

Table 8.3 Al_b Concentration (mmol Al/L) for Aged Samples from Experiment 34 by Two Different Procedures Using Al–Ferron Kinetics

	Procedure	
Aging Period (days)	This Work	Parker and Bertsch[10]
47	0.64	0.60
54	0.54	0.50
83	0.26	0.29

Preliminary titrations conducted at holding pH values other than pH 4.90 (pH 4.75, 5.00, 5.30, and 5.60) resulted in hydroxy-Al solutions in which, after about 3 to 5 months of aging, the Al_b levels decreased at rates comparable to the rates of decrease for solutions prepared at pH 4.90 during a similar time period, except for solutions prepared at pH 5.00 and pH 4.75 (Table 8.5). A combination of a smaller Al flux, a lower holding pH, and a lower ratio of total OH⁻ to total Al enhanced the temporal stability of Al_b. The respective Al fluxes for Experiment 24 (pH 5.00) and Experiment 29 (pH 4.75) were 0.056 and 0.23 µmol/min and the respective OH⁻/Al ratios were 1.44:1 and 0.71:1.

For Experiment 29 (pH 4.75) the original Al_b concentration in the freshly prepared hydroxy-Al solution was 0.15 mM. Over the next 227 days of aging, the Al_b level ranged from 0.13 mM to 0.19 mM; this was followed by a very slight decrease to 0.12 mM after 283 days (Figure 8.6). This decrease was accompanied by an increase in Al_a level from 0.26 mM to 0.29 mM. The Al_c

Table 8.4 Al_b Results for Filtered and Unfiltered, Aged, Partially Neutralized Aluminum Solutions

Expt.	Holding pH	Titration Temp. (°C)	Date of Prepn.	Date of Determin.	Aging Period (days)	pH	Al_b (µM Al)
13	5.20	25	Nov. 21 1986	Jun. 28 1990	1315	4.36	< 3
13[a]	5.20	25	Nov. 21 1986	Sep. 21 1990	1400	4.38	< 3
17	4.75	35	Feb. 20 1987	Jul. 23 1990	1249	4.41	21
17[a]	4.75	35	Feb. 20 1987	Sep. 14 1990	1302	4.40	23
18	5.00	10	Mar. 18 1987	Jun. 12 1990	1182	4.43	83
18[a]	5.00	10	Mar. 18 1987	Sep. 14 1990	1276	4.42	83
20	5.00	25	Apr. 24 1987	Jun. 19 1990	1152	4.19	23
20[a]	5.00	25	Apr. 24 1987	Sep. 21 1990	1246	4.19	22
23	7–9.5[b]	25	Dec. 2 1987	Jun. 21 1990	932	7.22	< 3
23[a]	7–9.5[b]	25	Dec. 2 1987	Sep. 12 1990	1015	7.20	< 3

[a] Filtered with Amicon PM10 ultrafiltration membrane, 25 Å nominal pore size (molecular weight cutoff, MWCO ~ 10 000).

[b] Held at pH 7.00, then at pH 8.00, and ending at pH 9.50.

Table 8.5 Change in Al_a, Al_b, and Al_c Concentrations (mM Al) with Several Months of Aging

Expt. (Anion)	pH	Al Flux (μmol/ min)	Aging Period (days)	Total Al	OH⁻/Al	Al_b	$-\Delta(Al_b)$ (%)	Al_a	Al_c
24 (Cl⁻)	5.00[a]	0.056	0	0.22	1.44	0.13	—	0.09	—
24 (Cl⁻)	4.70	—	122	0.22	1.44	0.11	15	0.07	0.04
25 (Cl⁻)	5.30[a]	0.056	0	0.23	2.49	0.091	—	0.05	0.09
25 (Cl⁻)	4.78	—	111	0.23	2.49	0.026	71	0.02	0.18
26 (Cl⁻)	5.30[a]	0.226	0	0.40	2.34	0.18	—	0.09	0.12
26 (Cl⁻)	4.52	—	132	0.40	2.34	0.03	83	0.06	0.31
27 (ClO₄⁻)	5.30[a]	0.226	0	0.37	2.53	0.10	—	0.06	0.21
27 (ClO₄⁻)	4.48	—	109	0.37	2.53	0.01	90	0.03	0.33
28 (ClO₄⁻)	5.60[a]	0.226	0	0.38	2.81	0.006	—	0.03	0.34
28 (ClO₄⁻)	4.73	—	125	0.38	2.81	< 0.003[b]	> 50	0.01	0.36
29 (ClO₄⁻)	4.75[a]	0.226	0	0.41	0.71	0.15	—	0.26	—
29 (ClO₄⁻)	4.43	—	134	0.41	0.71	0.13	13	0.25	0.03
30 (ClO₄⁻)	4.90[a]	0.226	0	0.39	1.48	0.24	—	0.15	—
30 (ClO₄⁻)	4.33	—	128	0.39	1.48	0.05	79	0.18	0.16
31 (ClO₄⁻)	4.90[a]	0.500	0	0.81	1.82	0.62	—	0.19	—
31 (ClO₄⁻)	4.22	—	144	0.81	1.82	0.07	89	0.25	0.49
32 (ClO₄⁻)	4.90[a]	0.125	0	0.81	2.19	0.66	—	0.11	0.03
32 (ClO₄⁻)	4.38	—	140	0.81	2.19	0.11	83	0.18	0.52
33 (ClO₄⁻)	4.90[a]	1.000	0	1.42	2.14	1.31	—	0.11	—
33 (ClO₄⁻)	4.26	—	117	1.42	2.14	0.19	79	0.30	0.93
34 (ClO₄⁻)	4.90[a]	0.250	0	1.27	2.03	1.13	—	0.14	—
34 (ClO₄⁻)	4.22	—	124	1.27	2.03	0.15	87	0.30	0.82

[a] Holding pH.
[b] Below detection limit.

fraction was barely perceptible after 134 days of aging. The scatter of Al–ferron data beyond 100 days of aging (Figure 8.6) could be due to errors introduced by sampling at different ambient temperatures or by not routinely shaking the sample container to ensure withdrawal of a representative subsample. After 10 days of aging, a qualitative NMR scan of the product solution showed a signal at 62.5 ppm, which was just above the spectral baseline, and a broad signal at 1 ppm. This was the first opportunity to use NMR spectrometry for analyzing the solutions described in Table 8.5. From this and subsequent scans of the product solutions of Experiments 30, 31, and 32, it was apparent that in order to investigate the time stability of the 62.5 ppm signal, the initial intensity of the signal would have to be increased by production of a larger concentration of metastable hydroxy-Al polymers. This was eventually achieved in Experiments 33 and 34.

Smith and Hem[26] had shown that the stability of hydroxy-Al polymers, synthesized by partial neutralization of acidic solutions, is related to the ratio of total OH⁻ added to the total amount of Al in solution. However, their

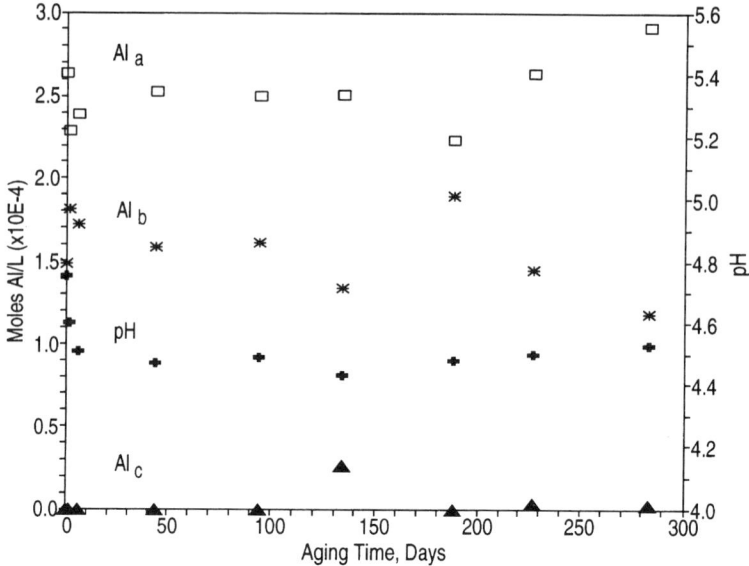

Figure 8.6 Al_a, Al_b, and Al_c concentrations and pH for the aged product solution of Experiment 29 prepared at pH 4.75; total Al concentration = 0.41 mM (as Al).

titration experiments were not conducted at constant pH, and therefore no effort was made to minimize any sites of local supersaturation with respect to gibbsite in the product solutions.

Previous investigators have examined rate constants for reaction of Al_b with ferron for further characterization of the polymeric species among batches of prepared solutions or in a particular solution at various times of aging.[9,21,33,34] Comparisons of rate constants among different studies cannot be made unless reagent formulations, procedural steps, instrument settings, and data treatments are similar. For this study, ranges of calculated psuedo-first-order rate constants for the development of the Al–ferron complex from Al_b are given in Table 8.6. During production of hydroxy-Al polymers by pH-stat titration, the ranges of rate constant values were more or less the same except for Experiment 28. The lower range of values from this experiment indicated that the Al_b fraction was slower to combine with ferron than the Al_b fractions produced in the other experiments. The more slowly reacting polymers of Experiment 28 were formed under the condition of a higher holding pH of 5.60 (Table 8.5) which allowed Al_c to form early in the titration process.

In the aged hydrolysate solutions the ranges of first-order rate constant values for reaction of Al_b with ferron resembled those found during the titrations, except with the solutions prepared in Experiments 25, 26, 27, and 28 (Table 8.6). The occurrence of smaller values with these latter solutions indicated that their Al_b fractions were slower to react with ferron than

Table 8.6 Ranges of Rate Constants (min^{-1}) for Development of the Al–Ferron Complex in Determinations of Al_b

Experiment	Rate Constant Range (min^{-1})	
	During titration	After titration
24	0.048–0.068	0.047–0.059
25	0.053–0.059	0.022–0.057
26	0.048–0.063	0.037–0.056
27	0.054–0.061	0.034–0.055
28	0.031–0.057	0.027–0.038
29	0.049–0.063	0.046–0.063
30	0.052–0.067	0.055–0.061
31	0.053–0.063	0.049–0.060
32	0.055–0.062	0.051–0.060
33	0.056–0.063	0.042–0.066
34	0.058–0.063	0.042–0.061

those of the other solutions. The more slowly reacting hydroxy-Al polymers developed upon aging as a consequence of the higher holding pH values of 5.30 and 5.60 employed during production of the hydrolysates (Table 8.5). As a group, the rate constants for interaction of ferron with Al_b in the aged solution from Experiment 28 were consistently low in comparison to other groups of rate constants in Table 8.6.

For Experiments 33 and 34 transmission electron microscopy (TEM) was used to examine the solid phase material from the Al_c fraction. The solid was collected with Amicon PM10 ultrafiltration membranes (25 Å nominal pore size). TEM images showed that the solid consisted of only hexagonal microcrystalline gibbsite platelets of diameters near 0.1 μm, as seen in Figure 8.7 for Experiment 34 and as reported earlier.[18] Analysis of the same material by XRD also confirmed the presence of microcrystalline gibbsite. No structures were detected in the solid, which could be interpreted as being evolved from the Al_{13} structure such as the tetrahedral crystals formed when freshly prepared Al hydrolysates were treated with sulfate at elevated temperatures[4,16] and when aged solutions were treated with sulfate at ambient temperatures.[33,34]

Formation of microcrystalline gibbsite is explained by edge-to-edge coalescence of hydroxy-Al hexamer units into larger polymeric hydroxy-Al fragments. As the process is repeated, solid-state behavior[18–20] is eventually exhibited by the larger polymer segments, but the generation and depletion of Al_{13} are not taken into account by this hexamer fragment model.

However, although the transient nature of Al_{13} is interpreted by the "cul-de-sac" model,[27] the process of how Al monomers, released by Al_{13} depolymerization, could initially form microcrystalline gibbsite before contributing to its subsequent crystal growth is not adequately addressed. The concepts of hexamer fragments and Al_{13} are complementary to each other in characterizing hydroxy-Al polymers,[35] yet each theory is exclusive of the other.

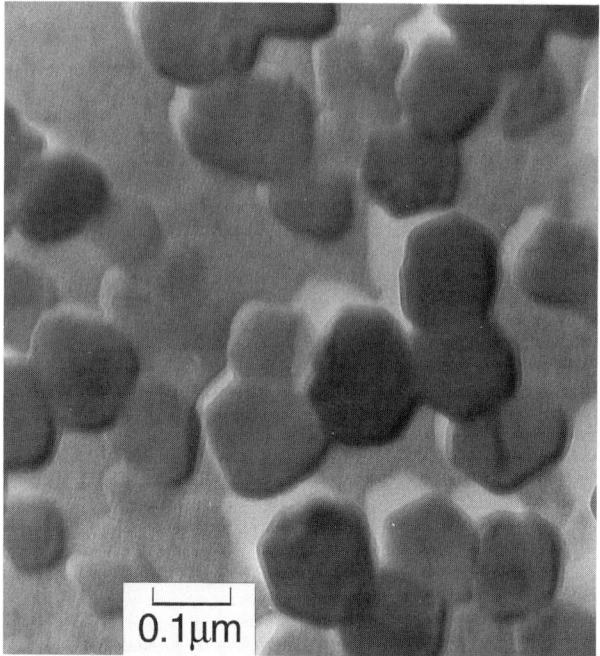

Figure 8.7 Electron micrograph of microcrystalline gibbsite recovered from the product solution of Experiment 34, aged 114 days.

Acknowledgments The helpful comments from Y.K. Kharaka of the US Geological Survey, D.Z. Denney and P.H. Hsu of Rutgers University, and one anonymous reviewer are gratefully acknowledged.

References

1. May, H., The hydrolysis of aluminum: conflicting models and the interpretation of aluminum geochemistry. In *Proc. 7th Int. Symp. on Water–Rock Interaction*, Vol. 1, Kharaka, Y. K., and Maest, A. N., (eds), A. A. Balkema, Rotterdam, 1992, p.13.
2. Akitt, J. W., Greenwood, N. N., Khandelwal, B. L., and Lester, G. D., ^{27}Al nuclear magnetic resonance studies of the hydrolysis and polymerisation of the hexa-aquo-aluminum(III) cation. *J. Chem. Soc., Dalton Trans.* 604, 1972.
3. Parthasarathy, N., and Buffle, J., Study of polymeric aluminum(III) hydroxide solutions for application in waste water treatment. Properties of the polymer and optimal conditions. *Water Res.* 19, 25, 1985.
4. Furrer, G., Ludwig, C., and Schindler, P. W., On the chemistry of the Keggin Al_{13} polymer. I. Acid–base properties. *J. Colloid Interface Sci.* 149, 56, 1992.
5. Bottero, J. Y., Cases, J. M., Fiessinger, F., and Poirer, J. E., Studies of hydrolyzed aluminum chloride solutions. 1. Nature of aluminum species and composition of aqueous solutions. *J. Phys. Chem.* 84, 2933, 1980.

6. Bottero, J. Y., Axelos, M., Tchoubar, D., Cases, J. M., Fripiat, J. J., and Fies-singer, F., Mechanism of formation of aluminum trihydroxide from Keggin Al_{13} polymers. *J. Colloid Interface Sci.* 117, 47, 1987.
7. Denney, D. Z., and Hsu, P. H., ^{27}Al nuclear magnetic resonance and ferron kinetic studies of partially neutralized $AlCl_3$ solutions. *Clays Clay Miner.* 34, 604, 1986.
8. Bertsch, P. M., Thomas, G. W., and Barnhisel, R. I., Characterization of hydroxy-aluminum solutions by aluminum-27 nuclear magnetic resonance spectroscopy. *Soil Sci. Soc. Am. J.* 50, 825, 1986.
9. Parker, D.R., Kinraide, T.B., and Zelazny, L.W., On the phytotoxicity of poly-nuclear hydroxy-aluminum complexes. *Soil Sci. Soc. Am. J.* 53, 789, 1989.
10. Parker, D. R., and Bertsch, P. M., Identification and quantification of the "Al_{13}" tridecameric polycation using ferron. *Environ. Sci. Technol.* 26, 908, 1992.
11. Hunter, D., and Ross, D. S., Evidence for a phytotoxic hydroxy-aluminum poly-mer in organic soil horizons. *Science* 251, 1056, 1991.
12. Teagarden, D. L., Hem, S. L., and White, J. L., Conversion of aluminum chlor-ohydrate to aluminum hydroxide. *J. Soc. Cosmet. Chem.* 33, 281, 1982.
13. Akitt, J. W., and Elders, J. M., Multinuclear magnetic resonance studies of the hydrolysis of aluminum(III). Part 8. Base hydrolysis monitored at very high mag-netic field. *J. Chem. Soc., Dalton Trans.* 1347, 1988.
14. Akitt, J. W., Elders, J. M., Fontaine, X. L. R., and Kundu, A. K., Multinuclear magnetic resonance studies of the hydrolysis of aluminum(III). Part 9. Prolonged hydrolysis with aluminum metal monitored at very high magnetic field. *J. Chem. Soc., Dalton Trans.*, 1889, 1989.
15. Keggin, J. F., The structure and formula of 12-phosphotungstic acid. *Proc. R. Soc. London Ser. A*, 144, 75, 1934.
16. Johansson, G., The crystal structure of a basic aluminum sulfate. *Arkiv. Kemi* 20, 403, 1963.
17. Bertsch, P. M., Barnhisel, R. I., Thomas, G. W., Layton, W. J., and Smith, S. L., Quantitative determination of aluminum-27 by high-resolution nuclear magnetic resonance spectrometry. *Anal. Chem.* 58, 2583, 1986.
18. Hem, J. D., and Roberson, C. E., *Form and Stability of Aluminum Hydroxide Complexes in Dilute Solution.* US Geological Survey Water Supply Paper 1827-A, US Government Printing Office, Washington DC, 1967.
19. Hem, J. D., and Roberson, C. E., Aluminum hydrolysis reactions and products in mildly acidic aqueous systems. In *Chemical Modeling in Aqueous Systems II*, Mel-chior, D. C., and Bassett, R. L. (eds), American Chemical Society, Washington DC, 1990, Chapter 10.
20. Hsu, P. H., and Bates T. F., Formation of X-ray amorphous and crystalline aluminum hydroxides. *Mineralog. Mag.* 33, 749, 1964.
21. Wang, W. Z., and Hsu, P. H., The nature of polynuclear OH–Al complexes in laboratory-hydrolyzed and commercial hydroxyaluminum solutions. *Clays Clay Miner.* 42, 356, 1994.
22. Smith, R. W., The state of Al(III) in aqueous solution and adsorption of hydro-lysis products on αAl_2O_3, Ph.D. Dissertation, Stanford University, Stanford, CA, 1969.
23. Smith, R. W., Relations among equilibrium and nonequilibrium aqueous species of aluminum hydroxy complexes. In *Nonequilibrium Systems in Natural Water Chemistry*, Hem, J.D. (ed.), American Chemical Society, Washington DC, 1971, Chapter 10.

24. Hem, J. D., *Study and Interpretation of the Chemical Characteristics of Natural Water*, 3rd edn., US Geological Survey Water Supply Paper 2254, US Government Printing Office, Washington DC, 1985, pp. 15–17.
25. Nordstrom, D. K., Plummer, L. N., Langmuir, D., Busenberg, E., May, H. M., Jones, B. F., and Parkhurst, D. L. Revised chemical equilibrium data for major water–mineral reactions and their limitations. In *Chemical Modeling in Aqueous Systems II*, Melchior, D. C., and Bassett, R. L. (eds), American Chemical Society, Washington DC, 1990, Chapter 31.
26. Smith, R. W., and Hem, J. D., *Effects of Aging on Aluminum Hydroxide Complexes in Dilute Solution*. US Geological Survey Water Supply Paper 1827-D, US Government Printing Office, Washington DC, 1972.
27. Bourrie, G., Two paths of Al hydroxide formation as function of behavior of polynuclear Al complexes: path in dilute conditions leading to gibbsite and boehmite, path in concentrated conditions leading to gels and bayerite. *C. R. Seances Acad. Sci., Ser. 2* 310, 1221, 1990.
28. Bourrie, G., Grimaldi, C., and Regeard, A., Monomeric versus mixed monomeric–polymeric models for aqueous aluminum species: Constraints from low-temperature natural waters in equilibrium with gibbsite under temperate and tropical climate. *Chem. Geol.* 76, 403, 1989.
29. Akitt, J. W., and Farthing, A., Aluminum-27 nuclear magnetic resonance studies of aluminum(III). Part 5. Slow hydrolysis using aluminum metal. *J. Chem. Soc., Dalton Trans.* 1624, 1981.
30. Changui, C., Stone, W. E. E., Vielvoye, L., and Dereppe, J., Characterization by nuclear magnetic resonance spectroscopy, ferron assay, and acidification of partially neutralized aluminum solutions. *J. Chem. Soc., Dalton Trans.* 1723, 1990.
31. Hsu, P. H., written communication.
32. Buffle, J., Parthasarathy, N., and Haerdi, W., Importance of speciation methods in analytical control of water treatment processes with application to fluoride removal from waste waters. *Water Res.* 19, 7, 1985.
33. Tsai, P. P., and Hsu, P. H., Aging of partially neutralized aluminum solutions of sodium hydroxide/aluminum molar ratio = 2.2. *Soil Sci. Soc. Am. J.* 49, 1060, 1985.
34. Tsai, P. P., and Hsu, P. H., Studies of aged OH–Al solutions using kinetics of Al–ferron reactions and sulfate precipitation. *Soil Sci. Soc. Am. J.* 48, 59, 1984.
35. Hsu, P. H., Aluminum hydroxides and oxyhydroxides. In *Minerals in Soil Environments*, 2nd edn., Soil Science Society of America, Madison, Wisconsin, 1989, Chapter 7.

9.

^{27}Al NMR Study of the Hydrolysis and Condensation of Organically Complexed Aluminum

FABIEN THOMAS, ARMAND MASION, JEAN YVES BOTTERO,

& PIOTR TEKELY

Environmental and Health Significance of Aluminum

Aluminum is the most abundant metal of the Earth's crust, of which it represents approximately 8%, ranking after oxygen and silicon. It exists mainly as oxides. In terrestrial environments, aluminum commonly exists as secondary (authigenic) hydroxide or aluminosilicate minerals, mainly clays. These minerals are highly insoluble at neutral pH. However, aluminum occurs in detectable amounts in natural waters, due to leaching of the soil minerals in acidic conditions. Soil acidity may have a natural origin, such as an acidic (silicic) mother rock, melted snow, dissolved carbonic acid,[1] or biologically generated organic acids.[2]

During the past two decades, it has been demonstrated that one of the major origins of increased aluminum mobilization and transport in forested soils is introduction of strong acid through atmospheric sulfur and nitrogen deposition.[3] It has also been shown that aqueous aluminum is the biogeochemical link between atmospheric pollution and damage caused to tree roots [4,5] and aquatic organisms such as plankton, crustaceans, insects, and fish.[6,7] Biological studies have shown that the different aluminum species exhibit various toxicities: the most toxic are the monomeric and the polynuclear species; complexation with organic acids results in low toxicity.[7–10]

The significance of aluminum to human health has long been regarded as negligible. There is a possible link between high-level aluminum contamination by renal dialysis or hemodialysis, and neurodegenerative health disorders such as Parkinson's or Alzheimer's diseases, but the part played by aluminum is not clear.[9] However, since aluminum salts are used on an industrial level as coagulants and flocculants in water treatment,[11,12] the

161

aluminum concentration and speciation in drinking water deserve careful monitoring.

Use of NMR Spectroscopy in Studying the Chemistry of Aluminum

Because of the specific toxicity of the aluminum species, there has been considerable concern in the past two decades over the speciation of aqueous aluminum present in soils and aquatic systems. To this end, several techniques have been developed in order to partition the aluminum species. The most common among them are chromatographic separation and categorization methods such as timed ferron reaction,[1,13] and computational methods derived from thermodynamic equilibrium constants.[14,15] However, significant discrepancies between the results have been noticed,[10] and attributed to the dramatic interference of organic and inorganic anions in the Al fractionation.[16,17] This interference becomes an advantage when nuclear magnetic resonance (NMR) is used, because this technique is sensitive to any structural change in the local environment of the aluminum nucleus. Aluminum is a favorable nucleus for NMR because of the 100% abundance of the ^{27}Al isotope, which possesses a 5/2 spin quantum number. These advantages have aroused numerous investigations into the chemistry of aluminum, as testified by the comprehensive reviews published on this topic in recent years.[18,19]

The most striking aspect of the chemistry of aluminum has been the discussion on the existence of a polynuclear precursor of aluminum hydroxide. Early studies proposed the aluminum hexamer and octamer, but these forms could not be identified by direct methods.[20-22] Later, ^{27}Al NMR proved the existence of the soluble tridecamer Al_{13}: $AlO_4Al_{12}(OH)_{24}(H_2O)_{12}^{7+}$,[13,23-27] previously described by Johansson.[28,29] More recently, theoretical approaches strongly supported the existence of this tridecamer.[30,31]

Aquatic media, in which the hydrolysis of aluminum takes place, contain organic molecules mostly bearing carboxylic acidity. These complexing molecules are able to modify or even block the succession of events leading from aluminum monomers to polymers. The stability constants of monomeric aluminum complexes with most of the low-molecular organic acids are well established[32] and efforts have been made to evaluate the speciation of Al as influenced by organic ligands.[15] However, new information about the mechanisms of aluminum hydrolysis in the presence of organic acids can be acquired on a molecular scale by ^{27}Al NMR, which is a powerful tool for investigating the molecular structure of the complexed aluminum species and their quantitative distribution.[33-38]

Accurate NMR measurements often require that the analyzed solutions be more concentrated in aluminum than they would be in natural waters. However, it has been established that Al_{13}, for instance, is always obtained in a wide range of total aluminum concentrations. A roughly constant proportion of Al_{13} is obtained at total aluminum concentrations ranging between 10^{-1} M and 10^{-4} M.[24,39] Thus, the general concepts of hydrolysis are valid even in the lowest concentrations encountered in natural waters. Similarly, the general trends concerning the evolution of organic acid–aluminum mixtures at various pH values may also be valid at low concentrations, at least qualitatively. Moreover microenvironments, such as soil interstitial pores and organic or mineral surfaces, may provide local conditions for the concentration of the percolating solutions, or for high local pH.[39-41] Thus, the phenomena occurring in these local conditions are correctly described by laboratory experiments at relatively high concentration.

The aim of the present work was to investigate the transition from soluble organic aluminum complexes in acidic conditions to the formation of precipitates at neutrality. This transition occurs through the formation of metastable, very reactive intermediate species such as dimers or trimers and tridecamers, and involves a competition between the hydroxyl groups and the organic ligands for the bonding sites of the aluminum atoms.[36-38] The acids studied, selected for their variety in chemical functionality and structure, are acetic (CH_3—$COOH$), oxalic ($HOOC$—$COOH$), lactic (CH_3—$CHOH$—$COOH$), tartaric ($COOH$—$CHOH$—$CHOH$—$COOH$), citric ($COOH$—CH_2—$CHOHCOOH$—CH_2—$COOH$) and salicylic (C_6H_4OH—$COOH$) acids. The speciation of aluminum in solution at pH values from 2 to 6 was studied qualitatively and quantitatively by liquid-state ^{27}Al NMR, and the aluminum species in dried precipitates were analyzed by solid-state ^{27}Al NMR.

Factors Affecting ^{27}Al NMR Spectra

The reference of the ^{27}Al NMR spectra is the resonance of the $Al(H_2O)_6^{3+}$ octahedron, which is assigned the value of 0 ppm. The resonance of the $Al(OH)_4^-$ tetrahedron is located at 80 ppm. These resonances are shifted by complexation, which induces a change in the electronic environment of the Al atom, i.e., a shielding of its local magnetic field. Thus inorganic anions, such as phosphate and sulfate, produce upfield shifts;[42] downfield shifts are generally observed with organic ligands. Nevertheless, in the case of aromatic ligands, the local field may be affected by the π electrons,[43,44] as exemplified below in the case of salicylate. The magnitude of the chemical shift is related to the number of coordinate bonds shared with the ligand.[45]

The drawback of the aluminum nucleus is that its quadrupole moment results in broad lines, to such a point that lines with widths of tens of hertz are regarded as narrow.[19] Quadrupole relaxation is strongly influenced by the surrounding electric field gradient. This means that the linewidth of an aluminum atom in a molecule is sensitive to the symmetry and to the motion of the molecule. For example, dissociation of a water molecule bound to aluminum, or its exchange by an organic ligand, is responsible for dissymmetry; formation of multiligand complexes or of aluminum polynuclear species larger than dimers reduces the mobility of the Al nuclei. Both cases result in line broadening. This is illustrated by the resonances of the tridecamer (Al_{13}), where only the 62.9 ppm resonance of the central Al tetrahedron is observed, the resonance of the 12 octahedra being so broad as to be undetectable.[19] Fast chemical exchange among the species in equilibrium also produces broad lines. If the exchange rate between several aluminum species is at least of the same order of magnitude as the NMR frequency being used, the corresponding resonances overlap and produce a broad line. This can be used to study the kinetics of dynamic equilibria.[46]

Experimental

Unhydrolyzed aluminum solutions at pH 3.3 were analyzed by liquid-state ^{27}Al NMR in order to obtain accurate information about the most probable complexes and their stoichiometry in the absence of hydrolysis. Stock solutions of 5×10^{-1} M $AlCl_3 \cdot 6H_2O$ and 5×10^{-1} M sodium acetate, oxalate, lactate, salicylate, citrate, or tartrate were prepared with deionized, 0.22 μm-filtered water. They were acidified with 5×10^{-1} M HCl. Volumes of acidified ligand were added to the aluminum solution, so that the aluminum concentration was 10^{-1} M, and the L/M (total ligand concentration/total metal concentration) ratio ranked between 0.1 and 3 according to the chelating power of the acid.

The NMR spectra were recorded within 2 h after the preparation of the samples. Spectra were obtained on a Bruker XWP 200 spectrometer at 52.1 MHz and 25 °C using a ^2H lock with D_2O. Typical experimental parameters included 10 μs pulses at $\pi/2$, recycle delays of 500 ms, 4096 transients, 8 kHz sweep width and 2 Hz line broadening. Samples were placed in a 10 mm coaxial cell. The central capillary contained a 5×10^{-2} M $Al(OH_4)^-$ solution used as a standard for the calibration of the peak areas, in order to derive absolute concentrations of the Al species.

Hydrolyzed samples analyzed by liquid-state ^{27}Al NMR were obtained by partial hydrolysis of the organic acid–aluminum mixtures (10^{-2} M ligand/2×10^{-2} M aluminum) up to the following pH values : 3.5, 4.2, 4.5, 5.0, 5.5, 6.0, and 6.5. The total aluminum concentration, 2×10^{-2} M, was suitable to record NMR spectra and required moderately concentrated NaOH (10^{-1} M) for titration. However, the chemical shifts on the NMR spectra were identical to those recorded with 10^{-1} M aluminum. The same complexes could then be taken into account in the following part of the study. At pH values above 4.2, the resonance of the central tetracoordinated Al^{IV} of Al_{13} was observed at 62.9 ppm (not shown). Preparation of 150 mL of partially hydrolyzed solutions was carried out at room temperature in a 200 mL vessel fitted with four baffles. The solution was vigorously stirred by means of a four-bladed paddle at a rate of 500 rpm. The titrant solution (1 M NaOH) was added with an automatic Tacussel Electroburex EBX 2 pipet. The injection speed was low, 0.04 mol NaOH min^{-1} per mol Al, in order to minimize local oversaturation of base.[39] The pH measurements were made with a Tacussel Titrimax TT 100 apparatus using a Tacussel XC 250 combined electrode.

Quantitative determination of the concentration of various aluminum species in solution was achieved by a novel technique[47] based on the analysis of the experimental free induction decays (FIDs). Calculated FIDs were obtained from a set of four adjustable parameters (linewidth, frequency, amplitude, and phase) for each presumable resonance, and were fitted to the experimental signal by a least-squares method. The NMR spectra yielded the proportions of the soluble aluminum species: hydrated monomeric aluminum (Alm) and complexed aluminum (Alc) from the resonances in the -10 ppm to $+40$ ppm region, and the tridecamer (Al_{13}) from the resonance of the central tetrahedron at 62.9 ppm. The results were expressed as percentages of the total aluminum. Experimental uncertainty in the calculated concentrations essentially originated from the signal/noise ratio in the NMR measurements, and was always lower than 10%.

An increasing part of the NMR signal of soluble aluminum was lost as hydrolysis progressed. This fraction of the aluminum was named Als ("solid" aluminum) and its proportion was calculated by difference between the total aluminum and the sum of the soluble species.

Dry samples for solid-state ^{27}Al NMR were obtained by centrifugation (15000 g, 30 min) of the precipitation product at pH values between 6.0 and 7.0, followed by freeze-drying of the residue. A reference solid phase was obtained by precipitating an Al_{13} solution with Na_2SO_4 following Johansson's procedure.[28]

Solid-state ^{27}Al magic angle spinning (MAS) NMR spectra were recorded on an XWP 300 spectrometer at 78.2 MHz. The pulse sequence was the same as for liquid-state NMR. The spinning frequency was 3 kHz in order to prevent excessive line broadening. The sidebands produced at that speed did not hide the resonance of tetrahedrally coordinated aluminum (62.9 ppm). An aluminum chloride solution (0.5 M) was used to calibrate the spectrometer: the $Al(H_2O)_6^{3+}$ octahedron resonance was taken as 0 ppm. Quantitative utilization of the solid-state NMR spectra was not possible, mainly because of the presence of the sidebands.

Organic Acid–Aluminum Complexes

The NMR spectra of acidic mixtures of organic acids and aluminum chloride (0.1 M) at various L/M ratios (Figures 9.1 to 9.6) display a sharp peak at 0 ppm originating from the hexacoordinated Al monomers and broad lines downfield of the monomer peak which were attributed to the aluminum complexed with the acids.

When L/M was increased, the aluminum monomer peak always decreased in intensity, and the downfield peaks increased in a reciprocal fashion. This indicates that in these experimental conditions exchange between the two species occurs which is slow with respect to the ^{27}Al chemical shift time scale.[45]

Acetate

Liquid–state ^{27}Al NMR spectra of acetate aluminum complexes at pH 3.3 and for L/M ratios of 0.5 and 1 are shown in Figure 9.1. Only a shoulder at 2.0 ppm downfield of the 0 ppm monomer peak was observed. It was attributed to the 1:1 acetate aluminum monodentate complex[36,48] since the carboxyl group of acetate is the only donor and gives a monodentate complex. Interaction with the Al octahedron occurs by exchange of a water molecule. This low chemical shift, corresponding to a weak distortion of the complexed Al octahedron, confirms the weakness of the acetate–aluminum bond. However, the broad shape of the 2 ppm line suggests the probable presence of other species, such as a 2:1 complex in fast exchange with the predominating 1:1 complex. This dimeric species gives a 4.2 ppm resonance, shifted to 3.9 ppm in the presence of acetate at high aluminum concentration (1 M) and temperature (80 °C).[48,50] However, in the experimental conditions of sample preparation (pH = 3.3, room temperature) the spectra in Figure 9.1 show that the 1:1 complex can be regarded as strongly predominant.

Oxalate

Liquid-state ^{27}Al NMR spectra of oxalate–aluminum complexes at pH values near 3, and for L/M ratios of 0.5 and 1 (Figure 9.2) displayed two distinct resonances from complexed aluminum: at L/M = 0.5, only one peak was

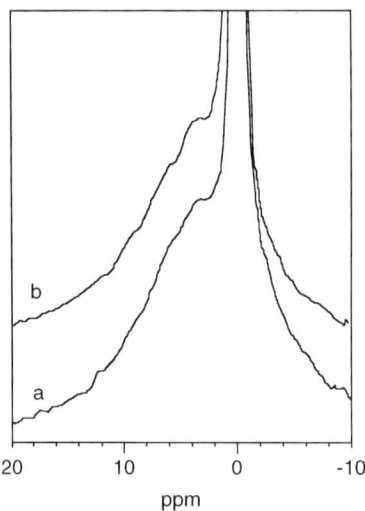

20   10   0   -10

ppm

Figure 9.1 Liquid-state ^{27}Al NMR spectra of aluminum chloride (0.1 M) with acetic acid (a) 0.05 M and (b) 0.1 M at pH 3.3 (reprinted with permission from Thomas et al., *Environ. Sci. Technol.* 25, 1553, 1991; Copyright 1994, American Chemical Society).

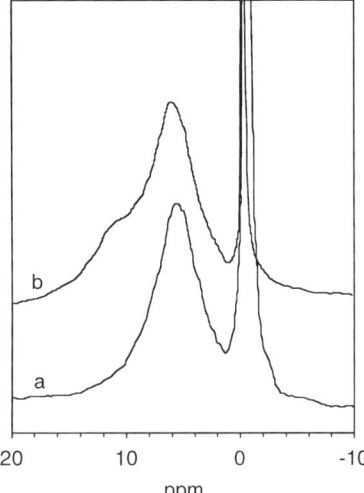

Figure 9.2 Liquid-state ^{27}Al NMR spectra of aluminum chloride (0.1 M) with oxalic acid (a) 0.05 M and (b) 0.1 M at pH 3.3 (reprinted with permission from Thomas et al., *Environ. Sci. Technol.* 25, 1553, 1991; Copyright 1994, American Chemical Society).

observed at 5.9 ppm (320 Hz width), and for L/M = 1, a second resonance was observed at 12.2 ppm (78 Hz width). Oxalate formed bidentate complexes, bearing up to three ligand molecules.[32] Thus, the first resonance observed on the NMR spectra at 5.9 ppm (Figure 9.2) was attributed to a 1:1 bidentate [AlL]$^+$; its large width (320 Hz) originates from the asymmetry of the complexed aluminum atom. The second resonance at 12.2 ppm may correspond to a 2:1 bidentate [AlL$_2$]$^-$, which is more symmetric than the 1:1 complex, as shown by the lower linewidth (78 Hz). Other structures such as [AlHL]$^{2+}$, [Al$_2$(OH)$_2$L$_4$]$^{4-}$, and Al$_3$(OH)$_3$L$_3$ have been described,[15] and may be formed, although they could not be detected by NMR in this work.

Lactate

Liquid-state ^{27}Al NMR spectra of lactate–aluminum complexes at pH 3.3 and for L/M ratios of 0.1, 1, 2, and 3 are shown in Figure 9.3. When L/M was increased from 0.1 to 13, the hexaaquo-aluminum monomer peak decreased in intensity, and the broad peak downfield increased in a reciprocal fashion and was progressively broadened by the contribution of resonances at 6, 9, 15, and 24 ppm. In accord with the proposed structural assignments of these peaks,[45] the 6 ppm resonance was assigned to the monodentate mononuclear complex [LM]$^{2+}$, the 9 ppm peak at L/M = 1 to the 1:1 bidentate [LM]$^+$, the 15 ppm shift to a 2:1 complex [L$_2$M]$^-$, and the 24 ppm shoulder to the emerging 3:1 complex [L$_3$M]$^{3-}$.[37] These complexes are in fast exchange,[32] as indicated by the widths of the NMR corresponding peak. This is supported by their narrow stability constants (Table 9.1).

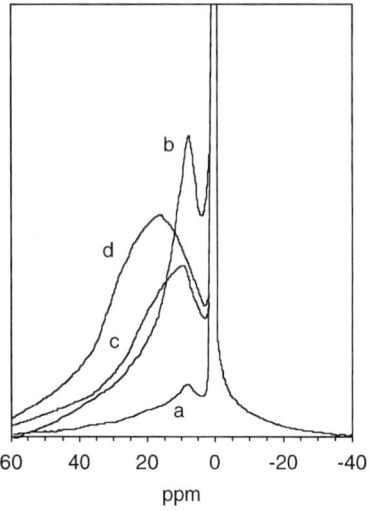

Figure 9.3 Liquid-state ^{27}Al NMR spectra of aluminum chloride (0.1 M) with lactic acid (a) 0.01 M,(b) 0.1 M, (c) 0.2 M, and (d) 0.3 M at pH 3.3 (reprinted with permission from Thomas et al., *Environ. Sci. Technol.* 25, 2551, 1993; Copyright 1994, American Chemical Society).

Salicylate

Liquid-state ^{27}Al NMR spectra of salicylate–aluminum complexes at pH 3.3 and for L/M ratios of 0.5 and 2 are shown in Figure 9.4. The broad line at 3 ppm was ascribed to the 1:1 bidentate [LM]$^+$ salicylate–aluminum complex,[37] since it is the most probable.[60] This chemical shift is surprisingly low for such strong complexes, since it would be expected to occur at 6 to 9 ppm as seen with oxalate and lactate[36,37] (Figure 9.3). Here the aluminum

Table 9.1 Association Constants of Aluminum Species, and Mononuclear Aluminum Complexes with Organic Acids[14,15,22,39,52]

Compound	log K	Compound	log K
Aluminum		Lactate–Aluminum	
Al(OH)$^{2+}$	5.02	AlL$^+$	2.38
Al(OH)$_2^+$	8.71	AlL$_2^-$	4.56
Al(OH)$_3^0$	10.4	AlL$_3^{3-}$	6.66
Al$_{13}$	97.6		
Al$_2$(OH)$_2$	6.27	Salicylate–Aluminum	
		AlL$^+$	12.9
Acetate–Aluminum		AlL$_2^-$	23.2
AlL^{2+}	1.51	AlL$_3^{3-}$	29.8
Oxalate–Aluminum		Citrate–Aluminum	
AlL$^+$	6.1/4.9	AlHL$^+$	2.68
AlL$_2^-$	11.1	AlL	4.92
AlL$_3^{2-}$	15.1	AlL$_2^{3-}$	12.53

atom is situated in a molecule where the p electrons of the benzene ring create a secondary magnetic induction which in this case is added to the magnetic field of the apparatus, and a shift upfield is added to the resonance of the complex.[51] The disymmetry in the 0–3 ppm resonance (Figure 9.4) toward high field, which increases at larger L/M values, could result from the formation of the highly stable 2:1 complex $[L_2M]^-$ (Table 9.1).

Citrate

Liquid-state ^{27}Al NMR spectra of citrate–aluminum complexes display a broad line at 9 ppm (Figure 9.5) for L/M = 0.5 and 1. It was considerably magnified and broadened when L/M was raised from 0.5 to 1. It shifted to 12 ppm at L/M = 2, and the monomer peak at 0 ppm totally disappeared. Similar resonances at 8, 10, and 12 ppm have been observed in citrate–aluminum solutions at pH 2 to 8.[45] By comparison with oxalate- and lactate–aluminum complexes, the 9 ppm line can be hypothetically attributed to a 1:1 bidentate. The broadening of the resonance can result from the formation of multiligand complexes,[52] or to several types of coordination involving changes in the number of chelate rings formed.[19]

Tartarate

Adding tartaric acid to aluminum chloride at pH near 3 and L/M ratios of 0.5, 1 and 2 resulted in liquid-state ^{27}Al NMR spectra (Figure 9.6) showing a very broad downfield line corresponding to numerous possible complexes.[53,54] This

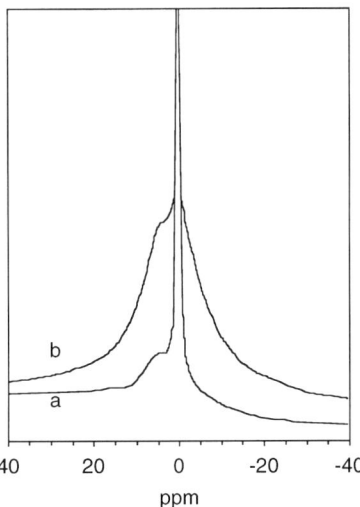

Figure 9.4 Liquid-state ^{27}Al NMR spectra of aluminum chloride (0.1 M) with salicylic acid (a) 0.05 M and (b) 0.2 M at pH 3.3 (reprinted with permission from Thomas et al., *Environ. Sci. Technol.* 25, 2511, 1993; Copyright 1994, American Chemical Society).

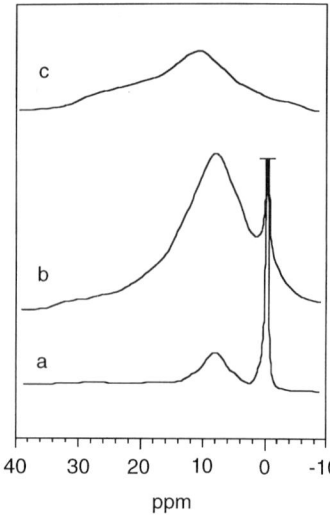

Figure 9.5   Liquid-state ^{27}Al NMR spectra of aluminum chloride (0.1 M) with citric acid (a) 0.05 M, (b) 0.1 M, and (c) 0.2 M at pH 3.3.

trend shows the limits of applicability of NMR in qualitative determination of the structure of the organic aluminum complexes.

NMR Data and Complexation Constants

The concentration of the free and complexed aluminum species calculated from the NMR spectra can be used to calculate complexation constants, or can be compared to those predicted by computer programs such as MINEQL. In general, good agreement has been found.[36,37] Only in the case of lactate,

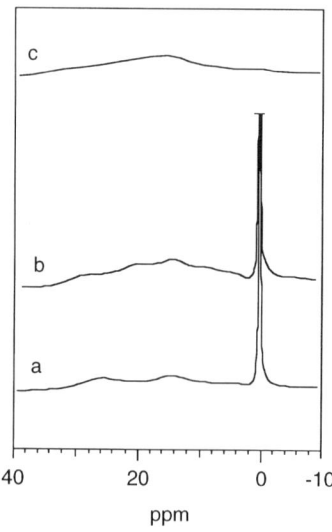

Figure 9.6   Liquid-state ^{27}Al NMR spectra of aluminum chloride (0.1 M) with tartaric acid (a) 0.05 M, (b) 0.1 M, and (c) 0.2 M at pH 3.3.

the measured concentrations of complexed aluminum were in disagreement with the predicted ones: large chemical shifts and peak areas were recorded by NMR, whereas the complex concentrations predicted from the complexation constants were negligible. Thus, re-examination of these constants has been proposed.[37]

Aluminium Species in Partially Hydrolyzed Organic Acid–Aluminum Solutions and Precipitates

Chloride

Hydrolysis of 2×10^{-2} M aluminum chloride (Figure 9.7) leads to the formation of Al_{13} at pH 4.2;[13,23,24] this involves 70 to 90% of the total aluminum, according to the preparation procedure.[34] At higher pH, aggregation of the tridecamers occurs,[24,25,55] and involves one-half of the total aluminum (Figure 9.7). When the hydrolysis ratio ($R = OH/Al$) approaches 3, the tridecamers are slowly converted into trihydroxides $Al(OH)_3$ without dissolution, and this transformation probably involves the detachment of the hexameric ring from the tridecamer.[25]

The mechanisms of the formation of the tridecamer are not yet well understood. Theoretical studies have shown that the precursor is the trimer.[30,31] The strongly nucleophilic central oxygen atom shared by the three aluminum atoms of the trimer (Figure 9.8) behaves as a nucleophilic ligand, and Al_{13} is formed by extremely fast condensation of four trimers around one monomer forced to adopt a tetrahedral configuration. However, the trimer has not been directly observed by ^{27}Al NMR, probably because it would give a very broad line. It was proposed from potentiometric titration and ^1H

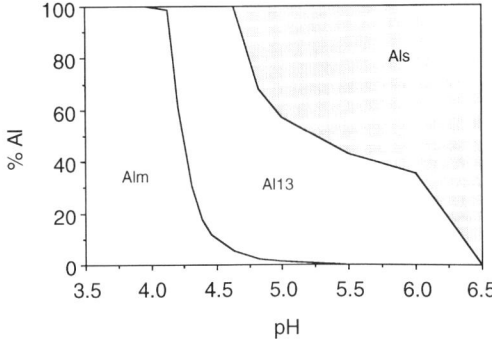

Figure 9.7 Aluminum speciation in partially hydrolyzed aluminum chloride solution (0.02 M), calculated from the fitted NMR spectra. Alm, monomers; Al13, tridecamer; Als, "solid" aluminum (reprinted with permission from Masion et al., *J. Non-Cryst. Solids* 171, 191, 1994).

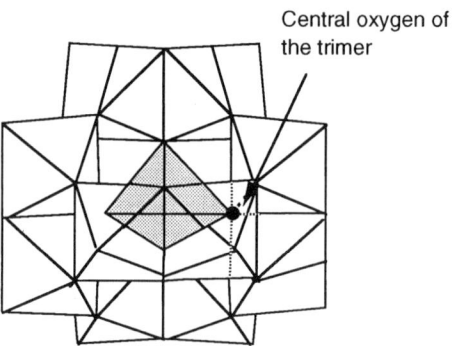

Figure 9.8   Structure of the Al$_{13}$ polymer showing the basic oxygen atom.

NMR,[34,49,52] and is also thought to exist in considerable amounts in hydro-lyzed mixtures of aluminum chloride and oxalate or citrate.[15,52] The dimer has been detected by ^{27}Al NMR in concentrated solutions and at 80 °C.[26,33] It has also been described in the presence of acetate.[48–50]

The nature of the precipitated phase was studied by solid-state NMR (Figure 9.9). The spectrum of pure Al$_{13}$ precipitated by sulfate displayed a line at 62.9 ppm from the central Al tetrahedron of the tridecamer and a broad peak near 6 ppm from the 12 octahedral Al atoms surrounding the central Al atom. Kirkpatrick et al.[56] and Bottero et al.[25] stated that shielding increases with increasing polymerization, and causes the chemical shift to reach 10 ppm in gibbsite or bayerite.

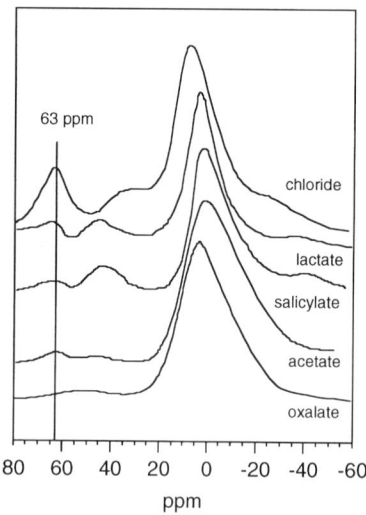

Figure 9.9   Solid-state ^{27}Al NMR spectra of freeze-dried precipitation products obtained by hydrolysis of aluminum (0.02 M)–organic acids (0.01 M) mixtures between pH 6 and 7 (reprinted with permission from Masion et al., *J. Non-Cryst. Solids* 171, 191, 1994.

Lactate

At the lowest pH value (pH 3.5), the aluminum was mainly in the monomeric form (Figure 9.10), the concentration of which decreased to the benefit of the 1:1 complex when titrant was added. More complex was formed as dissociation of the lactic acid occurred. NMR spectra carried out at pH 4.2 showed that one-half of the aluminum and all of the lactate were involved in a soluble bidentate complex. Al_{13} was formed without being hindered by lactate, in amounts close to those measured with aluminum chloride (Figure 9.7). However, in the pH range from 4.2 to 6.0, nearly 40% of the aluminum was involved in small clusters of the Als species (Figure 9.10), the size of which was of the order of 30 Å as derived from liquid-state NMR.[37] At pH > 6, the precipitation abruptly increased. The precipitate contained a relatively high amount of Al_{13} (Figure 9). The peak from the octahedral aluminum was only shifted by less than 6 ppm, and this was also observed for the other organic acids studied. It indicates that shielding from the neighbors around the octahedrally coordinated aluminum atoms is similar in the chloride and in the organic acid–aluminum precipitates. Aluminum polymerization is strongly limited, and the elementary particles of the precipitates are, besides Al_{13}, small aluminum oligomers, most probably monomers, dimers, and trimers.[57–59]

Acetate

At pH 3.5, 95% of the aluminum remains uncomplexed (Figure 9.11). A complex was progressively formed as the pH increased and the acetic acid dissociated. The proportion of the complex never exceeded 20%, as measured at pH 4.2. This is in agreement with its low complexation constant.[32]

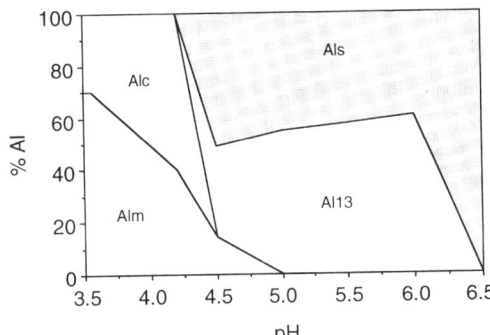

Figure 9.10 Aluminum speciation in partially hydrolyzed aluminum chloride (0.02 M) and lactate (0.01 M) solution, calculated from the fitted NMR spectra. Alm, monomers; Alc, complexes, A113, tridecamer; ALs "solid" aluminum (reprinted with permission from Masion et al., *J. Non-Cryst. Solids* 171, 191, 1994).

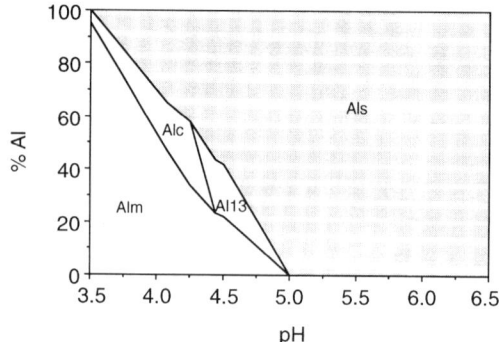

Figure 9.11 Aluminum speciation in partially hydrolyzed aluminum chloride (0.02 M) and acetate (0.01 M) solution, calculated from the fitted NMR spectra. Alm, monomers; Alc, complexes, A113, tridecamer; ALs "solid" aluminum (reprinted with permission from Masion et al., *J. Non-Cryst. Solids* 171, 191, 1994).

Although the stoichiometry of this complex is undoubtedly mononuclear at acidic pH, it is possible that a dinuclear complex forms at higher pH, according to the dissociation of acetic acid.[34,36,50] Paradoxically, despite the abundance of free monomers, less than 20% of Al_{13} was formed. Thus, the formation of Als from pH 3.5 suggests that complexed dimers are involved in the precipitation process. The precipitation of dimers then competes with the formation of Al_{13}, which is less abundant in the solid phase than with lactate (Figure 9.9).

Oxalate

The oxalate–aluminum bidentate complex which involved 47% of the aluminum and hence 94% of the oxalate was stable up to pH 4.5 (Figure 9.12). Since nearly all of the oxalate is involved in the monomeric complex, Al_{13} should be formed normally from the free monomers via condensation of trimers. However, the Al_{13} concentration in solution was less than 3% of the total aluminum, and no Al_{13} was found in the dry precipitate (Figure 9.9). The formation of Al_{13} may be hindered by complexation of the dimeric and trimeric precursors with oxalate. Complexes involving the aluminum oligomers, $[Al_2(OH)_2L_4]^{4-}$ and $Al_3(OH)_3L_3$, have been assumed by Sjöberg and Öhmann[15] to form at pH 4 to 7. These complexes could represent the major aluminum species at pH 4.5. They can be assumed to be undetected due to their oligomeric and multiligand structure, and to be included in the Als species. Therefore the precipitation of Als is probably due to the aggregation of small complexed oligomers as 1:3 mono- or bidentate trimers.[57–59]

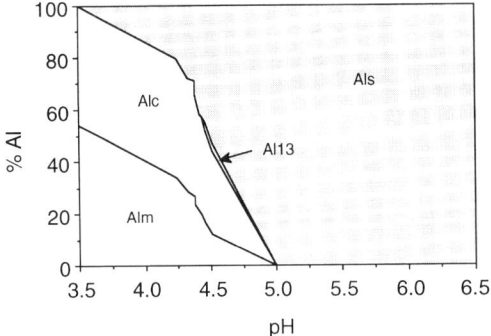

Figure 9.12 Aluminum speciation in partially hydrolyzed aluminum chloride (0.02 M) and oxalate (0.01 M) solution, calculated from the fitted NMR spectra. Alm, monomers; Alc, complexes, A113, tridecamer; Als "solid" aluminum (reprinted with permission from Masion et al., *J. Non-Cryst. Solids* 171, 191, 1994).

Salicylate

The bidentate salicylate–aluminum complex was the most stable among the complexes studied in this work. It involved 43% of the total aluminum and 86% of the ligand at pH 3.5 and was detected up to pH 5.5 (Figure 9.13). In the same pH range a large amount of free monomers (40% of the total aluminum) was available to form Al_{13}, but only 20% aluminum was finally involved within the tridecamers. Simultaneously, Als increased, and the precipitate contained a very small amount of Al_{13} and mainly octahedral aluminum involved within oligomers (Figure 9.9). The mechanism of hindrance of Al_{13} formation may be the same as with the other ligands, i.e., complexation of salicylate with the dimers or trimers. However, some Al_{13} was formed in solution, and precipitated probably through charge screening, since Al_{13} can accommodate up to six salicylate ligands.[60]

NMR and Aggregation

Due to quadrupole relaxation of the aluminum nucleus, the linewidth contains information on the symmetry of the local environment of the aluminum atom. This was exploited in the case of the hydrolysis of lactate– or salicylate–aluminum chloride mixtures, where the 0 ppm line was broadened according to pH but independently of the ligand nature or concentration. This led to the assumption that the hydrolysis of the free aluminum monomers occurs without perturbation by the organic ligands.[37] The linewidth also contains information about the mobility of the molecule in which the aluminum atom is involved.[19] Thus, it has been possible to calculate the hydrodynamic radius of small aggregates of Al_{13} from the broadening of the 63 ppm line due to the

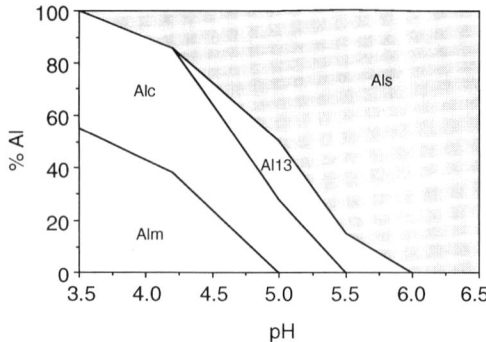

Figure 9.13 Aluminum speciation in partially hydrolyzed aluminum chloride (0.02 M) and salicylate (0.01 M) solution, calculated from the fitted NMR spectra. Alm, monomers; Alc, complexes; Al13, tridecamer; Als, "solid" aluminum (reprinted with permission from Masion et al., *J. Non-Cryst. Solids* 171, 191, 1994).

presence of lactate. The linewidth varies according to the correlation time (τ_c) of the species in solution:

$$\Delta\nu_{1/2} \propto \tau_c^{-1} \qquad (1)$$

and τ_c^{-1} varies according to the volume V or the hydrodynamic radius R of the solute.[43] Thus:

$$\Delta\nu_{1/2} \propto V \text{ or } R^3 \qquad (2)$$

Since the hydrodynamic radius of the bare Al_{13} in partially hydrolyzed solution is 12.6 Å,[25] and the corresponding 63 ppm NMR line is close to 12 Hz, the R value for $\Delta\nu_{1/2} = 150$ Hz is close to 30 Å. Such clusters may be formed of about five to seven tridecamers linked together by lactate molecules.

Conclusion

Numerous studies have proven that NMR is a powerful tool for the investigation of the chemistry of aluminum in the liquid and in the solid phase, and the data presented here support this statement. Beyond this, NMR can be used as a direct local probe of the aluminum nucleus in a situation where other methods such as titration or indirect chemical methods are inoperative, e.g., the metastable aluminum species formed during the transition from soluble complexes to precipitates in the presence of organic acids. As demonstrated by the above examples, appropriate exploitation of the NMR spectra can yield much information on the aluminum species. In the liquid state the structure and stoichiometry of the complexes, and the concentration of the various soluble species, as well as their symmetry, can be determined. The formation of precipitates with progressing hydrolysis has been quantified from the loss of signal. At last, solid-state NMR on the dried precipitates has yielded at least a qualitative speciation.

The data obtained in the present study show that the organic ligands hinder the formation of Al_{13} mainly at the stage of the intermediate oligomers. The oligomers form polynuclear complexes with the organic ligands, which thus hinder their aggregation to Al_{13}. The precipitates contain only poorly polymerized aluminum, and low amounts of tridecamer. When Al_{13} is formed in solution, it aggregates into small clusters through organic ligands.

The consequences of aluminum interaction with organic acids at near-neutral pH are of high environmental interest. The foregoing NMR results show the detoxifying power of organic acids toward aluminum. A proportion of 50 to 80% of the phytotoxic aluminum species (monomers and Al_{13}) is entrapped within less toxic complexes and organo-mineral precipitates. Monomers and oligomers are involved within organic complexes, the strength of which is related to the complexing capacity of the acid. Complexing acids strongly occupy the aluminum bonding sites, inhibiting the formation of hydroxo bridges between aluminum atoms and hence their aggregation to form Al_{13}. At near-neutral pH, a mixed organo-mineral phase is then precipitated. The pH at which this precipitation occurs is always lower in the presence than in the absence of organic acids.

Acknowledgments   This work was supported by the "Programme Dynamique et Bilans de la Terre 91-02" INSU-CNRS (Paper No. 630).

References

1. Driscoll, C. T., A procedure for the fractionation of aqueous aluminum in dilute acidic waters. *Int. J. Environ. Anal. Chem.* 16, 267, 1984.
2. Stevenson, F. J., and Fitch, A., Chemistry of complexation of metal ions with soil solution organics. In *Interactions of Soil Minerals with Natural Organics and Microbes,* Huang, P. M., and Schnitzer, M. (eds), Spec. Publ. No. 17, Soil Science Society of America, Madison, WI, 1986, Chapter 2.
3. Johnson, N. M., Acid rain: neutralization within the Hubbard Brook Ecosystem and regional implications. *Science* 204, 487, 1979.
4. Ulrich, B., Mayer, R., and Kanna, P. K., Chemical changes due to acid precipitation in a loess-derived soil in central Europe. *Soil Sci.* 130, 193, 1980.
5. Cronan, C.S., and Schofield, C.L., Relationships between aqueous aluminum in acidic deposition in forested watersheds of North America and northern Europe. *Environ. Sci. Technol.* 24, 1100, 1990.
6. Lawrence, G. B., and Driscoll, C. T., Aluminum chemistry of a whole-tree-harvested watershed. *Environ. Sci. Technol.* 22, 1293, 1988.
7. Gjessing, E. T., Alexander, J., and Rosseland, B. O., Acidification and aluminum-contamination of drinking water. In *Watershed 89,* Wheeler, D., Richardson, M.L., and Bridges, J. (eds), Pergamon, Oxford, 1989, Chapter 2.

8. Hue, N.V., Craddock, G.R., and Adams, F., Effect of organic acids on aluminum toxicity in soils. *Soil. Sci. Soc. Am. J.* 50, 28, 1986.

9. Bertsch, P. M., The hydrolytic products of aluminum and their biological significance. *Environ. Chem. Health* 12, 7, 1990.

10. Hodges, S. C., Aluminum speciation: a comparison of five methods. *Soil. Sci. Soc. Am. J.* 51, 57, 1987.

11. Bottero, J. Y., and Bersillon, J. L., Aluminum and Iron(III) chemistry. Some implications for organic substances removal. In *Aquatic Humic Substances*, Suffet, I. H., and MacCarthy, P. (eds), American Chemical Society, Washington DC, 1989, Chapter 26.

12. Bottero, J. Y., Coagulation–flocculation of minerals using Al, Fe(III) salts. What kind of flocs for what separation process? In *Influence and Removal of Organics in Drinking Water*, Mallevialle, J., Suffet, I. H., and Chan, U. S. (eds), Lewis Publishers, Boca Raton, FL, 1992, Chapter 9.

13. Bersillon, J. L., Hsu, P. H., and Fiessinger, F., Characterization of hydroxy-aluminum solutions. *Soil Sci. Soc. Am. J.* 44, 630, 1980.

14. Motekaitis, R. J., and Martell, A. E., Complexes of aluminum(III) with hydroxy carboxylic acids. *Inorg. Chem.* 23, 18, 1984.

15. Sjöberg, S., and Öhmann, L. O., Equilibrium and structural studies of silicon(IV) and aluminum(III) in aqueous solution. Part 13. A potentiometric and ^{27}Al nuclear magnetic resonance study of speciation and equilibria in the aluminum (III)–oxalic acid–hydroxide system,. *J. Chem. Soc., Dalton Trans.* 2665, 1985.

16. Alva, A. K., Sumner, M. E., Li, Y. C., and Miller, W. P., Evaluation of three aluminum assay techniques for excluding aluminum complexed with fluoride or sulfate. *Soil. Sci. Soc. Am. J.* 53, 38, 1989.

17. Jardine, P. M., and Zelazny, L. W., Influence of inorganic anions on the speciation of mononuclear and polynuclear aluminum by ferron. *Soil. Sci. Soc. Am. J.* 51, 889, 1987.

18. Wilson M. A., *NMR Techniques and Applications in Geochemistry and Soil Chemistry*, Pergamon, Oxford, 1988.

19. Akitt, J. W., Multinuclear studies of aluminum compounds. *Prog. NMR Spectrosc.* 21, 1, 1989.

20. Brosset, C., Biedermann, G., and Sillen, L. G., Studies on hydrolysis of metal ions. *Acta Chem. Scand. Ser. A* 8, 1917, 1954.

21. Matijevic, E., Mathai, K.G., Ottewill, R.H., and Kerker, M., Detection of metal ion hydrolysis by coagulation. III. Aluminum. *J. Phys. Chem.* 65, 826, 1961.

22. Hsu, B. H., and Bates T. F., Formation of X-ray amorphous and crystalline aluminum hydroxides. *Miner. Mag.* 33, 749, 1964.

23. Bottero J. Y., Cases, J. M., Fiessinger, F., and Poirier, J. E., Studies of hydrolyzed aluminum chloride solutions. 1. Nature of aluminum species and composition of aqueous solutions. *J. Phys. Chem.* 84, 2933, 1980.

24. Bottero, J. Y., Marchal, J. P., Poirier, J. E., Cases, J. M., and Fiessinger, F., Étude, par RMN, de l'aluminium-27, des solutions diluées de chlorure d'aluminium partiellement hyrolysées. *Bull. Soc. Chim. Fr.* 11–12, I 439, 1982.

25. Bottero, J. Y., Axelos M., Tchoubar, D., Cases, J. M., Fripiat, J. J., and Fiessinger, F., Mechanisms of formation of aluminum trihydroxide from Keggin Al_{13} polymers. *J. Colloid Interface Sci.* 117, 47, 1987.

26. Kloprogge, J. T., Seykens, D., Geus, J. W., and Jansen, J. H. B., Temperature influence on the Al_{13} complex in partially neutralized aluminum solutions: a [27]Al nuclear magnetic resonance study. *J. Non-Cryst. Solids* 142, 87, 1992.

27. Kloprogge, J. T., Seykens, D., Jansen, J. H. B., and Geus, J. W., A [27]Al nuclear magnetic resonance study on the optimization of the development of the Al_{13} polymer. *J. Non-Cryst. Solids* 142, 94, 1992.

28. Johansson, G., On the crystal structures of some basic aluminum salts. *Acta Chem. Scand. Ser. A* 14, 769, 1960.

29. Johansson, G., On the crystal structure of a basic aluminum sulfate and the corresponding selenate. *Acta Chem. Scand. Ser. A* 14, 772, 1962.

30. Brinker, C. J., and Sherer, G. W., *Sol Gel Science*, Academic Press, New York, NY, 1989.

31. Henry, M., Jolivet, J. P., and Livage, J., Aqueous chemistry of metal cations: hydrolysis, condensation and complexation. In *Structure and Bonding*, Springer, Berlin, 1992, p. 155.

32. Nordstrom, D. K., and May, H. M., Aqueous equilibrium data for mononuclear aluminum species. In *The Environmental Chemistry of Aluminum*, Sposito, G. (ed.), CRC Press, Boca Raton, FL, 1989, Chapter 2.

33. Akitt, J. W., and Farthing, A., Aluminium-27 nuclear magnetic resonance studies of the hydrolysis of aluminium(III). Part 2. Gel-permeation chromatography. *J. Chem. Soc., Dalton Trans.* 1606, 1981.

34. Akitt, J. W., Elders, J. M., Fontaine, X. L. R., and Kundu, A. K., Multinuclear magnetic resonance studies of the hydrolysis of aluminium(III). Part 10. Proton, carbon-13, and aluminium-27 spectra of aluminium acetate at very high magnetic field. *J. Chem. Soc., Dalton Trans.* 1897, 1989.

35. Öhman, L. O., and Forsling, W., Equilibrium and structural studies of silicon (IV) and aluminum (III) in aqueous solution. III. A potentiometric study of aluminum (III) hydrolysis and aluminum(III) hydroxo carbonates in 0.6 M NaCl. *Acta Chem. Scand, Ser. A* 35, 795, 1981.

36. Thomas, F., Masion, A., Bottero, J. Y., Rouiller, J., Genévrier, F., and Boudot, D., Aluminum(III) speciation with acetate and oxalate. A potentiometric and ^{27}Al NMR study. *Environ. Sci. Technol.* 25, 1553, 1991.

37. Thomas, F., Masion, A., Bottero, J. Y., Rouiller, J., Montigny, F., and Genévrier, F., Aluminum(III) speciation with hydroxy-carboxylic acids. ^{27}Al NMR study. *Environ. Sci. Technol.* 27, 2511, 1993.

38. Masion, A., Thomas, F., Bottero, J. Y., Tchoubar, D., and Tekely, P., Formation of amorphous precipitates from aluminum–organic ligands solutions: macroscopic and molecular study. *J. Non-Cryst. Solids* 171, 191, 1994.

39. Parker, D. R., and Bertsch, P. M., Formation of the "Al_{13}" tridecameric polycation under diverse synthesis conditions. *Environ. Sci. Technol.* 26, 914, 1992.

40. Furrer, G., Trusch, B., and Müller, C., The formation of polynuclear Al_{13} under simulated natural conditions. *Geochim. Cosmochim. Acta* 56, 3831, 1992.

41. Furrer, G., Ludwig, C., and Schindler, P.W., On the chemistry of the Keggin Al_{13} polymer. I. Acid–base properties. *J. Colloid Interface Sci.* 149, 56, 1992.

42. Akitt, J. W., Greenwood, N. N., Khandelwal, B. L., and Lester, G. D., [27]Al nuclear magnetic resonance studies of the hydrolysis and polymerization of the hexa-aquo-aluminum(III) cation. *J. Chem. Soc., Dalton Trans.* 604, 1972.

43. Farrar, T. C., and Becker, E. D., *Pulse and Fourier Transform NMR: Introduction to Theory and Methods*, Academic Press, London, 1971.

44. Delpuech, J. J., Khaddar, M. R., Peguy, A., and Rubini, P., Octahedral and tetrahedral solvates of the aluminum cation. A study of the exchange of free and bound organophosphorous ligands by Nuclear Magnetic Resonance spectroscopy. *J. Am. Chem. Soc.* 97, 3373, 1975.

45. Karlik, S. J., Tarien, E., Elgavish, G. A., and Eichhorn, G. L., Aluminum-27 Nuclear Magnetic Resonance study of aluminum(III) interactions with carboxylate ligands. *Inorg. Chem.* 22, 525, 1983.

46. Ichikawa, K., and Matsumoto, T., An aluminum-27 study of chemical exchange and NMR line broadening in molten butylpyridinium chloride + AlCl$_3$. II. *J. Magn. Reson.* 63, 445, 1985.

47. Montigny, F., Brondeau, J., and Canet, D., Analysis of time domain NMR data by standard non linear least squares. *Chem. Phys. Lett.* 170, (3), 175, 1990.

48. Akitt, J. W., and Mann, B. E., ^{27}Al NMR spectroscopy at 104.2 MHz. *J. Magn. Reson.* 44, 584, 1981.

49. Hiraishi, M., Harada, S., Uchida, Y., Kuo, H. L., and Yasunaga, T., Aluminum monoacetate complex formation studied by the pressure-jump method. *Int. J. Chem. Kinetics* 12, 387, 1980.

50. Akitt, J. W., and Millic, N. B., Aluminum-27 nuclear magnetic resonance studies of the hydrolysis of aluminum(III). Part 6. Hydrolysis with sodium acetate. *J. Chem. Soc., Dalton Trans.* 981, 1984.

51. Canet, D., *La RMN, Concepts et Méthodes*, Inter Editions, Paris, 1991.

52. Öhman, L. O., Equilibrium and structural studies of silicon(IV) and aluminum(III) in aqueous solution. 17. Stable and metastable complexes in the system H$^+$–Al^{3+}–citric acid. *Inorg. Chem.* 27, 2565, 1988.

53. Greenaway, F. T., Aluminum 27 NMR Study of aluminum(III) interactions with hydroxy carboxylic acids. *Inorg. Chim. Acta* 16, 21, 1986.

54. Venema, F. R., Peters, J. A., and Van Beckkum, H., Multinuclear magnetic resonance study of the coordination of aluminum(III) with tartaric acid in aqueous solution. *Inorg. Chim. Acta* 191, 261, 1992.

55. Axelos, M. A. V., Tchoubar, D., Bottero, J. Y., and Fiessinger, F., Détermination, par D.P.A.X., de la structure fractale d'agrégats obtenus par collage d'amas. Étude de deux solutions d'hydrolyse d'aluminium (AlOH)$_x$ avec $x = 2,5$ et 3. *J. Phys.* 46, 1587, 1985.

56. Kirkpatrick, R. J., Smith, K. A., Schramm, S., Turner, G., and Yang, W. H., Solid-state nuclear magnetic resonance spectroscopy of minerals. *Annu. Rev. Earth Planet. Sci.* 13, 29, 1985.

57. Masion, A., Thomas, F., Villiéras, F., Tchoubar, D., Bottero, J. Y., and Tekely, P., Chemistry and structure of Al(OH)/organics precipitates. A small angle X-ray scattering study. I. *Langmuir* 10, 4344, 1994.

58. Masion, A., Thomas, F., Tchoubar, D., Bottero, J. Y., and Tekely, P., Chemistry and structure of Al(OH)/organics precipitates. A small angle X-ray scattering study. II. *Langmuir* 10, 4349, 1994.

59. Masion, A., Thomas, F., Tchoubar, D., Bottero, J. Y., and Tekely, P., Chemistry and structure of Al(OH)/organics precipitates. A small angle X-ray scattering study. III. *Langmuir* 10, 4353, 1994.

60. Rakotonarivo, E., Tondre, C., Bottero, J. Y., and Mallevialle, J., Polymerized and hydrolyzed aluminum(III) complexation by salicylate ions. Kinetic and thermodynamic study. *Water Res.* 9, 1337, 1989.

10.

Cation and Water Interactions in the Interlamellae of a Smectite Clay

ANDREA LABOURIAU, CLIFF T. JOHNSTON, & WILLIAM L. EARL

Advances in NMR instrumentation and availability have led to increased application to mineral systems and to environmental problems. The sensitivity of high-field NMR systems is nearly sufficient to work at real environmental concentrations. Even with limited sensitivity, the amount of chemical information obtained through NMR spectroscopy makes it a very valuable technique in many model systems. The application of NMR spectroscopy in mineral systems has been primarily limited to studies of the structural metals aluminum and silicon. However, in recent years there have been several publications on mobile cations in minerals, including work on the exchangeable cations in clays. Our interests lie in understanding the sorption of cations in clays, the structural sites available for that sorption, and the role of water in cation–clay interactions. Our goal is to eventually understand the molecular interactions that determine the adsorption and diffusion of cations in clays and, thus, the role of clays in determining cation transport through the geosphere.

This fundamental understanding has applications in the fate of heavy metals, radionuclides, and even the mobility of nutrients for plants. It is well known that there are very strong interactions between metals and humic materials and these are also strong contributors to cation mobility. However, for simplicity, we have chosen to focus on the interactions of mobile metal ions with well-characterized clays. An NMR-based approach to this problem can take two complementary directions: first, studies of the structural components of clays such as ^{29}Si and ^{27}Al NMR as a function of cation or hydration; second, NMR studies of probe molecules—which in this case are the cations themselves. High magnetic field, multinuclear NMR spectrometers make it quite possible to study various "uncommon" nuclei with relative ease. It is our experience that using the cations as probe nuclei for studying sorption phenomena yields more information than studies of structural nuclei. This chapter is basically a report of work in progress on several systems that are starting to yield interesting results, which it is hoped will lead to a general understanding of these complex systems.

Because the field of NMR in clays (phyllosilicates) is somewhat specialized, we will start with a short, general description of the clays themselves and their adsorption and hydration properties which have been determined by other methods. This will give an overall picture of the systems studied. We will then summarize some of the excellent NMR studies in clay systems which are precursors to this work. This is followed by the experimental details, results, and finally a discussion of those results. Since this work is still in its infancy, there are no strong conclusions but rather general impressions of the directions and utility of NMR in understanding these important soil components.

Clay Structures

The structural and dynamic features of clays have been studied for many years, as is evidenced by the countless monographs and several research journals dedicated to reporting the results of academic studies of clay structure and properties. Clays are complexes of silicon, aluminum, and magnesium oxides. They are sheet structures much like a slightly disordered deck of cards, as depicted in Figure 10.1. Each sheet in the superstructure is itself composed of layers. There are various ways of categorizing clays which divide them into successively more homogeneous groups. A division according to the structure of the sheets recognizes 1:1 and 2:1 clays. In 1:1 clays each sheet is composed of one tetrahedral layer and one octahedral layer. The tetrahedral layer is primarily silicon oxide with Si in the center of the tetrahedron and the octahedral layer is either aluminum or magnesium oxide with the metal in the octahedral environment. Since the work reported here does not cover 1:1 clays they will not be described in any more detail. The 2:1 clays are composed of two layers of tetrahedral SiO_2 sandwiching an octahedral layer of magnesium oxide or aluminum oxide. Figure 10.2 is an idealized structure of one sheet of a 2:1 clay. Further, clays are categorized as dioctahedral or trioctahedral. In an ideal structure, each half unit cell ($\frac{1}{2}$ u.c.) contains three octahedral sites. If all three sites are filled with a metal with a formal charge of $+2$ such as Mg^{2+}, charge is balanced and this is referred to as a trioctahedral clay. If, instead, the octahedral sites are filled with a formally trivalent metal like Al^{3+}, charge balance requires that only two sites be occupied and the third be vacant. This is a dioctahedral clay. The clays studied in this work are montmorillonites or smectites, which are 2:1 dioctahedral clays.

Charge is developed in the structural layers of clays by substitution of metals of lower valence. For example, a single Mg^{2+} can replace a single Al^{3+} in a dioctahedral clay. This gives a net charge of -1 in the octahedral sheet which must be balanced by an exchangeable cation between the layers.

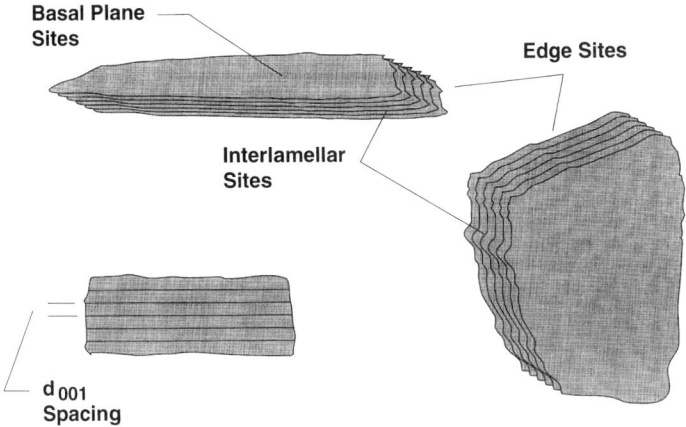

Basal Plane Sites

Edge Sites

Interlamellar Sites

d_{001} Spacing

Figure 10.1 Diagrammatic representation of a macroscopic clay structure.

Additionally, Al^{3+} can substitute for Si^{4+} in the tetrahedral layer, again giving a net -1 charge. There are numerous other substitutions including Fe^{3+} and Li^{+} in the octahedral layers. The distribution of charge is different when the substitution is in the tetrahedral layer (where the charge is quite close to the interlayer) or in the octahedral layer (where the charge is more diffuse). These substitutions change the electrical charge properties of the clay and thus establish the differences between clays found in nature. More highly charged clays require more exchangeable cations and thus more water of hydration. There are several monographs which give a more detailed structure of different clays, e.g., refs. 1–3.

For the purpose of illustration, we give a few examples of different clays and some of their properties. Vermiculite is a highly charged trioctahedral

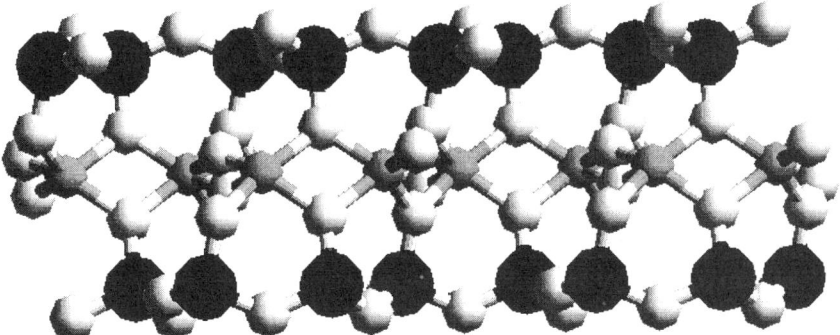

Figure 10.2 The structure of a single clay sheet, with the two layers of tetrahedral metals sandwiching an octahedral layer. All oxygens are shown, demonstrating the shared oxygen atoms within and between layers.

clay with considerable substitution of Al for Si in the tetrahedral sheets. This results in a system with high charge adjacent to the exchangeable cations. This charge, with the corresponding exchangeable cations, holds the sheets tightly together, giving vermiculites an ordered structure. This tight layering makes vermiculites difficult to hydrate, requiring special solvents or conditions. Hectorite is a trioctahedral clay with Li substitution for the Mg in the octahedral sheet. It has a fairly high charge but this charge is in the octahedral layer so is more diffuse than that in vermiculite. Hectorite is much less ordered and easier to hydrate than vermiculite. These two clays are the most common in NMR studies because they are very low in iron or other paramagnetic metals[3] so they give excellent high-resolution results. Montmorillonites are a class of dioctahedral clays whose layer charge is primarily due to substitution of Mg for Al in the octahedral sheet. Besides the charge due to substitution in the layers, clay sheets are not infinite in dimension so there are edges and defects which result in locations where charge is necessarily imbalanced. This results in "edge sites", depicted in Figure 10.1, which require cations to balance the charge. These sites are much less structured than interlayer sites and are relatively difficult to quantify.

The total layer charge of a given clay is measured through ion exchange experiments and is usually reported as the cation exchange capacity (CEC) in equivalents/gram. The clays used in this work are Clay Minerals Society reference clays so the CEC has been measured by several groups and reported in the literature.[3-6] The CEC is an overall, bulk measurement and thus does not describe the detailed nature or exact location of the charge. There are several modifications of the CEC measurement and chemical tests that are used to distinguish the octahedral charge from the tetrahedral charge.[5,6] These measurements are reported for the samples used in this work, so we have a qualitative idea of the total charge distribution.[5]

Cations in the interlayer bring waters of hydration with them. The distance between the layers varies, depending upon the exact clay structure, but more significantly it depends on the exchangeable cation and the extent of hydration.[7,8] There is invariably some sheet-to-sheet disorder. In other words, diffraction techniques give no indication of the alignment of one sheet with the sheet above or below it, i.e., clays are oriented but not crystalline materials. The sensitivity of NMR to local structure or short-range order makes it an ideal complement to diffraction for determining structures and dynamics in clays.

Our eventual goal is to determine the detailed location of exchangeable cations. A frequently proposed position for the cation is in a site referred to as the oxygen ditrigonal cavity. The ditrigonal cavity is a hexagonal arrangement of oxygen atoms that occurs on the tetrahedral surface of all clays. Placing a cation in this location minimizes the distance between the positive cation charge and the negative charge on the oxygen atoms in the clay. A cation site in this cavity is reasonable if the cation is small and has no waters of hydration, but it becomes more difficult when the cation is large and the waters of hydration are tightly bound. In a 2:1 clay, the ditrigonal cavities of one layer may orient directly over those of the adjacent layer, giving a pocket with 12-fold coordination, or they may not line up at all, giving lower symmetry to this "site."

There are defects and dislocations in the layers, so this picture of a structured "deck of cards" is overly idealized. A realistic structure contains large sheets interspersed with some smaller ones making an imperfect crystallite. There are crystallites of different sizes which pack together in more or less random fashion in most preparations. We note that it is relatively easy to prepare oriented clay samples which have most of the sheets parallel to one another (turbostratic structure). In our work the clays are ground with a mortar and pestle and packed into MAS rotors, so they are probably nearly random powders. There is no evidence for ordering in any of our experiments.

Most spectroscopic studies of clays start with ion exchanged, homoionic samples. The average d_{001} spacing can then be controlled through control of hydration.[9] There are a number of water adsorption isotherms reported in the literature. The easiest experiment to visualize is one in which a clay is placed in an atmosphere of controlled relative humidity, the hydration level is allowed to equilibrate and then the d_{001} spacing is measured with X-ray diffraction (XRD).[4] This can be repeated with different relative humidities, resulting in a plot such as that in Figure 10.3. The data in Figure 10.3 are taken from work by Sato et al.[4] for two of the clays used below.

NMR in Clays

For the purposes of this chapter we assume that the reader is familiar with the fundamentals of NMR but not necessarily with the applications and history of NMR in clay systems. NMR studies of solid clays can be separated into three types: studies of water using proton and deuteron NMR, structural studies using ^{27}Al and ^{29}Si, and, most recently, studies of the adsorbed cations using the NMR resonance of the ion of interest.

Early studies of the proton and deuteron in hydrated clay systems strove to determine the location, structure, and extent of hydration. Hougardy et al. investigated the proton NMR of well-ordered vermiculite sedimented from solution, from which they could infer the structure of the water molecule relative to the exchangeable cations.[10] They continued their proton NMR in different hydrates of lithium- and sodium-exchanged vermiculite and were able to correlate the NMR data with the water exchange isotherm.[11] These studies give considerable insight into the structure of the water but suffer the disadvantage that vermiculite is one of the most structured of the clays, so they establish a limit in a very structured system. Woessner and coworkers also published several studies of water in clays, primarily using deuterium NMR.[12] They were also able to obtain relatively detailed structural information about the water in the interlayer of clay. Although the state of water in the interlayer is a critical problem in clay properties, there are

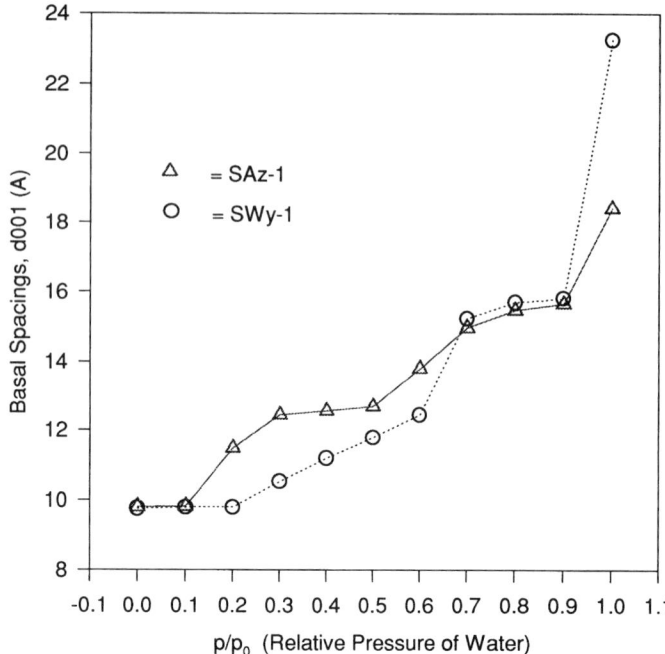

Figure 10.3 A plot of the d_{001} spacing as a function of p/p_0 for the Na^+-exchanged forms of SAz-1 and SWy-1. The transition between relatively well-defined hydration states for SAz-1 can be seen at p/p_0 of about 20 and 60. The transitions are less clear in SWy-1.

intrinsic difficulties with these studies. First, there are large numbers of hydro-xyl groups in the structure and in defects of clays which exchange with the protons and deuterons studied. Additionally, the limited chemical shift range of hydrogen isotopes limits the chemical information available. The water molecules are highly fluxional which, combined with the intrinsic disorder in clays, results in a dispersion of sites. It is only through very careful work on well-prepared samples that these researchers were able to add significantly to our understanding of how water binds and orients in clays. The role of water in metal ion adsorption and transport in clays is a critical issue. As seen below, our immediate approach to understanding the role of water in clays is to study the cations as a function of hydration.

By far the most extensive use of NMR in clays and other minerals has been the use of ^{27}Al and ^{29}Si NMR to study the short-range structure. The excellent book by Engelhardt and Michel[13] summarizes much of the work up to the date of its publication. It is clear that the use of NMR in clay systems derives much of its history from work in zeolites. The reasons are primarily the economic importance of zeolites as industrial catalysts and the fact that most zeolites can be prepared in very clean, microcrystalline form. This makes them far easier to study than clays which are usually of geological origin and consequently are not as "regular" in chemistry or structure. There are several ^{27}Al NMR studies of Al substitution in the tetrahedral sheet in phyllosilicates. This type of information is unavailable from X-ray measurements because of both disorder in the substitution and the fact that the diffraction cross sections of Al and Si are similar. Herrero and coworkers published an excellent study of the extent of ordering of Al in the tetrahedral sheet of vermiculites which is relevant to any clay studies that are sensitive to layer charge.[14,15] They conclude that Al substitution follows Lowenstein's rule. Further, their results are consistent with a model in which each hexagon of tetrahedral metals (i.e., silicon in the tetrahedral sheet) contains one or two Al atoms and there are no hexagons with zero or three Al atoms. They also conclude that there is no long-range structural order, as had been postulated in other studies.[16] Since vermiculite is one of the most ordered clays, the implication is that montmorillonites are also unlikely to have long-range order in the tetrahedral layer. Thus, we expect all of the ditrigonal cavities to be similar and there are no clearly distinct structural sites in the interlayer from charge ordering in the tetrahedral sheet.

Rather than attempting an exhaustive review of the literature, we point out that a number of research groups have attempted to use [29]Si chemical shifts as a means of determining bond lengths and bond angles in the tetrahedral sheets of clays.[17–21] These structural studies are useful in establishing an overall picture of the structure of clays. Upon examination of the many NMR studies of clay structure, it is clear that samples were carefully selected to avoid any with significant paramagnetic interferences. Iron commonly substitutes for Al in the octahedral sheet and then can undergo oxidation and reduction and electron spin exchange, severely perturbing the NMR experiments.

Recently several research groups have started using the NMR of the cations themselves to investigate sorption in clays. This work benefited from developments in multinuclear NMR spectrometers, improvements in MAS, and other solid-state techniques. The earliest cation NMR was work done by Conard[22] using ^{7}Li NMR in montmorillonite to attempt to obtain information about water structure. Lithium is a particularly interesting cation for study because of its small size and ability to orient water. Conard's work did not have the advantage of magic angle spinning (MAS) so the spectral resolution is not very high, but it demonstrated the potential of using NMR of the metal ions in the interlayer. Bank and coworkers[23] made a relatively complete study of the [113]Cd NMR of several exchanged clays. They interpreted their results in terms of mobility of the cation and interactions with paramagnetic iron in the clays studies. More recently, Tinet et al. investigated Cd^{2+} adsorption on a montmorillonite and hectorite[27] using static chemical shift tensor analysis. Their analysis of the static spectra disagrees with some of Bank's conclusions about the source of different shifts and linewidths in these clays. They conclude that the different resonances are due to hydrated

Cd^{2+} both inside and outside the layers. Weiss et al. made a fairly complete study of Cs on hectorite[24] followed by a survey of Cs^+ on a large suite of well-characterized clays.[25] The use of a relatively large series of clays allowed these workers to draw conclusions about the effects of changes in the clay structure. In our work reported below, Cs^+ has given us some problems because of the low charge to size ratio for this cation and the low hydration energy. Prost and coworkers have made the largest number of different contributions in this field.[26–28] They studied Na^+, Cd^{2+}, and Cs^+ adsorption on a Texas vermiculite.[26] This system is ideal for such studies because of the high layer charge, low iron substitution, and straightforward structural changes as a function of hydration. Their work also relies on prior structure–hydration studies by van Olphen.[7,8] van Olphen defined the amount of water adsorbed at different hydration levels, which makes a good correlation with the NMR studies of the cations as a function of hydration. They were able to relate chemical shifts to structural changes in the vermiculite with hydration— except for Cs^+, where the state of water is apparently poorly defined. Most recently, Lambert et al. demonstrated that ^{39}K NMR has potential as a probe nucleus for studying hydratable potassium in minerals.[28] They point out that for such low-γ nuclei (γ = magnetogyric ratio) and for quadrupolar systems, very high magnetic fields are a distinct advantage.

Finally, the interest in "pillared clays" has led to publications on adsorption of cationic amines in clays directed towards understanding both the pillaring process and cation sorption.[29]

Experimental

The clays used in this work are Source Clays obtained from the Clay Mineral Society. They are readily available and so have been characterized by a number of authors over the years. Their chemical formulas and cation exchange capacities are given in Table 10.1. It can be seen that SAz-1 has considerably more total charge than the other two clays but the charge in the tetrahedral sheet is nearly equivalent to that in SWy-1. On the other hand, the total charge of STx-1 and SWy-1 are nearly the same, but STx-1 has a much lower charge in the tetrahedral sheet. The clays were fractionated by sedimentation and the $<2\,\mu m$ fraction was used for most experiments. They were exhaustively exchanged with Na^+ by treating a small amount of clay with a large excess of 1 M NaCl solution followed by centrifugation and another treatment with 1 M NaCl. Cesium-substituted samples were prepared from the Na-SAz-1 by exchanging the sample twice with 1 M CsCl solutions. Lithium-substituted samples were prepared by exchanging the natural clay with two washes in 1 M LiCl solutions. In all cases the exchange steps were followed by several washes in deionized water until the supernatant tested negative to Cl^-. The clays were then air-dried at room temperature for about 36 h, ground in a mortar and pestle, and exposed to different relative humidities. The controlled relative humidity atmospheres were prepared by making

Table 10.1 Properties of the Clay Samples Used

Clay	Formula	Source	Ref.
SAz-1	$(Si_{7.86}Al_{0.14})(Fe^{3+}_{0.20}Al_{2.80}Mg_{1.00})O_{20}(OH)_4 X^+_{1.14}$	Apache County, Arizona	30
STx-1		Gonzales County, Texas	
SWy-1	$(Si_{7.84}Al_{0.16})(Fe^{3+}_{0.26}Al_{3.22}Mg_{0.40})O_{20}(OH)_4 X^+_{0.68}$	Crook County, Wyoming	31

Cation Exchange Capacities[a]

Clay	Total CEC	Tetrahedral Charge	Octahedral Charge
SAz-1	125	19.1	106
STx-1	84.2	9.5	74.7
SWy-1	79.8	19.7	60.1

[a] The cation exchange capacities are taken directly from work by Jaynes and Bigham.[5] We have taken their MgEC value as the charge in the tetrahedral sheet, and the difference between the total CEC and MgEC as the charge in the octahedral sheet.

saturated solutions of several salts and placing those solutions in a tightly sealed container with the air-dried clay for at least 60 h. The hydration levels of the clays are reported as the ratio of the partial pressure of water to the partial pressure of an atmosphere saturated with water vapor, p/p_0. Pure water and P_2O_5 were used to produce atmospheres with p/p_0 of 1.0 and 0.0, respectively. The equilibrated clays were then quickly removed from the controlled atmosphere and packed into zirconia rotors which were sealed with a cap containing a double o-ring. In our experience, these caps effectively exclude air for periods of many days.

NMR experiments were performed using a Varian Unity-400 spectrometer. The probe was a standard Varian Variable Temperature MAS probe with the Jacobsen design stator and rotor. Spinning speeds were approximately 7000 Hz for all samples. The nominal radio frequency was 105.8 MHz for ^{23}Na, 155.4 MHz for ^{7}Li, and 52.5 MHz for ^{133}Cs. No proton decoupling was used.

Results and Discussion

The level of hydration is one of the most important parameters determining the interactions of cations and clays. As noted above, several groups have successfully used ^{1}H NMR techniques to investigate directly the state of water. Our rather preliminary investigations of the ^{17}O NMR of clay–cation systems have not given particularly interesting results, although we believe that more thorough experiments have some promise. In this work, we pursue the effect of water by investigating the ^{23}Na NMR of the samples as a function of hydration. Figure 10.4 contains the ^{23}Na spectrum of the Arizona montmorillonite at four hydration levels. Within this set of spectra, the easiest to understand is the one at $p/p_0 = 1.0$. In this case, the Na$^+$ is highly hydrated, very mobile, and gives a spectrum with a chemical shift equal to that in aqueous solution. The linewidth can be attributed to a number of possible interactions, including interactions with the slight paramagnetism in the clay, chemical shift dispersions from different regions in the clay, and diffusion of the Na$^+$ between regions with differing bulk susceptibilities. As the clay is dried, the resonance shifts to higher field, and in the very dry sample there are clearly two overlapping peaks with distinct spinning sidebands. We interpret this spectrum to be indicative of at least two distinct sites for the Na$^+$. If there is any exchange between the sites, the lifetime is longer than about 2×10^{-4}s because the separation between the peaks is about 1500 Hz. This spectrum is highly reproducible. We have dried Na$^+$-substituted SAz-1 by heating to various temperatures (as high as 400 °C) and by equilibrating it over P_2O_5: the spectra for all samples are virtually identical.

The interpretation of the chemical shifts at intermediate levels of hydration is less straightforward. One is tempted to view this as a chemically exchanging cation, with the two extremes being represented by the fully dried and fully hydrated samples. The rate of exchange and the populations

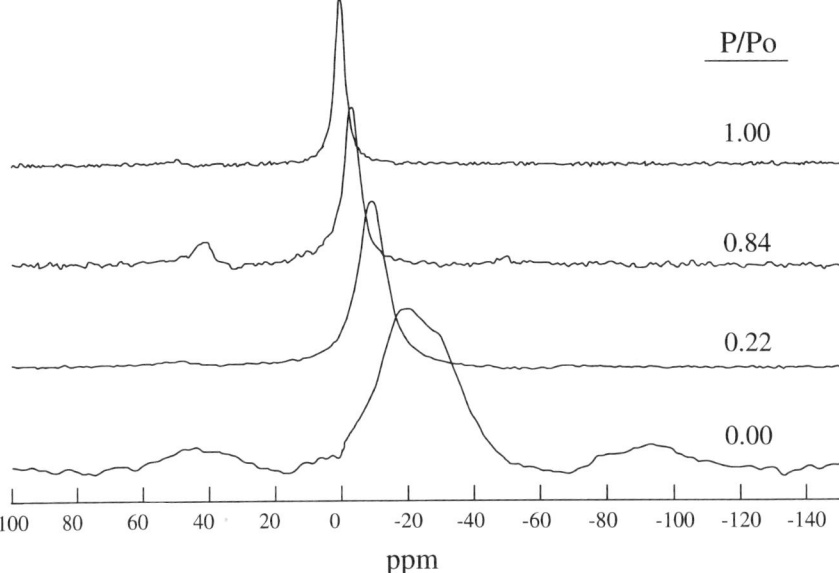

Figure 10.4   The ^{23}Na spectra of SAz-1 equilibrated at different relative humidities, p/p_0.

of the two sites are all functions of the hydration level, and this would provide an explanation of the spectra observed. If such is the case, variable-temperature NMR for one of the hydrated samples will slow the exchange and eventually we should obtain two peaks. Figure 10.5 contains the spectra of the Na-SAz-1 at three hydration levels, $p/p_0 = 0.0$, 0.22, and 1.0, all at very low temperatures. These spectra were not acquired at exactly the same temperature but nevertheless comparisons can be made. The dry sample shows only some broadening and little chemical shift as a function of temperature, even down to $-140\,^{\circ}$C. The sample at $p/p_0 = 1.0$ also only broadens as the temperature is lowered. The sample at intermediate hydration, $p/p_0 = 0.22$, is starting to show evidence of a splitting into a peak with a resonance near that of the dry sample and another one near the ^{23}Na resonance of the completely hydrated clay. It is known that when a fluid is constrained to small spaces such as pores or the interlamellar region of a clay, the freezing point is depressed. We believe that, even at $-105\,^{\circ}$C, there is still mobility of water and probably of the Na^{+} in this system.

As the hydration is increased, the d_{001} spacing increases; this changes the electric field at the sodium nucleus, in principle producing a combination of a chemical shift and a second-order quadrupole shift. Since we have not assigned the shifts to a definite structure, it is difficult to give an unequivocal interpretation of the chemical shift results. Our present interpretation is that there are different types of Na^{+} in these clay samples—two structurally

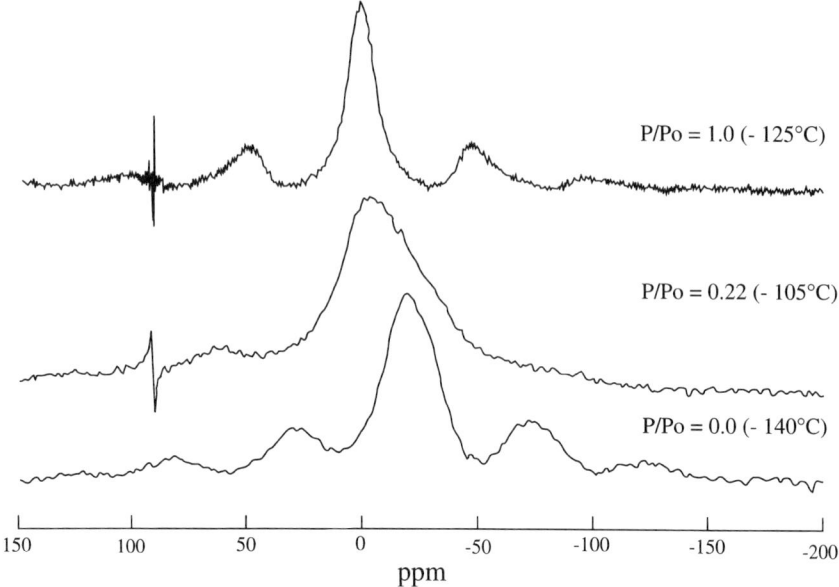

P/Po = 1.0 (- 125°C)

P/Po = 0.22 (- 105°C)

P/Po = 0.0 (- 140°C)

ppm

Figure 10.5 The ^{23}Na spectra of SAz-1 equilibrated at different relative humidities. These spectra were all taken at temperatures lower than $-100\,^{\circ}$C.

different sites in a dry SAz-1, and a completely hydrated Na$^+$ which is mobile in the interlayer when the sample is well hydrated. As the well-hydrated sample is cooled, the cation stays hydrated in the interlayer. This is much like cooling and freezing of aqueous sodium ion in a bulk solution. We believe that at intermediate hydration levels there is chemical exchange between fully hydrated Na$^+$ and Na$^+$ more tightly bound to the siloxane surface. We are continuing to work on this problem with variable-temperature and two-dimensional NMR, with the goals of confirming our assumption that the spectra reflect chemical exchange between sites and of measuring the rate of exchange.

The two peaks seen in the spectrum of dry Na-SAz-1 must be related to different structural sites in the clay. What are the two sites? Laperche et al.[26] investigated ^{23}Na NMR of vermiculite and saw only a single shift for Na in that work. Vermiculite is a more highly charged clay than montmorillonite and has considerably more order. Thus, the two peaks seen in our work may result from disorder in the dried system. A proposal suggested by Weiss and coworkers[24,25] is that there are sites of 12-fold coordination generated when the clay layers line up with two ditrigonal cavities directly opposite each other. If the clay layers are less ordered, then a site of lower symmetry is generated. They suggest nine-fold symmetry as an alternative, i.e., six oxygen atoms from one tetrahedral sheet and three from the other.

To investigate structural possibilities for cation sites, we have obtained ^{23}Na NMR spectra of several different Na$^+$-exchanged montmorillonite clays. The clays chosen are standard clays from the Source Clay Repository but they have slightly different charge distributions. We expected that changes in the total layer charge and location of the charge (octahedral or tetrahedral) will change the NMR spectra in an interpretable fashion. Figure 10.6 contains ^{23}Na spectra of the three clays after equilibration at $p/p_0 = 0.0$. The distribution of layer charges is given in Table 10.1. The spectra all have chemical shifts in the same region but significantly different linewidths and peak intensities for the two peaks in each spectrum. A priori, one might expect the spectrum of STx-1 to be different from the other two because it has a very small tetrahedral charge. In fact, the shoulders of SAz-1 and STx-1 at about -20 and -30 ppm nearly line up, indicating that there is some similarity between the sodium sites on these two clays. The relative intensity ratios and linewidths are quite different and require further work for clarification. The ^{23}Na spectrum of the SWy-1 sample is shifted from the other two. This spectrum has a peak at about -25 ppm, intermediate between the two peaks in STx-1, and a poorly defined shoulder at about -15 ppm. This clay is fairly difficult to prepare because it forms gels in aqueous suspensions of low ionic strength, so there may be some anomalies in this particular spectrum due to errors in sample preparation. These ^{23}Na chemical shifts in different homoionic clays can only be considered tantalizingly not yet indicative of structural effects on the ^{23}Na spectra.

We are also pursuing the effects of particle size on clay samples. There are edge sites and basal plane sites that may well have different chemical shifts from those in the interlayer. We expect the relative concentrations of these to change with clay particle size, as can be seen from the cartoon in Figure 10.1. In a single trial, we have seen slight changes in the intensity of the two peaks in the ^{23}Na spectrum of STx-1 at $p/p_0 = 0.0$.

We believe that the shifts seen in the ^{23}Na spectra of Figure 10.6 are due to a combination of chemical shifts and second-order quadrupolar shifts rather than quadrupole splitting, although we have no direct evidence. Our arguments are as follows. First, the quadrupole moment of ^{23}Na is relatively large, so even small electric field gradients would produce large quadrupolar shifts and we therefore expect the satellite lines to be outside the range swept in these experiments. Laperche et al. make similar arguments regarding ^{23}Na spectra obtained in vermiculite. Secondly, we have obtained data on ^{7}Li- and ^{133}Cs-exchanged clays (see above); the ^{133}Cs spectra of dried clays show two peaks with splittings greater than those in the ^{23}Na spectra, in spite of the fact that the electric quadrupole moment of ^{133}Cs is almost two orders of magnitude smaller than that of ^{23}Na.

Finally, we thought that it would be interesting to investigate the NMR spectra of different cations, with different hydration properties, in a single clay. Figure 10.7 contains the ^{133}Cs, ^{23}Na, and ^{7}Li NMR spectra of homoionically prepared SAz-1. We have discovered several interesting things about

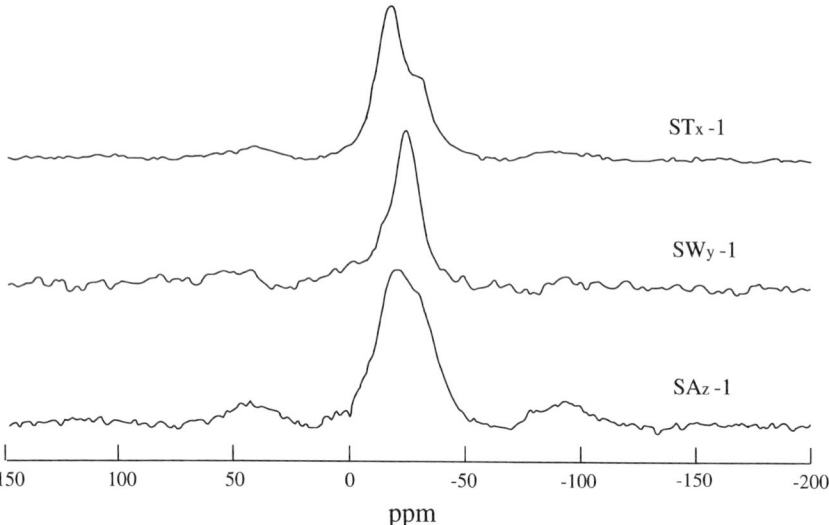

Figure 10.6 The ^{23}Na spectra of Na$^+$-substituted SAz-1, SWy-1, and STx-1. All spectra were taken at ambient temperature ($\approx 30\,°C$) and the samples were equilibrated at 0% relative humidity.

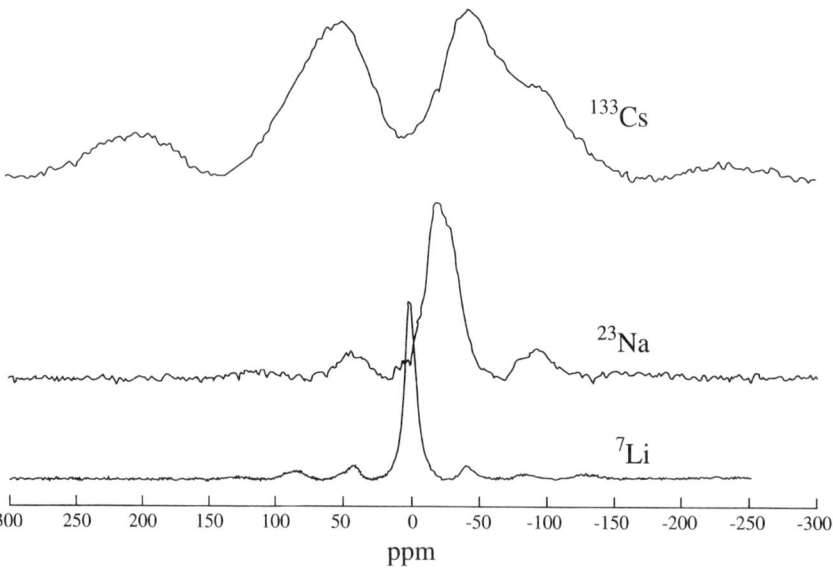

Figure 10.7 The ^{23}Na, ^{133}Cs, and ^{7}Li spectra of SAz-1 which was substituted with the cations. The zero for the chemical shift scale for all three spectra is set at the position of the resonance of 1 M aqueous solutions of LiCl, NaCl, and CsCl.

194

these systems, none of which has yet led us to any novel conclusions about the interactions of these cations with water in the clay interlayer.

First, we find that ^{7}Li has a very small chemical shift range and thus poor resolution. Presumably this small shift range is due to the fact that Li$^+$ has few electrons (a total of three) so the effects of subtle changes in bonding are small. This is analogous to the small chemical shift range of the proton. The electric quadrupole moment of ^{7}Li is intermediate between deuterium and ^{23}Na, resulting in a linewidth of several hundred hertz in samples that we have investigated. Consequently the ratio of the chemical shift range to line-width is small, and we cannot discriminate between different "types" of Li$^+$ in our samples. On the other hand, ^{133}Cs has a large number of electrons (a large chemical shift range) and the quadrupole moment is quite small. The ^{133}Cs spectrum in Figure 10.7 contains two clearly discernible peaks with shoulders from the MAS spinning sidebands. We again interpret this spectrum to be indicative of two sites for Cs$^+$ in the dry clay. The problem with this ion is that it is quite large and has a small hydration energy. When we attempt to study the behavior of the ^{133}Cs peaks as a function of hydration and temperature we obtain irreproducible results. The details of the spectrum are a function of the hydration history rather than simply of the p/p_0 at which the sample is equilibrated. This is probably due to the fact that Cs$^+$ is a very large ion with low hydration energy. Upon drying, the clay sheets may collapse on the ion. It then becomes very difficult to rehydrate the Cs$^+$ in the interlamellar space. The spectrum in Figure 10.7 is interesting because there are two well-separated peaks for the dry sample, again indicating that there are two sites for cation adsorption. The linewidths of these peaks are much larger than would be expected for ^{133}Cs, which has a rather small quadrupole moment. The immediate explanation for the linewidths is a dispersion of shifts from regions of the clay with differing hydration levels. Laperche et al. found similarly odd NMR spectra in a Cs$^+$-exchanged vermiculite.[26] They attributed their results to Cs$^+$ in micropores created by the packing of the clay particles. We are continuing to pursue ^{133}Cs NMR in several clays as a function of hydration. We are combining powder X-ray diffraction measurements with the NMR spectroscopy to determine the d_{001} spacing for the same samples as are used for NMR.

Conclusions

We find multinuclear NMR experiments of cations to be useful in elucidating the structure and dynamics of alkali-metal ions sorbed on clays. Our present interpretation, based primarily on ^{23}Na results, is that the cation is fully hydrated and very mobile in a well-hydrated clay. This environment is similar to a cation in a bulk aqueous solution. Under very dry conditions, there are at least two, chemically shifted, sites for cations. We have not yet assigned those sites to definite locations in the structure but are continuing work towards that end. At intermediate hydrations, it appears that the spectra are representative of chemical exchange. However, the exchange is not between sites that correspond to the sites in fully hydrated and dry clays. It is more probably an exchange between two partially hydrated sites, or a partially hydrated site and one similar to the sites in a dry clay. These arguments are based on chemical shift measurements which are difficult to quantify. The use of multiple, well-defined clays and multiple cation resonances holds promise for returning information on the details of cation binding and mobility in clay minerals.

References

1. Brindley, G. W., and Brown, G., *Crystal Structures of Clay Minerals and Their X-Ray Identification*, The Mineralogical Society, London, 1980.
2. Grim, R. E., *Clay Mineralogy*, McGraw-Hill, New York, NY, 1978.
3. van Olphen, H., and Fripiat, J. J., *Data Handbook for Clay Materials and other Non-metallic Minerals*, Pergamon, Oxford, 1979.
4. Sato, T., Watanabe, T., and Otsuka, R., Effects of layer charge, charge location, and energy change on expansion properties of dioctahedral smectites. *Clays Clay Miner.* 40, 103–113, 1992.
5. Jaynes, W. F., and Bigham, J. M., Charge reduction, octahedral charge, and lithium retention in heated, Li-saturated smectites. *Clays Clay Miner.* 35, 440–448, 1987.
6. Stul, M. S., and Mortier, W. J., The heterogeneity of the charge density in montmorillonites. *Clays Clay Miner.* 22, 391–396, 1974.
7. van Olphen, H., Thermodynamics of interlayer adsorption of water in clays. *J. Colloid Sci.* 20, 822–937, 1965.
8. van Olphen, H., Thermodynamics of interlayer adsorption of water in clays, II. Magnesium vermiculite. In *Proc. Int. Clay Conf.*, Heller, L. (ed.), Israel Universities Press, Jerusalem, 1969, pp. 649–657.
9. Johnston, C. T., Sposito, G., and Erickson, C., Vibrational probe studies of water interactions with montmorillonite. *Clays Clay Miner.* 40, 722–730, 1992.
10. Hougardy, J., Stone, W. E. E., and Fripiat, J. J., NMR study of adsorbed water. I. Molecular orientation and protonic motions in the two-layer hydrate of a Na vermiculite. *J. Chem. Phys.* 64, 3840–3851, 1976.
11. Hougardy, J., Stone, W. E. E., and Fripiat, J. J., Complex proton NMR spectra in some ordered hydrates of vermiculites. *J. Magn. Reson.* 25, 563–567, 1977.
12. Woessner, D. E., and Snowden, Jr., B. S., Study of the orientation of adsorbed water molecules on montmorillonite clays by pulsed nuclear magnetic resonance. *J. Colloid Interface Sci.* 30, 54–68, 1969.
13. Engelhardt, G., and Michel, D., *High-resolution Solid-state NMR of Silicates and Zeolites*, John Wiley, New York, NY, 1989.
14. Herrero, C. P., Sanz, J., and Serratosa, J. M., Si, Al distribution in micas: analysis by high-resolution ^{29}Si NMR spectroscopy. *J. Phys. C: Solid State Phys.* 18, 13–22, 1985.
15. Herrero, C. P., Sanz, J., and Serratosa, J. M., Dispersion of charge deficits in the tetrahedral sheet of phyllosilicates. Analysis of ^{29}Si NMR spectra. *J. Phys. Chem.* 93, 4311–4315, 1989.
16. Gatineau, L., *Bull. Soc. Fr. Miner. Cristallogr.* 87, 321–355, 1964.

17. Sanz, J., and Serratosa, J. M., ^{29}Si and ^{27}Al high-resolution MAS–NMR spectra of phyllosilicates. *J. Am. Chem. Soc.* 106, 4790–4793, 1984.

18. Weiss, C. A., Altaner, S. P., and Kirkpatrick, R. J., High resolution ^{29}Si NMR spectroscopy of 2:1 layer silicates: correlations among chemical shift, structural distortions, and chemical variations. *Am. Miner.* 72, 935–942, 1987.

19. Lipsicas, M., Raythatha, R. H., Pinnavaia, T. J., Johnson, I. D., Giese, Jr., R. F., Costanzo, P. M., and Robert, J.-L., Silicon and aluminum site distributions in 2:1 layered silicate clays. *Nature (London)* 309, 604–607, 1984.

20. Barron, P. F., Slade, P., and Frost, R. L., Ordering of aluminum in tetrahedral sites in mixed-layer 2:1 phyllosilicates by solid-state high-resolution NMR. *J. Phys. Chem.* 89, 3880–3885, 1985.

21. Woessner, D. A., Characterization of clay minerals by ^{27}Al nuclear magnetic resonance spectroscopy. *Am. Miner.* 74, 203–215, 1989.

22. Conard, J., Structure of water and hydrogen bonding on clays studied by ^{7}Li and ^{1}H NMR. In *Magnetic Resonance in Colloid and Interface Science*, Resing, H. A., and Wade, C. G., (eds.), American Chemical Society, Washington, DC, 1976, pp. 85–93.

23. Bank, S., Bank, J. F., and Ellis, P. D., Solid-state ^{113}Cd nuclear magnetic resonance study of exchanged montmorillonites. *J. Phys. Chem.* 93, 4847–4855, 1989.

24. Weiss, C. A., Kirkpatrick, R. J., and Altaner, S. P., The structural environments of cations adsorbed onto clays: ^{133}Cs variable-temperature MAS NMR spectroscopic study of hectorite. *Geochim. Cosmochim. Acta* 54, 1655–1669, 1990.

25. Weiss, C. A., Kirkpatrick, R. J., and Altaner, S. P., Variations in interlayer cation sites of clay minerals as studied by ^{133}Cs MAS nuclear magnetic resonance spectroscopy. *Am. Miner.* 75, 970–982, 1990.

26. Laperche, V., Lambert, J. F., Prost, R., and Fripiat, J. J., High-resolution solid-state NMR of exchangeable cations in the interlayer surface of a swelling mica: ^{23}Na, ^{111}Cd, and ^{133}Cs vermiculites. *J. Phys. Chem.* 94, 8821–8831, 1990.

27. Tinet, D., Faugere, A. M., and Prost, R., ^{113}Cd NMR chemical shift tensor analysis of cadmium-exchanged clays and clay gels. *J. Phys. Chem.* 95, 8804–8807, 1991.

28. Lambert, J. F., Prost, R., and Smith, M. E., ^{39}K solid-state NMR studies of potassium tecto- and phyllosilicates: the in situ detection of hydratable K^+ in smectites. *Clays Clay Miner.* 40, 253–261, 1992.

29. Pratum, T. K., A solid-state ^{13}C NMR study of tetraalkylammonium/clay complexes. *J. Phys. Chem.* 96, 4567–4571, 1992.

30. Knechtel, M. M., and Patterson, S. H., U.S. Geol. Survey Bull. 1082-M, pp. 957–958, 1962.

31. Grim, R. E., and Guven, N., *Developments of Sedimentology*, Vol. 24, *Bentonites: Geology, Mineralogy, Properties, and Uses,* Elsevier, Amsterdam, 1978.

11.

^{2}H NMR and Gel Formation of the Ultrafine Solids Fraction Associated with the Athabasca Oil Sands Fine Tails

JOHN A. RIPMEESTER, L. S. KOTLYAR, B. D. SPARKS, & R. SCHUTTE

The two oil sands plants operated by Syncrude Canada Ltd. and Suncor Canada Ltd. near Fort MacMurray, Alberta, use a hot water process for the separation of bitumen from oil sands. In brief, hot water and oil sands, with caustic soda as dispersing agent, are mixed thoroughly, and bitumen is floated to the top of the resulting slurry by streams of air. After secondary bitumen recovery, the remaining tailings are carried to ponds, where the coarse sands are used to form dikes, the fine tails are left to settle, and freed water is recycled. Typical production figures for the Syncrude plant are 390 000 barrels of diluted bitumen per day produced from 325 000 tonnes of oil sand.

One complicating factor is that the fine tails dewater only to a solids content of ~30%, requiring ponds of ever increasing size (the Syncrude pond is 22 km^2) to store the resulting sludge. As the ponded material is toxic to wildlife, it poses a considerable local environmental hazard. In addition, there is the potential hazard of contamination of surface water and a major river system as a result of seepage or potential dike failure.

The work reported here was carried out as part of a major project initiated to address the problem of the existing tailings ponds, and also to modify the currently used separation process so as not to produce sludge.

Starting with the recognition that the very stable fine tails, consisting of water, silt, clay and residual bitumen, have gel-like properties,[1,2] we employed the strategy of fractionating the fine tails with the hope of identifying a specific fraction which might show gel-forming propensity.[3] This was done by breaking the gel, and collecting fractions according to sedimentation behavior during centrifugation. Fractions consisting of the coarser solids (> 0.5 μm) settled rapidly, whereas fractions with smaller particle sizes (termed ultrafines) gave suspensions which set into stiff, thixotropic gels on standing.

198

Gel formation and the sol–gel transition in colloidal clay suspensions are classical problems which have received much attention over the years; however, much remains to be learned.[4,5] NMR techniques have shown considerable promise in understanding clay–water interactions at a microscopic level. Early work[6–8] concentrated on clay–water systems with relatively high solids contents, but recent studies[9–12] have shown that even relatively dilute systems (1%) are amenable to study. This great sensitivity is due to the fact that in suspensions of smectite clays the clay platelets are strongly oriented by the magnetic field, which in turn causes the bound water layer to be oriented, and this net orientation effect is transmitted to the bulk water by fast exchange between surface and bulk water. This is seen most easily by observing the residual quadrupolar splitting of the ^{2}H NMR signal in suspensions containing ^{2}H$_2$O, although it may also be monitored by observing the ^{17}O NMR signal.

In this chapter we show that ^{2}H NMR methods have contributed a great deal towards defining a model which accounts for sludge properties. In a more general way, the methods described may be used to good advantage to study the sol–gel transformation in colloidal clay suspensions at a microscopic level.

Theory

The theory which accounts for quadrupolar splittings of solvent molecules in heterogeneous systems[13] has been presented previously in the context of smectite suspensions,[10–12] and its salient features will be reviewed. A description of the orienting effect of a magnetic field on single, anisotropic, charged particles undergoing Brownian motion has been given in terms of the anisotropic magnetic susceptibility associated with the particle and the rotational diffusion constants.[14] There is not as yet a theory for the case where the particles are allowed to interact in the presence of an electrolyte; however, for our purposes all we need to know is that orienting effects are expected. The ordering of particles in the magnetic field can be described by the Euler angles Ω_{WLD}. The resulting quadrupolar splittings Δ of surface-bound water molecules are given by

$$\Delta = \frac{3\chi}{4I(2I-1)} \langle 3\cos^2\theta_{LD} - 1 \rangle A \tag{1}$$

where χ is the quadrupole coupling constant e^2qQ/h, A is a residual anisotropy with values between $+1$ and -1, and θ_{LD} is the angle between the vector perpendicular to the surface of the oriented platelet (known as the director $\boldsymbol{D}$) and the magnetic field $\boldsymbol{B}_0$ (Fig. 11.1). The residual anisotropies for ^{2}H$_2$O are

$$A(^2\text{H}) = \frac{0.05}{2}\langle 3\cos^2\theta_{DM} - 1 \rangle - \frac{0.41\sqrt{6}}{2}\langle \sin^2\theta_{DM}\cos 2\phi_{DM} \rangle$$

$$A(^{17}\text{O}) = \frac{0.03}{2}\langle 3\cos^2\theta_{DM} - 1 \rangle + \frac{0.80\sqrt{6}}{2}\langle \sin^2\theta_{DM}\cos 2\phi_{DM} \rangle \tag{2}$$

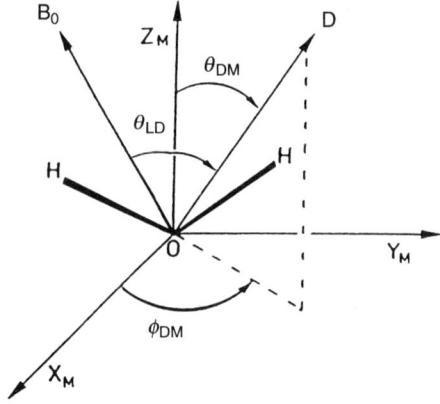

Figure 11.1 Diagram illustrating the angles defined in equation (1); angle θ_{LD} between the magnetic field $\boldsymbol{B_0}$ and the director $\boldsymbol{D}$ (perpendicular to the clay platelet surface), angles θ_{DM} and ϕ_{DM} which define the orientation of the water molecule with respect to the director $\boldsymbol{D}$.

where the Euler angles $\Omega_{DM}(\theta_{DM}, \phi_{DM})$ describe the orientation of the water molecules at the clay surface (Fig. 11.1). It should be noted that no splittings are observed when either equation (1) or (2) is zero, i.e., when there is random reorientation of either the water molecules or the platelets. Under normal circumstances, the surface-bound water in suspensions exchanges rapidly with bulk water, which has $\Delta = 0$, and the observed splitting is then a weighted average,

$$\Delta = \Sigma_i \rho_i \Delta_i \tag{3}$$

where the ρ_i are the fractions of water molecules with splitting Δ_i. This description is sufficient for dilute suspensions; however, a number of observations on the systems studied here indicate that aggregates can be formed in random as well as oriented fashions. Since the aggregates are probed by mobile water molecules, a length scale is introduced quite naturally and the concept of domain structures becomes useful, and indeed necessary, as will be seen later on.

Experimental

Samples

The fine tails samples were supplied by the Research Department of Syncrude Ltd. The samples were taken from a small containment pond which had been filled with fine tails pumped from the 17 m level of the main tailings pond. The samples contained 26% solids and 1% bitumen, the remainder being water. The procedure for separation of fine tails into different fractions was as follows: a sample was agitated for 15 min using a high-intensity mixer and then centrifuged for 1 h at 200 g. The following layers could be identified (top to bottom): a bitumen layer with associated solids (BS), an aqueous colloidal suspension layer (A) and a compact layer of residual (bulk) solids. The layer A, accounting for about 4% of the total tails, was fractionated again on the basis of particle size by centrifugation at 500 g, 1500 g, and 91000 g. These subfractions were designated A-500, A-1500, and A-91000. Concentrated suspensions of these subfractions were diluted with pond water in order to obtain a number of different concentrations for study. As water chemistry is extremely important, the ion concentrations (ppm) in the pond water were: Na 430, Mg 7.0, Ca 5.4, K 22, Si 12, A 4.7, B 4.1.

NMR Experiments

For the NMR experiments, about 7% ^2H_2O (99.9 atom% ^2H; MSD Isotopes) was added to each sample. ^2H and ^{17}O NMR spectra were recorded on a Bruker MSL-300 NMR spectrometer at frequencies of 46.07 MHz and 40.68 MHz, respectively. Typically, for ^2H, 4–16 transients were co-added in 8K datum points at a sweep width of 2000 Hz and zero-filled to 16 K before Fourier Transformation. For ^{17}O, 20 000–40 000 transients were co-added in 1K datum points at a sweep width of 20 kHz. Spectra were run in a 10mm broadband high-resolution probe. Spinning at low frequencies (10 Hz) did not affect the spectra. The 90° pulse lengths were 25–30 µs, and the delay times 0.5–2 s.

Some experiments were carried out in a solenoid (5 mm) probe fitted with a goniometer device to allow recording of the spectrum of oriented samples as a function of rotation angle. The 90° pulse lengths were 2–3 µs in this case.

Results and Discussion

The Nature of the Solids

The complete characterization of the solids fractions proved to be a complex problem.[15] Results of transmission electron microscopy showed that the particles are anisotropic, some of hexagonal but mostly of irregular morphology, and very thin (down to single aluminosilicate sheets). Particle sizes decreased from 200 nm for the A-500 fraction down to 20 nm for the A-91000 fraction. X-ray diffraction analysis performed on powder specimens in random mounts (for whole mineral composition) and in preferred orientation (for characterization of phyllosilicates) indicates that the main inorganic components are kaolinite and mica, with smectite-type minerals present only in trace quantities. Results of X-ray photoelectron spectroscopy (XPS) indicate that iron accounted for less than 0.2 atom% of the particle surfaces. XPS also shows that there is a significant amount of organic carbon on some of the particle surfaces. Cation exchange capacities were 11 and 20 meq/100 g for the A-500 and A-1500 fractions, respectively.[16]

Initial NMR Experimental Protocol

Initially, suspensions of different concentrations prepared from each subfraction were examined by NMR without much attention being paid to the time period between sample preparation and the actual recording of the spectra. The samples varied in consistency from free-flowing suspensions to stiff gels. The ^{2}H NMR spectra for the sol samples showed a characteristic doublet splitting. The gel samples showed no splitting at all, even though the samples were left in the field for a prolonged time (several hours). This contrasts with results reported for smectite gels, which gave time-independent splittings after several hours of residence time in the field.[12] With this procedure there was no systematic variation of the doublet splitting with concentration of the solid, as is observed for the smectites.

Some experiments were then carried out in order to establish a protocol for the study of such samples. Figure 11.2(a) shows ^{2}H and ^{17}O NMR spectra for a sample containing 5% solids which has been left to stand for about a week to give a highly viscous material. The ^{2}H splitting is about 3 Hz, and there is no fine structure evident for the ^{17}O NMR line. These spectral features do not change on leaving it in the field for a day or so. If the sample is then shaken, it becomes free-flowing. The ^{2}H and ^{17}O NMR spectra for this sample are shown in Figure 11.2(b). The ^{2}H NMR doublet splitting Δ is now 40 Hz, and the ^{17}O spectrum shows a broadened quintuplet with a characteristic splitting of 200 Hz. When this sample is now left in the field, gel formation takes place, and some subtle changes occur in the ^{2}H NMR spectrum (Figure 11.2(c)). The ^{2}H quadrupole splitting is essentially invariant; however, the change in linewidth of the doublet components indicates a change in relaxation mechanism.[12]

From observation of the sample viscosity, it is clear that over a short term the equilibrium state of the sample is the gel, and that the gel is thixotropic, i.e., the structure can be broken quite easily by shaking the sample. The NMR observations show that the gel can be produced in oriented as well as in random forms. The oriented form is produced by placing a freshly shaken sample in the magnetic field to give a suspension of oriented particles with the characteristic signature shown in Figure 11.2(b). The gel structure apparently can form from the sol without a net loss of oriented surface. This is an important conclusion, as it precludes the presence of the classical cardhouse structures which are cubic,[4] and instead favors structures where the main features are ribbons or sheets, with platelets arranged in edge-to-edge or face-to-face patterns. It is worth noting the edge-to-edge ribbon models proposed by M'Ewen to explain optical and X-ray scattering measurements.[17,18]

The observation of the random gel, where very small or no ^{2}H splittings are observed, can be explained in terms of the presence of domains.

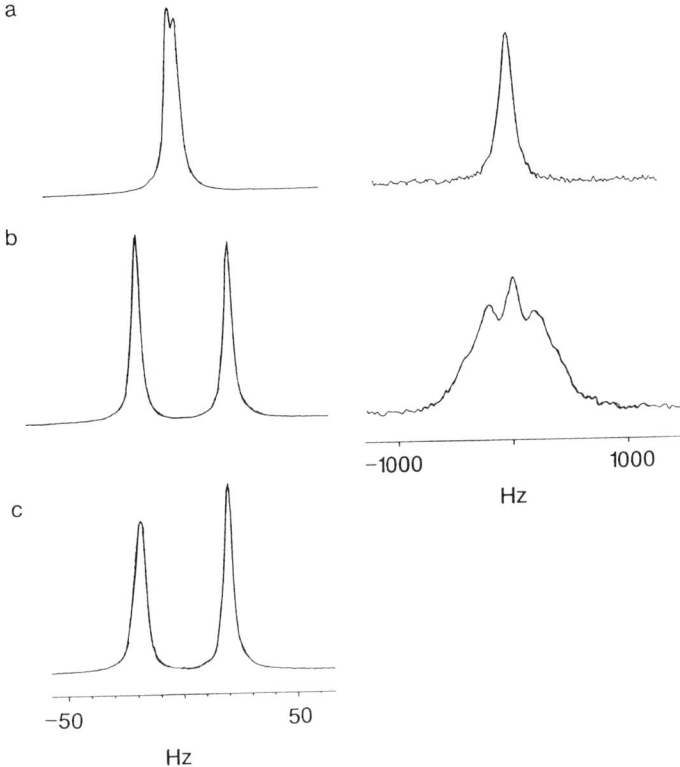

Figure 11.2 (a) ^{2}H and ^{17}O NMR spectra of 5% solids sample after standing for a week; (b) ^{2}H and ^{17}O NMR spectra of same sample after shaking; (c) ^{2}H NMR spectra of sample (b) after residence time of 2 days in magnetic field.

Woessner[7] has provided a detailed analysis of oriented water in domain structures, and we will follow his analysis. It is clear that in the sample under study, aggregates of platelets grow to fill the entire sample volume, and that the platelets are locked in the gel structure so that they can no longer orient in the magnetic field (Figure 11.3). In the presence of ribbon or sheet structures, one should expect local regions where the aggregated platelets have a common orientation with respect to the magnetic field, and such a region may be termed a domain. The other variable that now becomes important is the diffusion length of the water molecules. A correlation time τ_c can be assigned to interdomain motions of the water molecules. If the water molecules stay within a domain so that $\Delta\tau_c \gg 1$, and the domains are randomly oriented, a ^{2}H NMR powder pattern is expected. If the domains have a common orientation, a "single crystal" doublet is expected. On the other hand, if the water molecules visit a sufficiently large number of randomly oriented domains so that $\Delta\tau_c \ll 1$, an effective motional averaging takes

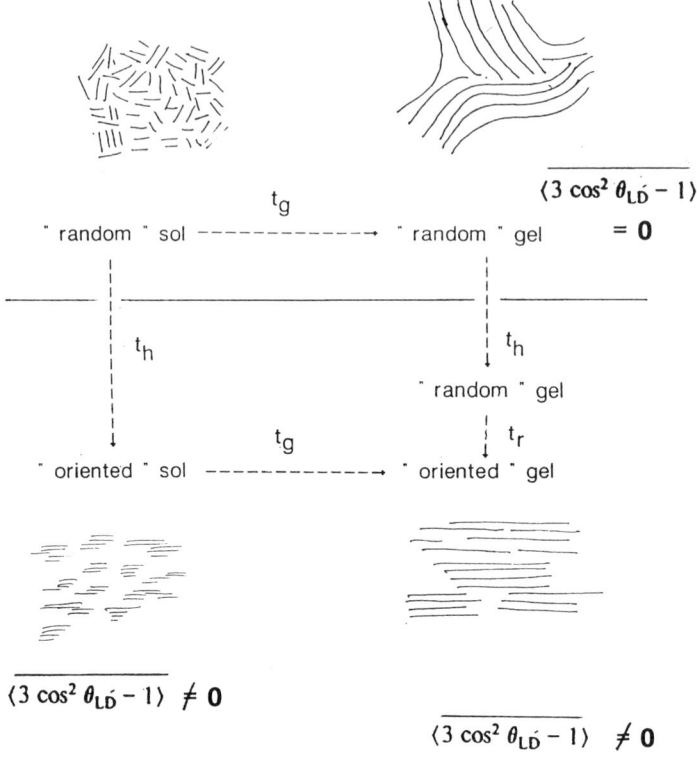

Figure 11.3 Schematic presentation of the time dependence observed in the spectra of sol and gel samples shown in Figure 11.2. Gelation takes place in a characteristic time t_g; vertical arrows labeled t_h represent the placing of the sample in the magnetic field in a time t_h (typically seconds). The time t_r is a hypothetical time for the reorganization of a random to an oriented gel. For a fully set gel, this time is very long.

place which reduces the doublet to a single line. In terms of the formalism presented in the theory section, in the presence of sufficiently fast interdomain diffusion ($\Delta \tau_c \ll 1$) of water molecules, $\langle 3\cos^2\theta_{LD} - 1\rangle$ in equation (1) can be replaced by its time average, which is equal to 0 for the random gel (Figure 11.3). The local ordering of the water molecules with respect to the platelet surfaces, as expressed by the anisotropies A in equation (2), are not expected to change on going from the suspension to either the oriented or the random gel.

In order to confirm that the gel can be produced in macroscopically anisotropic fashion, oriented gels were prepared by allowing sol samples to set to a gel inside the magnetic field. Subsequently, spectra were recorded as a function of rotation angle, both for a sample rotated about an axis at right angles to the orienting field, and for a sample rotated about an axis parallel to the orienting field (Figure 11.4). In the former case, the spectral splitting

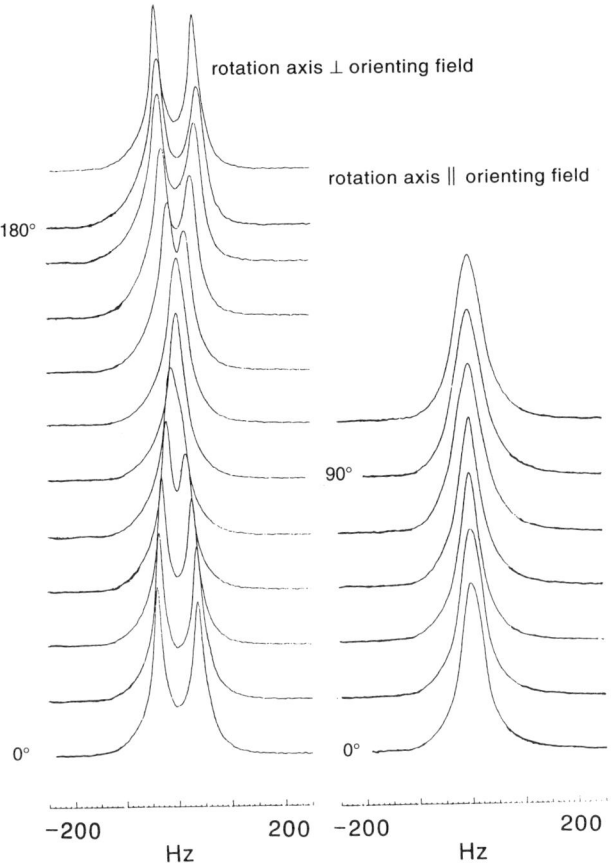

rotation axis ⊥ orienting field

rotation axis ∥ orienting field

180°

90°

0°

0°

−200   200 −200   200

Hz   Hz

Figure 11.4 ^{2}H NMR spectrum of oriented gel samples as a function of rotation angle (18° steps) for samples rotated about axes perpendicular and parallel to the orienting field.

changes continuously with rotation angle, the splitting disappearing at 90°, and returning to the 0° value when the rotation angle is 180°. In the latter case, the spectra show no splitting at any angle. The results are consistent with a macroscopically aligned sample with axial symmetry about the orienting field, and with two-fold symmetry along the field.

The Sol–Gel Transition

The experiments described above suggest that consistent results can be obtained as long as one considers that the suspensions often are in a non-equilibrium state, and that a time reference point must be established. This should be the last time the gel structure was broken.

With such a protocol for obtaining spectra it was shown that indeed the splittings vary systematically with concentration.[19] To a first approximation, at low concentrations the splitting is relatively independent of particle size, and the splitting can be written as

$$\Delta_s = aC$$

It is now possible to study the sol–gel transition itself. After the gel structure has been broken, the doublet splitting can be measured as a function of time. Of course, the sol–gel transition must be allowed to take place outside the magnetic field so that the platelets aggregate to form domain structures without a net orientation. The equilibrium random gel structure produced will then give zero doublet splitting. Of course, in order to measure the progress of aggregate formation, the sample must be inserted into the magnetic field to record the spectrum. If we use a simple two-state model, the observed doublet splitting $\Delta(t)$ is given by

$$\Delta(t) = f_s(t)\Delta_s + (1 - F_s(t))\Delta_r$$

where $f_s(t)$ is the fraction of water molecules associated with the sol, $(1 - f_s(t))$ is the fraction of water molecules associated with the gel, and Δ_r is the splitting for the random gel. Since the splitting $\Delta_r = 0$, $\Delta(t) = f_s(t)\Delta_s$, and a plot of $\Delta(t)/\Delta_s$ directly gives the time-dependent fraction of water molecules associated with the sol.

Typical results are shown in Figure 11.5, where we show normalized splittings $\Delta(t)/\Delta_s$ as a function of time for one of the fractions, for different concentrations. The observation of a single doublet again also implies that the system is homogeneous on an NMR time scale and that the water can exchange rapidly between both gel and sol fractions. The fast exchange observation certainly puts a limit on the size of sol and gel domains. Woessner and Snowden[22] have shown how domain sizes can be measured from the ^{2}H splitting, the spin–spin relaxation time T_2, and diffusion constant measurements, and it may be useful to consider measurements of this kind in the future.

The plot (Figure 11.5) for a 6.9% concentration of solids in the sol fraction shows that $\Delta(t)/\Delta_s$ is constant at 1.0 for a considerable length of time, then drops sharply as the gel forms initially, and more slowly as all of the particles are incorporated in the gel structure. Some of these features resemble to a marked degree calculations modeling the crosslinking of polymer gels.[20] Conceptually, the description of the sol–gel transition as an ordering transition from a randomly disordered state to a colloidal mesophase is quite attractive,[21] although it becomes rather important to obtain additional information on the nature of the ordered phase. We note also that water chemistry is critical, as gel formation does not occur, for instance, in distilled water, and the ^{2}H NMR splittings are essentially time-independent.

Not only can the ^{2}H NMR results be used to obtain independent estimates of the time for gel formation based on microscopic criteria, but also estimates of the minimum concentration required for gel formation can be

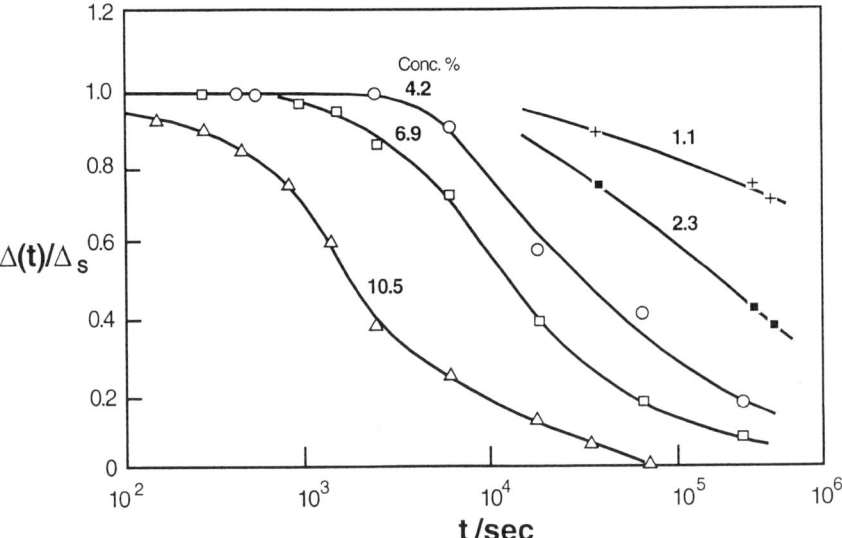

Figure 11.5 Normalized ^{2}H quadrupole splittings as a function of time for different concentrations.

made. Figure 11.6 shows plots of $\Delta_0 - \Delta_\infty$, the difference between splitting at zero time and the splitting nine days later, plotted as a function of solids concentration. In the absence of gel formation, the splittings will be the same; if gel formation has occurred, the long-term splitting will be very small, and the difference large. If the linear portions of the curves at low and high concentrations are extrapolated, the crossing point gives a good indication of the minimum solids concentration required for gel formation. From the results shown it is clear that gel formation takes place at a much lower concentration for the fraction with the smaller particle sizes. Of course, care must be taken that Δ_∞ indeed has reached its limiting long-time value.

Conclusions

It is clear that considerable information on gel structure and the sol–gel transition can be obtained at the *microscopic* level with NMR techniques, and that it may be possible to develop some useful models. Of course, it will be necessary to link the microscopic information from NMR, which has limited structural information but a strong kinetic component, to other techniques which give direct information on structure, and finally to macroscopic techniques such as viscometry.

Specifically with regard to Athabasca Oil Sands tailings, a simplified picture would seem to be one where settling of solids takes place upon discharge of the tailings until the ultrafines undergo the sol–gel transition, and

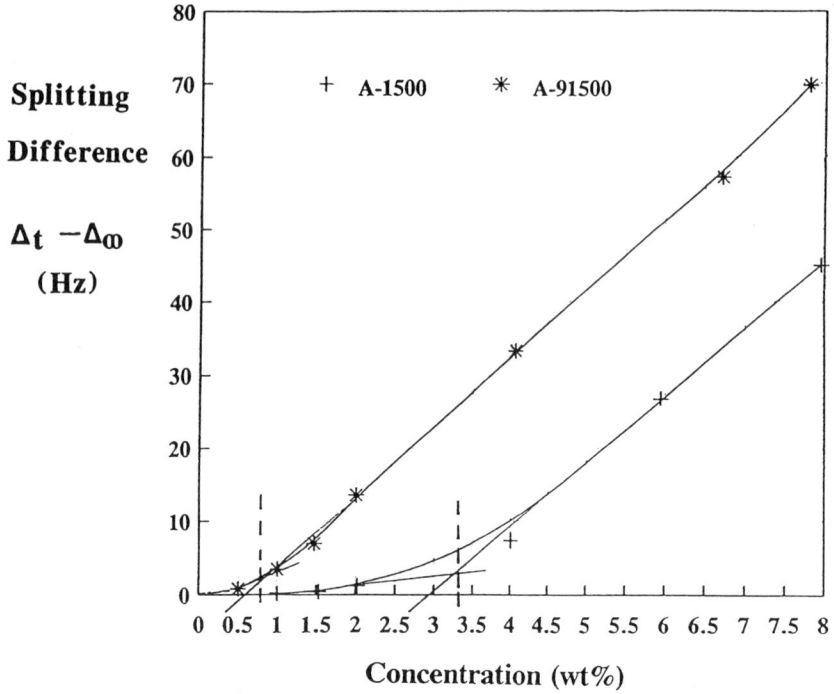

Figure 11.6 Plot of $\Delta_0 - \Delta_\infty$, the difference between initial and final splitting, as a function of concentration for two size fractions. The intercepts of the extrapolated linear portions of the initial and final slopes give a measure of the minimum concentration required for gel formation.

coarser particles and bitumen are trapped when the gel becomes sufficiently viscous. Recent experiments[23] have shown that water chemistry, particle size, and the presence of a biwetted surface-active solids fraction all contribute to gel formation. One especially important result of this work is the observation that the ultrafine fraction responsible for gel formation contains little or none of the swelling clays traditionally held responsible for gel formation, unless these are present in a delaminated state.

Solutions to the sludge formation problem naturally require economic as well as scientific consideration. Possible solutions include selective mining so as to avoid high fines feed, modification of the water chemistry so that the sol–gel transition does not take place during settling, or separate ponding of high fines tails. Dealing with the existing tailings ponds is a more difficult problem; however, methods based on the breaking of gel structure, either temporarily by mechanical means or more permanently by freeze–thaw cycling, so that additional settling can take place, are possibilities. NMR methods can play a role in providing microscopic information on gel formation, or the breaking of gel structures, in most of the cases mentioned above.

Acknowledgments This work was performed under the Sludge Fundamentals Consortium Agreement. Participants: Alberta Energy; Alberta Oil Sands Technology and Research Authority; Alberta Research Council; Energy Mines and Resources Canada (CANMET); The National Research Council of Canada; Suncor, Inc.; and Syncrude Canada Ltd.
The authors thank M. Lynds for making some of the measurements reported in this work.
Published as NRCC no. 39132.

References

1. Scott, J. D., Dusseault, M. B., and Carrier, III, W. D., Behaviour of the clay/bitumen/water system from oil sands extraction plants. *Appl. Clay Sci.* 1, 207, 1985

2. Pierre, A. C., Zou, J., and Barker, C., Structure comparison of an oil sands tailings sludge with a montmorillonite gel model. *Fuel* 71, 1373, 1992.

3. Sparks, B. D., Kotlyar, L. S., Ripmeester, J. A., Woods, J., and Schutte, R., *Colloidal Properties of Ultrafine Solids from Suncor and Syncrude Fine Tails.* Institute for Environmental Chemistry Report EC-1222-91S, 1991.

4. van Olphen, H., *An Introduction to Clay Colloid Chemistry*, Interscience, New York, NY, 1963.

5. Forslind, E., and Jacobsen, A., Clay–water systems. In *Water. A Comprehensive Treatise*, Vol. 5, Franks, F. (ed.), Plenum Press, London, 1975, Chapter 4.

6. Woessner, D. E., and Snowden, B. S., NMR doublet splitting in aqueous montmorillonite gels. *J. Chem. Phys.* 50, 1516, 1969.

7. Woessner, D. E., NMR and structure in aqueous heterogeneous systems. *Mol. Phys.* 34, 899, 1977.

8. Woessner, D. E., An NMR investigation of the range of the surface effect on the rotation of water molecules. *J. Magn. Reson.* 39, 297, 1980.

9. Fripiat, J. J., Surface activities of clays. In *Spectroscopic Characterization of Minerals and their Surfaces*, ACS Symp. Ser. No. 415, American Chemical Society, Washington DC, 1989, p. 360.

10. Grandjean, J., and Laszlo, P., Multinuclear and pulsed gradient magnetic resonance studies of sodium cations and of water reorientation at the interface of a clay. *J. Magn. Reson.* 83, 128, 1989.

11. Grandjean, J., and Laszlo, P., Deuterium NMR studies of water molecules restrained by their proximity to a clay surface. *Clays Clay Miner.* 37, 403, 1989.

12. Delville, A., Grandjean, J., and Laszlo, P., Order acquisition by clay platelets in a magnetic field. NMR study of the structure and microdynamics of the adsorbed clay layer. *J. Phys. Chem.* 95, 1383, 1991.

13. Halle, B., and Wennerstrom, H., Interpretation of magnetic resonance data from water nuclei in heterogeneous systems. *J. Chem. Phys.* 75, 1928, 1981.

14. van de Ven, T. G. M., *Colloidal Hydrodynamics*, Academic Press, London, 1989.

15. Kotlyar, L. S., Sparks, B. D., Kodama, H., Deslandes, Y., Ripmeester, J. A., Woods, J., and Schutte, R., *Separation of Syncrude and Suncor Fine Tails Based on Particle Size and Hydrophobicity Using a Modified CWAT Procedure.* Institute for Environmental Chemistry Report EC-1223-91S, 1991.

16. Mercier, L., Detellier, C., and Ripmeester, J. A., *Characterization of Gel-forming Fractions from the Athabasca Tailings Pond Fine Tails.* Institute for Environmental Chemistry Report EC-1250-92S, 1992.

17. M'Ewen, M. B., and Pratt, M. I., The gelation of montmorillonite, Part 1. *Trans. Faraday Soc.* 53, 535, 1957.
18. M'Ewen, M. B., and Pratt, M. I., The gelation of montmorillonite, Part 2, *Trans. Faraday Soc.* 53, 549, 1957.
19. Ripmeester, J. A., Kotlyar, L. S., and Sparks, B. D., ^{2}H NMR and the sol–gel transition in suspensions of colloidal clays. *Colloids and Surfaces,* 78, 57, 1993.
20. Ziman, J. M., *Models of Disorder,* Cambridge University Press, Cambridge, 1979.
21. Lekkerkerker, H. N. W., Crystalline and liquid crystalline order in concentrated colloidal dispersions: an overview. *NATO ASI Ser. B* 211, 165, 1989.
22. Woessner, D. E., and Snowden, Jr., B. S., A pulsed NMR study of dynamics and ordering of water molecules in interfacial systems. *Ann. N. Y. Acad. Sci.* 204, 113, 1973.
23. Kotlyar, L. S., Kumar, A., Ripmeester, J. A., Schutte, R., and Sparks, B. D., Ultrafines gelation and fine tails structure formation. In *Oil Sands—Our Petroleum Future*, Conference Proceedings, April 4–7, 1993, Edmonton, Paper F7.

NUTRIENT AND NATURAL ORGANIC MATTER CYCLING IN THE ENVIRONMENT

12.

Characterization of Natural Organic Matter by Nuclear Magnetic Resonance Spectroscopy

JERRY A. LEENHEER

Natural organic matter (NOM) is a major intermediate in the global carbon, nitrogen, sulfur, and phosphorus cycles. NOM is also the environmental matrix that frequently controls binding, transport, degradation, and toxicity of many organic and inorganic contaminants. Despite its importance, NOM is poorly understood at the structural chemistry level because of its molecular complexity and heterogeniety. Nuclear magnetic resonance (NMR) spectroscopy is one of the most useful spectrometric methods used to investigate NOM structure because qualitative and quantitative organic structure information for certain organic elements can be generated by NMR for NOM in both the solution and solid states under nondegradative conditions. However, NMR spectroscopy is not as sensitive as infrared or ultraviolet–visible spectroscopy; it is not at present applicable to organic oxygen and sulfur, and quantification of NMR spectra is difficult under certain conditions. The purpose of this overview is to present briefly the "state of the art" of NMR characterization of NOM, and to suggest future directions for NMR research into NOM. More comprehensive texts concerning the practice of NMR spectroscopy and its application to NOM in various environments have been produced by Wilson[1] and by Wershaw and Mikita.[2]

Elemental Composition Considerations

Carbon, hydrogen, and oxygen are the major elements of NOM; together they comprise about 90% of the mass. The minor elements that constitute the remainder are nitrogen, sulfur, phosphorus, and trace amounts of the various halogen elements. With the exception of coal, in which carbon is the most abundant element, the order of relative abundance in NOM on an

213

atomic basis is H > C > O > N > S > P = halogens. The optimum NMR-active nuclei for these elements are ^{1}H, ^{13}C, ^{17}O, ^{15}N, ^{33}S, ^{31}P, and ^{19}F. The natural abundances and receptivities of these nuclei relative to 1H are given in Table 12.1.[3] Quadrupolar effects for ^{17}O, ^{33}S, and halogen elements other than ^{19}F lead to line broadening that greatly limits resolution in NMR studies of these elements in NOM.[3]

The first application of NMR spectroscopy to NOM characterization was by Oka et al.,[4] who determined the solution-state proton NMR of peat humic acids. This and other early attempts to obtain useful NMR spectra of NOM by continuous wave NMR spectroscopy yielded poorly resolved spectra that contained little useful structural information.[5,6] The advent of pulse Fourier Transform NMR spectrometers with higher magnetic fields greatly increased the sensitivity and resolution of both ^{1}H and ^{13}C NMR spectroscopy of NOM in the solution state.[7,8] Early attempts to obtain ^{13}C NMR spectra of solids provided little useful structural information on NOM because of line broadening due to C–H dipolar interactions and chemical shift anisotropy. These problems were minimized by high-power ^{1}H decoupling and by cross polarization combined with magic angle spinning.[9] Now, ^{1}H and ^{13}C NMR spectroscopy of NOM in both the solution and solid states are readily conducted.[1,2]

^{15}N NMR spectroscopy can be performed on ^{15}N-enriched samples,[10–13] but its low relative abundance and relative receptivity (Table 12.1) greatly limit natural-abundance ^{15}N NMR studies of NOM. One of the first reports of natural-abundance ^{15}N NMR of NOM is given by Knicker et al.[14] Knicker, Fründ and Lüdemann describe this work in Chapter 15 of this book.

^{31}P NMR of NOM is limited only by the low percentage of ^{31}P in NOM. Extensive concentrations and extractions of organic ^{31}P in NOM are necessary, as explained in Chapters 13 and 14. Additional problems arising from extractions and concentration procedures are molecular aggregation with other dissolved constituents and hydrolysis of phosphate esters.

^{19}F NMR has been applied to fluorinated derivatives of humic substances to determine functional group composition,[15] but there are no known applications of ^{19}F NMR to natural-abundance NOM.

Natural Organic Matter Phase Considerations

NOM characteristics determine whether NMR spectra are measured in the solution or solid-state phase. Dissolved aquatic organic matter and soil and sediment extracts can generally have NMR spectra measured in a suitable solvent, but many types of NOM are not soluble in any solvent, and solid-state NMR spectroscopy is the only option. The analyst's choice of the NOM phase for NMR studies also depends on the resolution desired and the sensitivity of the sample.

Table 12.1 Natural Abundance and Relative Receptivity of NMR-active Nuclei in Natural Organic Matter[a]

Nucleus	Natural Abundance (%)	Relative Receptivity[b]
^{1}H	99.985	1
^{13}C	1.108	1.76×10^{-4}
^{17}O	0.037	1.08×10^{-5}
^{15}N	0.36	3.85×10^{-6}
^{33}S	0.76	1.71×10^{-5}
^{31}P	100	6.63×10^{-2}
^{19}F	100	0.8328
^{35}Cl	75.53	3.55×10^{-3}
^{81}Br	49.46	4.87×10^{-2}
^{127}I	100	9.34×10^{-2}

[a] Data from Harris and Mann.[3]
[b] Receptivities are calculated relative to the ^{1}H nucleus.

Solution NMR

Applications and reviews of solution NMR of humic substances are given by Preston[16] and Steelink et al.[17] Solution NMR is used primarily for ^{1}H, ^{13}C, ^{31}P, and ^{15}N NMR spectroscopic studies of NOM.

The major advantage of solution NMR is its superior resolution compared to solid-state NMR; however, with the broad peak shapes obtained for most NOM samples, this advantage is not generally significant. A much larger data base of chemical standards exists for solution NMR than for solid-state NMR because of its application in chemistry and biochemistry, because of the superior resolution advantage, and because solution NMR predates solid-state NMR.[18,19]

Another significant advantage of the solution phase for ^{1}H NMR is that exchangeable protons in NOM can be differentiated from nonexchangeable protons or removed by deuterium exchange (Chapter 16). In solid-state ^{1}H Combined Rotation and Multiple Pulse (CRAMPS) NMR, linewidths of proton resonances in pure standards are broad and inclusion of exchangeable protons over a wide range of chemical shifts tends to smear out the spectra so that little useful information is obtained for NOM samples where the ratio of exchangeable to nonexchangeable protons is large. ^{1}H-CRAMPS has been used to estimate aromatic and aliphatic protons in coal.[1]

A final advantage of solution NMR is that NOM samples remain in the solution phase and avoid possible chemical changes that occur during drying. Lactone ester structures in fulvic acid form during drying of certain hydroxy acids.[20]

Disadvantages of solution NMR are that nucleus relaxation times are longer than in the cross-polarization magic angle spinning (CPMAS) solid-state NMR. Because the sample has to be dissolved, nucleus concentrations are lower than in the solid state. There is also frequent uncertainty about whether NOM is truly dissolved. Molecular aggregations that do not result in precipitates can still lead to changes in NMR spectra (broad lines, quantitation differences) that are difficult to interpret (Chapter 13). These aggregations may occur between NOM macromolecules or with inorganic metals, borates, carbonates, or polysilicates. Filtration or centrifugation may remove large aggregates and precipitates. Solvents have to be selected so that solvent peaks do not interfere with NMR spectra of NOM. Finally, solvent effects on observed chemical shifts must be taken into account when comparisons are made with NMR spectra run in different solvents.

Solid-state NMR

Solid-state NMR spectroscopy has become the favored spectroscopic method for NOM characterization in almost any kind of matrix, such as humic substances,[21-23] wood,[24] oil shale,[25] whole soil, peat, lignite, and coal,[1] and stream sediments (Chapter 17 in this book). Polarization transfer from ^1H to ^{13}C, ^{31}P, or ^{15}N (cross polarization) increases sensitivity for solid-state NMR compared to solution NMR. However, the effects of cross-polarization dynamics in CPMAS solid-state NMR must be considered and compensated for to obtain a quantitative determination of various organic structures in NOM.

The rate of the proton cross-polarization process is dependent on the proximity of structural protons to the nuclei being studied by NMR spectroscopy, and on molecular motion.[1] For example, a ^{13}C-labeled methylene group is efficiently cross-polarized, but ^{13}C in graphite is not cross-polarized because of the virtual absence of structural protons in graphite. As a consequence, the various carbon atoms in NOM have differential responses in solid-state ^{13}C NMR.[26] The same is true for other nuclei. With carefully chosen cross-polarization contact times and pulse delay times, quantitative ^{13}C spectra can be obtained with most types of NOM.[27] Certain coals, with variable maceral structures that approach graphite at one extreme and saturated aliphatic hydrocarbons at the other, present unique challenges in obtaining quantitative ^{13}C NMR spectra.

The nature of the phase also affects quantitation in ^{13}C NMR. Crystalline substances are not as efficiently cross polarized as amorphous materials, but amorphous NOM with liquid domains in its structure, such as liquid water or liquid hydrocarbons, results in molecular mobility that adversely affects the cross-polarization process which in its turn affects quantitation.[1,28] Hatcher and Wilson[29] found that the fulvic acids studied by Earl[28] were hygroscopic and that as little as 2% of adsorbed water caused significant changes in spin lattice relaxation times and cross-polarization contact times that affected quantitation. Drying the fulvic acid samples immediately before running the ^{13}C NMR spectrum was recommended.

Paramagnetic elements and free radicals shorten nucleus relaxation times. In some cases, this results in sensitivity losses and line broadening which may limit the usefulness of the NMR spectra. Soils and sediments with high levels of iron and manganese are difficult phases from which to obtain useful NMR spectra.

The primary effort towards obtaining NMR spectra of these types of samples has been to remove the iron and manganese by chemical or physical means (chapter 17). Reduction of iron and manganese by sodium dithionite[1] was used by Oades et al.[30] to improve the ^{13}C NMR spectra of red Australian soils, and removal of iron and manganese by successive treatments with pyrophosphate, oxalate, and dilute HF was used to improve the ^{13}C NMR of Mississippi River sediments (Chapter 17). Organic matter can be separated from clay containing paramagnetic elements by sequential extraction followed by ^{13}C NMR characterization of the extracts,[31] but significant portions of soil organic matter are not extractable by any solvent. The limitations of chemical treatments are the organic matter losses and chemical changes such as hydrolysis and functional group reduction or oxidation that occur upon treatment.

Physical treatments to remove iron and manganese have focused on particle size[30,32] and density separations.[33–35] In addition to fractionating out paramagnetic elements, these physical separation methods have become essential to the characterization of soil organic matter with respect to understanding the diagenetic processes in the formation and selective preservation of NOM in soil.

The quantitative limitations of solid-state NMR are not well understood, and care in addressing these limitations with appropriate spectral parameters can maximize the quantitative significance of NOM studies. The improvement in NMR instrumentation and software has increased the level of sensitivity for ^{13}C NMR from a carbon content of 3% of the matrix[1] to about 1.0%,[36] providing there is no significant cross polarization or paramagnetic element limitations. Even lower levels of carbon than 1% in soils and sediments can be studied by ^{13}C NMR spectroscopy if this carbon is enriched by extraction, mineral dissolution, and size or density fractionations. Similar considerations with regard to quantitation and sensitivity apply to ^{31}P and ^{15}N CPMAS solid-state NMR.

Future Directions

The trend toward increased sensitivity for various NMR-active nuclei in NOM will undoubtedly continue. Large-volume sample spinners for high-sensitivity solid-state NMR studies have been developed,[37] but there have been few applications of these large-volume spinners toward better characterization of NOM in environments low in organic matter, such as aquifer sediments. Field instruments with improved sensitivity are also becoming available for both solution and solid-state NMR.

As this sensitivity increases, ^{17}O NMR studies of NOM might be possible; the initial use of ^{17}O-enriched NOM may be followed by NMR studies of unenriched samples. There is a critical need for ^{17}O NMR spectroscopy of NOM, because oxygen is one of the major elements in most NOM samples, and ^{13}C NMR and infrared spectroscopy are not able to supply quantitative structural information on all oxygen-containing structures, such as structural ethers. Despite the problem with quadrupolar broadening of linewidths, paramagnetic ionic interactions with oxygen functional groups, and low isotopic abundance of ^{17}O (Table 12.1), the broad chemical shift range over which oxygen functional-group structures occur, the abundance of oxygen in NOM, and the existing data base for correlating chemical shifts with structure,[3] make ^{17}O NMR studies of NOM worthy of serious consideration.

Better techniques of sample preparation for NMR studies will continue to lead towards better understanding of NOM structure and properties. Isotopic enrichment of NOM by chemical derivatization, growth and decay in isotopically enriched media,[33] enzyme-catalyzed binding studies,[38] or isotopic exchange reactions results in greatly increased sensitivity for nuclei with low isotopic abundances and low relative receptivities (^{13}C, ^{15}N, and ^{17}O). Fractionation of heterogeneous NOM to more homogeneous fractions followed by NMR characterization will continue to yield more interpretable spectra. Much research needs to be done on separation of paramagnetic elements from NOM samples (Chapter 17), and on the use of lanthanide shift reagents to increase resolution[39] to obtain better NMR characterization of NOM.

There is a need to integrate quantitative spectroscopic methods (^{13}C NMR, ^{1}H NMR, ^{15}N NMR, and infrared spectroscopy) with molecular weight distributions, elemental composition, and titrimetric data to derive structural models of NOM based on quantitative data.[40] While such models are controversial because they present discrete chemical structures for heterogeneous NOM, they are valuable because they require the investigator to apply a variety of analytical constraints to the characterization of NOM such that the structural models reflect the predominant structures and properties of NOM. Significant differences between allochthonous and autochthonous dissolved organic matter were illustrated by the quantitative structural model approach,[41] and the chemical structure of strong acids in a stream of fulvic acid was determined by structural models of fulvic acid altered by fractionations and chemical derivatizations and degradations.[20]

Finally, most types of NOM are chemically and physically bound to clay minerals and their sesquioxide coatings.[42] Extending the scope of NOM characterization to the inorganic matrix may be possible by conducting ^{27}Al, ^{29}Si, and perhaps even ^{17}O NMR studies of clay mineral–organic complexes or enriched preparations that simulate surface–complex interactions.

References

1. Wilson, M. A., *NMR Techniques and Applications in Geochemistry and Soil Chemistry*, Pergamon, Sydney, Australia, 1987.
2. Wershaw, R. L., and Mikita, M. A. (eds), *NMR of Humic Substances and Coal*, Lewis Publishers, Chelsea, MI, 1987.
3. Harris, R. K., and Mann, B. E. (eds), *NMR and the Periodic Table*, Academic Press, New York, NY, 1978.
4. Oka, H., Sasaki, M., Itho, M., and Suzuku, A., *Nenryo Kyokai Shi (J. Japanese Fuel Soc.)* 48, 295–302, 1969.
5. Schnitzer, M., and Khan, S. U., *Soil Organic Matter*, Elsevier Scientific, Amsterdam, 1978.
6. Stevenson, F. J., *Humus Chemistry, Genesis, Composition, Reactions*, John Wiley, New York, NY, 1982.
7. Gonzãles-Vila, F. J. G., Lentz, H., and Lüdemann, H. D., *Biochem. Biophys. Res. Commun.* 72, 1063–1069, 1976.
8. Lentz, H., Lüdemann, H. D., and Ziechmann, W., *Geoderma* 18, 325–328, 1977.
9. Hatcher, P. G., Breger, I. A., Dennis, L. W., and Maciel, G. E., in *Aquatic and Terrestrial Humic Materials*, Christman, R. F., and Gjessing, E. T. (eds), Ann Arbor Science, Ann Arbor, MI, 1983, pp. 37–82.
10. Benzing-Purdie, L., Ripmeester, J. A., and Preston, C. M., *J. Agric. Food Chem.* 31, 913–915, 1983.
11. Preston, C. M., Rauthan, B. S., Rodger, C., and Ripmeester, J. A., *Soil Sci.* 134, 277–293, 1982.
12. Steelink, C., in *Humic Substances in the Global Environment and Implications on Human Health*, Senesi, N., and Maino, T. M. (eds), Elsevier Science, Amsterdam, 1994, pp. 405–426.
13. Thorn, K. A., Arterburn, J. B., and Mikita, M. A., *Environ. Sci. Technol.* 26, 107–116, 1991.
14. Knicker, H., Fründ, R., and Lüdemann, H.-D., *Naturwissenschaften* 80, 219–221, 1993.
15. Leenheer, J. A., and Noyes, T. I., in *Humic Substances II. In Search of Structure*, Hayes, M. H. B., MacCarthy, P., Malcolm, R. L., and Swift, R. S. (eds), John Wiley, New York, NY, 1989, pp. 257–280.
16. Preston, C. M., in *NMR of Humic Substances and Coals*, Wershaw, R. L., and Mikita, M. A. (eds), Lewis Publishers, Chelsea, MI, 1987, pp. 3–32.
17. Steelink, C., Wershaw, R. L., Thorn, K. A., and Wilson, M. A., in *Humic Substances II. In Search of Structure*, Hayes, M. H. B., MacCarthy, P., Malcom, R. L., and Swift, R. S. (eds), John Wiley, New York, NY, 1989, pp. 281–308.
18. Simons, W. W., *The Sadtler Handbook of Proton NMR Spectra*, Sadtler Research Laboratories, Philadelphia, PA, 1978.
19. Simons, W. W., *The Sadtler Guide to Carbon-13 NMR Spectra*, Sadtler Research Laboratories, Philadelphia, PA, 1983.
20. Leenheer, J. A., Reddy, M. M., and Wershaw, R. L., *Environ. Sci. Technol.* 29, 399–405, 1995.
21. Frye, J. S., Bronnimann, C. E., and Maciel, G. E., in *NMR of Humic Substances and Coals*, Wershaw, R. L., and Mikita, M. A. (eds), Lewis Publishers, Chelsea, MI, 1987, pp. 33–46.

22. Malcolm, R. L., in *Humic Substances II. In Search of Structure*, Hayes, M. H. B., MacCarthy, P., Malcolm, R. L., and Swift, R. S. (eds), John Wiley, New York, NY, 1989, pp. 33–67.
23. Wilson, M. A., Basic concepts and techniques. In *Humic Substances II. In Search of Structure*, Hayes, M. H. B., MacCarthy, P., Malcolm, R. L., and Swift, R. S. (eds), John Wiley, New York, NY, 1989, pp. 309–338.
24. Hatcher, P. G., and Breger, I. A., *Org. Geochem.* 3, 49–55, 1981.
25. Miknis, F. P., Netzel, D. A., Smith, J. W., Mast, M. A., and Maciel, G. E., *Geochim. Cosmochim. Acta* 46, 977–984, 1982.
26. Vassallo, A. M., in *NMR of Humic Substances and Coals*, Wershaw, R. L., and Mikita, M. A. (eds), Lewis Publishers, Chelsea, MI, 1987, pp. 211–224.
27. Fründ, R., and Lüdemann, H.-D., *Sci. Total Environ.* 81–82, 157–168, 1989.
28. Earl, W. L., in *NMR of Humic Substances and Coals*, Wershaw, R. L., and Mikita, M. A. (eds), Lewis Publishers, Chelsea, MI, 1987, pp. 167–188.
29. Hatcher, P. G., and Wilson, M. A., *Org. Geochem.* 17, 293–299, 1991.
30. Oades, J. M., Vassallo, A. M., Waters, A. G., and Wilson, M. A., *Aust. J. Soil Res.* 25, 71–82, 1987.
31. Schnitzer, M., Ripmeester, J. A., and Kodama, H., *Soil Sci.* 145, 448–454, 1988.
32. Preston, C. M., Newman, R. H., and Rother, P., *Soil Sci.* 157, 26–35, 1994.
33. Baldock, J. A., Oades, J. M., Vassallo, A. M., and Wilson, M. A., *Aust. J. Soil Res.* 28, 193–212, 1990.
34. Beudert, G., Kögel-Knabner, I., and Zech, W., *Sci. Total Environ.* 81–82, 401–408, 1989.
35. Golchin, A., Oades, J. M., Hemstad, J. O., and Clarke, P., *Aust. J. Soil Res.* 32, 285–309, 1994.
36. Baldock, J. A., Oades, J. M., Waters, A. G., Peng, X., Vassallo, A. M., and Wilson, M. A., *Biogeochemistry* 16, 1–42, 1992.
37. Zhang, M., and Maciel, G. E., *Anal. Chem.* 62, 633–638, 1990.
38. Bortiatynski, J. M., Hatcher, P. G., Minard, R. D., Dec, J., and Bollag, J.-M., in *Humic Substances in the Global Environment and Implications on Human Health*, Senesi, N., and Maino, T. M. (eds), Elsevier Science, Amsterdam, 1994, pp. 1091–1098.
39. Nanny, M. A., and Minear, R. A., *Environ. Sci. Technol.* 28, 1521–1527, 1994.
40. Leenheer, J. A., McKnight, D. M., Thurman, E. M., and MacCarthy, P., in *Humic Substances in the Suwannee River, Georgia: Interactions, Properties, and Proposed Structures*, Averett, R. C., Leenheer, J. A., McKnight, D. M., and Thorn, K. A. (eds), US Geological Survey Water Supply Paper 2373, 1994, pp. 195–211.
41. Leenheer, J. A., in *Environmental Chemistry of Lakes and Reservoirs*, Baker, L. A. (ed.), Adv. Chem. Ser. No. 237, American Chemical Society, Washington DC, 1994, pp. 195–222.
42. Leenheer, J. A., in *Organic Substances and Sediments in Water I. Humics and Soils*, Baker, R. A. (ed.), Lewis Publishers, Chelsea, MI, 1991, pp. 3–22.

13.

^{31}P FT-NMR of Concentrated Lake Water Samples

MARK A. NANNY & ROGER A. MINEAR

The use of phosphorus-31 Fourier Transform nuclear magnetic resonance (^{31}P FT-NMR) spectroscopy for the study of dissolved organic phosphorus (DOP) in fresh water has been recently established by Nanny and Minear.[1] The fact that NMR is an element-specific technique, is nondestructive, and has the ability to differentiate between similar phosphorus compounds makes it invaluable for the identification and characterization of DOP. Such information regarding DOP is required in order to understand aquatic nutrient cycling.

The difficulty with using ^{31}P FT-NMR spectroscopy for such studies is the extremely low DOP concentration; usually ranging from $< 1\,\mu g$ P/L in oligotrophic lakes to approximately $100\,\mu g$ P/L for eutrophic systems. Nanny and Minear[1] raised the DOP concentration into the NMR detection range, which is on the order of milligrams of phosphorus/liter, by concentrating large volumes of lake water with ultrafiltration (UF) and reverse osmosis (RO) membranes. Volume concentration factors of several ten thousand fold provided DOP concentrations of up to $60\,mg$ P/L. Other DOP concentration methods such as anion exchange,[2] lanthanum hydroxide precipitation,[3] and lyophilization[4] require severe chemical and/or physical transformations of the sample and/or they need long processing times, all of which increase the risk of DOP hydrolysis. Sample concentration with UF and RO membranes does not require the sample to undergo these major changes and is also a relatively rapid concentration method. In addition to these concentration capabilities, the use of ultrafiltration and reverse osmosis membranes permitted fractionation of the DOP samples according to molecular size. Nanny and Minear[1] used three membranes in series with decreasing pore size: 30 kDa (kilodaltons), 1 kDa, and RO (95% NaCl rejection) to separate the high-molecular-weight, intermediate-molecular-weight, and low-molecular-weight DOP species.

In the intermediate-molecular-weight fraction, Nanny and Minear[1] observed the presence of monoester and diester phosphates. Spectra from ten samples collected over a year typically consisted of a large broad signal in the monoester phosphate region spanning from a chemical shift of 2.00 ppm to −0.50 ppm. The maximum of this signal was usually in the range of 1.00 to 1.50 ppm. This broad signal had a shoulder in the diester phosphate region which sometimes was intense enough to appear as an individual signal. In addition to the large broad signal, a second, smaller signal was usually present in the monoester phosphate region at a chemical shift position between 4.00 and 5.00 ppm. The height of this smaller signal relative to that of the large broad signal varied with season, and the ratio could be correlated with visual observations of lake water quality. Orthophosphate was also occasionally detected in these samples and was attributed to the hydrolysis of DOP compounds.

Attempts to identify the DOP species giving rise to the large broad signal indicated that the signal was probably due to DOP species incorporated into or adsorbed to an aggregate or micelle structure. These structures could be formed during the sample concentration procedure from surfactant-type materials, humic and fulvic acids, hydrophobic compounds such as lipids and phospholipids, and silicates which can condense into polymeric silicates. Spiking the concentrated samples with known organic phosphorus compounds that had ^{31}P FT-NMR signals at chemical shift positions different from that of the region of the large broad signal, only caused an increase in intensity of the large broad signal rather than the appearance of a new signal arising from the added DOP species. This change in the chemical shift position of the added DOP compounds indicated that the DOP was being adsorbed or incorporated into a structure, rather than remaining dissolved in solution.

Further experimental evidence for DOP incorporation into some kind of structure was provided by examining the ^{31}P FT-NMR spectrum of a concentrated sample at various pH values. Each time a phosphate group gains or loses a proton, the electronic density around the phosphorus nucleus changes, causing the chemical shift position of the signal to change. Nanny and Minear[1] observed this for several monoester phosphate compounds. When the sample pH was varied, the position of the large broad signal surprisingly did not change. This indicated that the phosphorus causing this signal was isolated or protected from the solvent environment; the phosphorus was incorporated into some kind of a structure and did not interact with the solvent. The chemical shift position of the smaller signal (4.00 to 5.00 ppm) was a function of pH. Nanny and Minear hypothesized that this smaller signal could represent freely dissolved DOP, or DOP that is present on the surface of the aggregate or micelle and could interact with the solvent. A similar situation exists with the bilayer micelles formed from L-dipalmitoyl-α-lecithin in the presence of low concentrations of Pr^{3+}.[5] When no Pr^{3+} is present, the ^{31}P FT-NMR signals from the inner and outer phosphate groups of the bilayer micelle appear as a single, broad signal. Addition of Pr^{3+}

causes two signals to appear, one in the original position, which are due to the inner phosphate groups, and a second broad signal at a higher frequency, which is due to the outer phosphate groups in contact with the Pr^{3+} ions. Because this behavior is similar to that of the large broad signal position as a function of pH, it was thought that phospholipids, with a major portion of the phosphate groups enclosed inside the bilayer micelle, might be contributing to the intensity of the large broad signal. Many phospholipids also produce ^{31}P FT-NMR signals in this chemical shift region,[6] which gave further support to this hypothesis. The presence of phospholipids with new algal growth would not be unexpected, since phospholipids are major components of cellular membranes.

In light of the work by Nanny and Minear, it is realized that numerous problems need to be overcome for effective identification and characterization of DOP with ^{31}P FT-NMR spectroscopy. Despite the fact that the concentrated lake water samples contain adequate phosphorus concentrations, that a large number of spectral acquisitions are obtained, and that sufficient T_1 relaxation times are used, low sensitivity and poor resolution still hinder analysis. Also, if ^{31}P FT-NMR spectra are to be properly understood, questions regarding the possible interaction of DOP with aggregates or micelles during or after the concentration procedure must be addressed. In addition, if a complete understanding of phosphorus cycling is to be obtained, ^{31}P FT-NMR spectra of the high-molecular-weight and the low-molecular-weight fractions need to be examined over several seasons.

This chapter presents ^{31}P FT-NMR spectra of the high-, intermediate-, and low-molecular-weight samples collected at different seasons and discusses the characterization of the DOP in each sample. It also provides possible reasons for changes in the spectral patterns and signal quality with different seasons. Because the ability to identify individual DOP species conclusively is hindered by the low sensitivity and poor resolution in the ^{31}P FT-NMR spectra, and also in order to gain further understanding of the possible interactions of DOP with aggregates or micelles, several extraction, isolation, and degradation techniques were examined. These techniques were the addition of chelating agents to complex paramagnetic ions, the use of XAD-8 resin to remove dissolved hydrophobic humic and fulvic acids, organic solvent extractions to remove hydrophobic materials, freeze-drying the sample prior to various extraction schemes, diafiltration of the sample to remove dissolved salts, alkaline bromination degradation to oxidize any aggregate or micelle structures, and the use of lanthanide shift reagents to separate overlapping signals.

Experimental

Sampling

Pelagic lake water samples were collected at Crystal Lake, a man-made mesotrophic lake in Champaign County, Illinois. It has an average depth of 10 ft

(3 m) with a maximum depth of 13 ft (4 m) and is fed by a 200 ft (61 m)-deep well. The only other water source is surface runoff from the surrounding area, which is wooded with deciduous trees. No storm sewers flow into the lake. For each sample, 250 to 500 L of water were filtered on-site with a plankton net and stored in acid-washed, 55 L polyethylene containers until processed in the laboratory approximately 30 min later. At the laboratory, the samples are stored and processed at 4 °C in a refrigerated room.

Concentration system

The detailed description of the ultrafiltration and reverse osmosis concentration system is presented elsewhere.[7] A brief account will be given here for clarity. The water was first filtered with a 0.2 μm tangential flow filter (Millipore Corp.) to remove algal cells, bacteria, and colloidal and suspended solids. The water then passed through an Na^+ cation exchange column containing 6.2 L of 50–100-mesh Dowex 50X8 in the sodium form to remove Ca^{2+} and Mg^{2+}, and then through a second tangential flow filtration unit, consisting of a 30 kDa polysulfone membrane (Millipore Corp.). The 30 kDa membrane was used because it was expected to remove enzymes that would hydrolyze organic phosphorus in the lower-molecular-weight fractions. The retentate was continuously recycled to the second tangential flow filter while the filtrate passed to a third ultrafiltration tangential flow unit, consisting of a 1 kDa cellulose acetate membrane (Millipore Corp.). The 1 kDa membrane was used because it is the UF tangential flow membrane with the smallest pore size available. The retentate from the 1 kDa membrane was continuously recycled through the 1 kDa membrane and the filtrate was passed onto a Millipore Lab Pro Reverse Osmosis (RO) Bench Top unit that contains a spiral-wound polyamide RO membrane, rated at 99% NaCl rejection (Millipore Corp.). The retentate from this third filtration was recycled back to the spiral-wound membrane and the filtrate was discarded. The final volume of the units is fixed because of the internal holdup volumes. The holdup volume of the tangential flow unit was usually 0.8 L, and 1.5 L for the spiral-wound RO unit. The spiral-wound RO retentate was passed through a second, smaller cation exchange column containing the same resin as the first cation exchanger to remove any remaining Ca^{2+} and Mg^{2+}. Samples were further concentrated to a final volume of 8 mL with a series, in order of decreasing size, of batch pressure ultrafiltration units (Amicon) containing cellulose acetate filters (Amicon) with an effective pore size of either 1000 or 500 Da. The batch membrane cutoff selected depended upon the molecular weight selectivity of the antecedent continuous flow system.

NMR Spectrometers

All ^{31}P FT-NMR spectra were collected with a GN 300 Narrow Bore or a GN 300 Wide Bore nuclear magnetic resonance spectrometer at the University of

Illinois, School of Chemical Sciences Molecular Spectroscopy Laboratory. Crystal Lake samples were scanned for 24 to 48 h using the GN 300 Wide Bore NMR, or over several 12 h periods with the GN 300 Narrow Bore NMR, followed by compiling the individual free induction decay patterns (FIDs) into a single FID. A Fourier Transformation was then performed upon this composite FID to obtain the final spectrum. All ^{31}P FT-NMR spectra, obtained at 121.648 MHz, were generated by a pulse width of 20 to 24 μs with a pulse delay of 3 to 6 s, depending upon the sample. All spectra were proton decoupled and the spectral width was 10 000 Hz. Magnet shimming and signal phasing were done automatically by computer, and all chemical shift measurements were measured relative to 85% H_3PO_4. Samples were placed in 10 mm glass NMR tubes with coaxial inserts (Wilmad Corporation) containing deuterium oxide (Sigma).

Methods and Techniques

TSP and SRP Concentration Measurements

Total soluble phosphorus (TSP) and soluble reactive phosphorus (SRP) concentrations were measured using the ascorbic acid–molybdate method.[8] Organic phosphorus was oxidized for TSP measurements using potassium persulfate oxidation.[8]

Addition of Chelating Molecules

The effect of *trans*-1,2-diaminocyclohexane-N, N, N', N'-tetraacetic acid (CDTA) (Sigma) was examined by comparing the ^{31}P FT-NMR spectra of inositol hexaphosphate (Sigma) at a concentration of 20 mg P/L, dissolved in a concentrated lake water matrix, before and after the addition of CDTA at a final concentration of 0.013 M. This concentration provided a CDTA/P ratio of ≈200. The effect of tetrasodium ethylenediaminetetraacetate (EDTA) was examined by dissolving enough Na_4EDTA in a concentrated sample to provide an EDTA concentration of approximately 0.2 M and examining the ^{31}P FT-NMR spectrum before and after· the addition of EDTA.

XAD-8 Research

The procedure employed for the removal of hydrophobic humic and fulvic acids was similar to that used by Leenheer.[9] The concentrated sample was acidified with 6 M HCl to a final pH of 2 and then passed through an XAD-8 resin at a flow rate of approximately 1 mL/min. The filtrate was neutralized with 0.1 M NaOH. The adsorbed material on the XAD-8 column was removed by flushing the column with three times the column volume of 0.1 M NaOH. This fraction was neutralized with 0.1 M HCl.

Organic Solvent Extractions

The methanol–chloroform extractions were slight modifications of the Bligh and Dyer method for phospholipid extraction.[10] Thin layer chromatography for the detection of lipids used activated silica on glass plates. Lipids were detected using 2,7-dichlorofluoroscein and phospholipids were further characterized using an ascorbic acid–molybdate spray.

Freeze-drying Samples

Samples were freeze-dried and then solvent extractions were performed by mixing 3 mL of solvent with 0.1 g of freeze-dried sample in a test tube. For the ammonia extraction 0.1 M NH_4OH was used, and for acetonitrile extractions a 75% acetonitrile (aq.) solution containing tetrabutylammonium bromide (TBA) (Sigma) at a concentration of 0.16 M Acetonitrile extractions were also performed by adding the TBA to the sample before freeze-drying and then extracting the dried sample with 75% acetonitrile (aq.).

Diafiltration

Diafiltration experiments were performed by placing 5 mL of concentrated sample in a 12 mL pressure batch concentrator, diluting the sample with 5 mL of Milli-Q treated water and reconcentrating the sample to 5 mL. The membrane used for the high-molecular-weight sample was a 1 kDa membrane and a 500 Da membrane was used for the low-molecular-weight samples. This procedure was repeated seven times for each sample. The effectiveness of the diafiltration procedure was monitored by measuring the conductivity and total organic carbon in the filtrate.

Alkaline Bromination

Alkaline bromination procedures were modified from the procedure used by Irving and Cosgrove.[11] A 3 mL sample was placed in a 10 mL Erlenmeyer flask, cooled in an ice bath, and 0.1 g of NaOH and 0.3 mL of pure bromine (Br_2) were added. The solution was allowed to sit at room temperature for 1.5 h, after which it was boiled until the excess bromine boiled off and the solution had a light yellow color, which took about 5 min. Then 1 mL of conc. NH_4OH and 0.1 g of sodium metabisulfate were added. Approximately 0.5 g of Na_4EDTA was added to the sample, which was then filtered through a prewashed paper filter (Whatman).

Lanthanide Shift Reagents

The lanthanide shift reagent, praseodymium ethylenediaminetetraacetate (PrEDTA), was synthesized by adding dropwise 50 mL of 0.2 M praseodymium trichloride (Johnson Matthey) aqueous solution to 50 mL of a 0.2 M Na_4EDTA solution. After addition, the solution was stirred for 45 to 60 min at a temperature of approximately 80 °C. After 24 h, a light green precipitate formed, which was filtered and allowed to dry overnight in a vacuum desiccator. Elemental analysis showed that the molecular formula was $PrEDTA \cdot 9H_2O$. A 0.05 M PrEDTA solution with a 10-fold excess of EDTA to Pr was made. The amount of this solution added to concentrated samples was dependent upon the TSP concentration of the sample and the Pr/P ratio desired.

Results and discussion

^{31}P FT-NMR Spectra of High-molecular-weight Samples

The high-molecular-weight samples consisted of dissolved compounds that were in the molecular size range $0.2 \mu m > x > 30 kDa$. The TSP concentrations for these samples ranged from 12 to 22 mg P/L. Figure 13.1 contains five high-molecular-weight ^{31}P FT-NMR spectra collected over several seasons. Despite similar TSP concentrations, two different spectral patterns are seen. The most common pattern consists of a large broad signal with a maximum at a chemical shift position of approximately 1.00 ppm, indicative of monoester phosphates. These broad signals have shoulders or overlap into the diester phosphate region of 0.00 to −2.00 ppm. High-molecular-weight diester phosphates that previously have been detected in fresh water include DNA[12] and RNA.[13] In addition to monoester and diester phosphates, signals in the region ascribed to phosphonates (20 to 30 ppm) are present in the May 1992 and October 1992 samples and possibly present in the February 1993 sample (Figure 13.2). This is the first time that phosphonates have been detected in lake water. Spiking the May 1992 sample with phosphonoacetic acid and phosphonoformic acid produced signals at 17.0 ppm and 30.0 ppm respectively. Phosphonates originate from cellular membranes of algal cells.[14] The spectrum of the May 1992 spring sample also shows a small signal at −10.0 ppm which, if a signal at all, could indicate the presence of pyrophosphate. ^{31}P FT-NMR signals of pyrophosphate appear in the region of −9.0 to −11.0 ppm, depending upon the sample pH.[15] No orthophosphate has been detected in any of these spectra, which is expected because the SRP concentration was below the ^{31}P FT-NMR detection limit.

The second type of high-molecular-weight spectral pattern has much lower sensitivity, despite similar TSP concentrations with samples exhibiting the large broad signal. These spectra comprise several overlapping signals in the monoester and diester phosphate region. None of these spectra contain signals in the phosphonate or polyphosphate regions. The July 1992 spectrum does contain an orthophosphate signal, which is believed to arise from the hydrolysis of labile organic phosphorus. Previous experimental results[7] illustrate that orthophosphate is not retained by the 30 kDa membrane; therefore, orthophosphate would not be expected to present in any appreciable amount unless by DOP hydrolysis after sample concentration. It is currently unclear

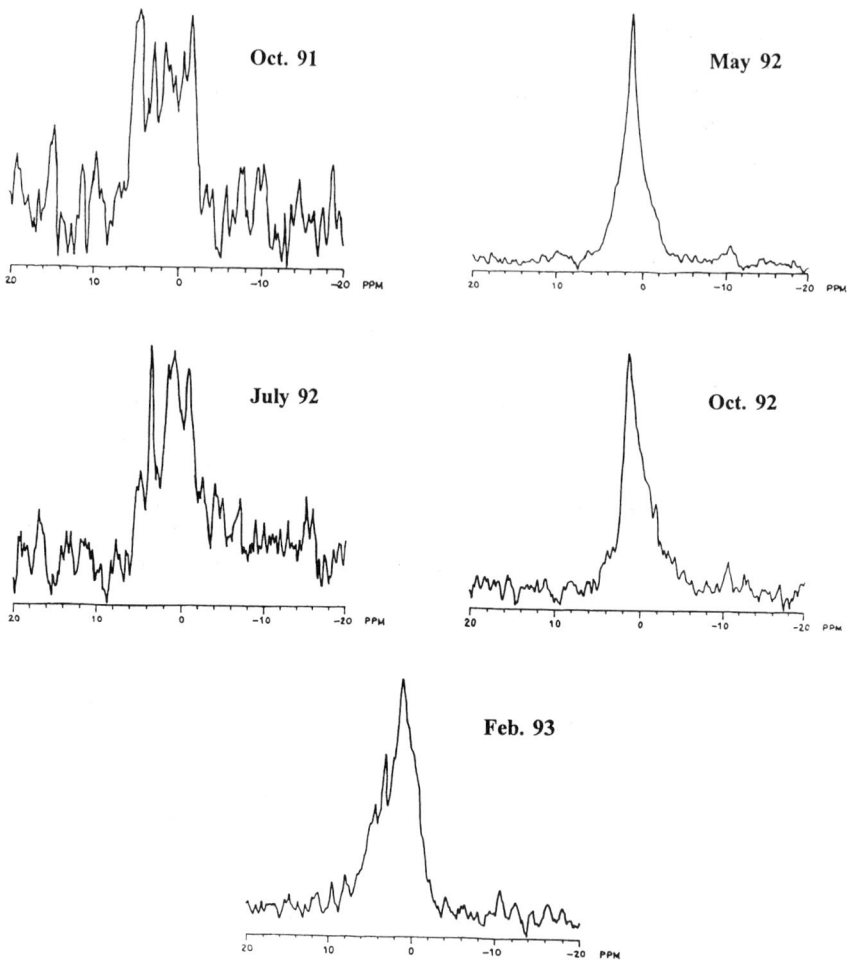

Figure 13.1 Seasonal ^{31}P FT-NMR spectra of high-molecular-weight fractions $(0.2\,\mu m < x < 30\,kDa)$.

why there is a difference in the patterns and sensitivity of the two types of high-molecular-weight spectra.

Intermediate-molecular-weight Samples

Spectra for the intermediate-molecular-weight samples have signals in both the diester and the monoester phosphate region, although the signals in the diester phosphate region are not as intense as in the high-molecular-weight samples (Figure 13.3). Common to all samples is a large broad signal at 1.0 to 1.5 ppm. This is not due to orthophosphate because when orthophosphate is

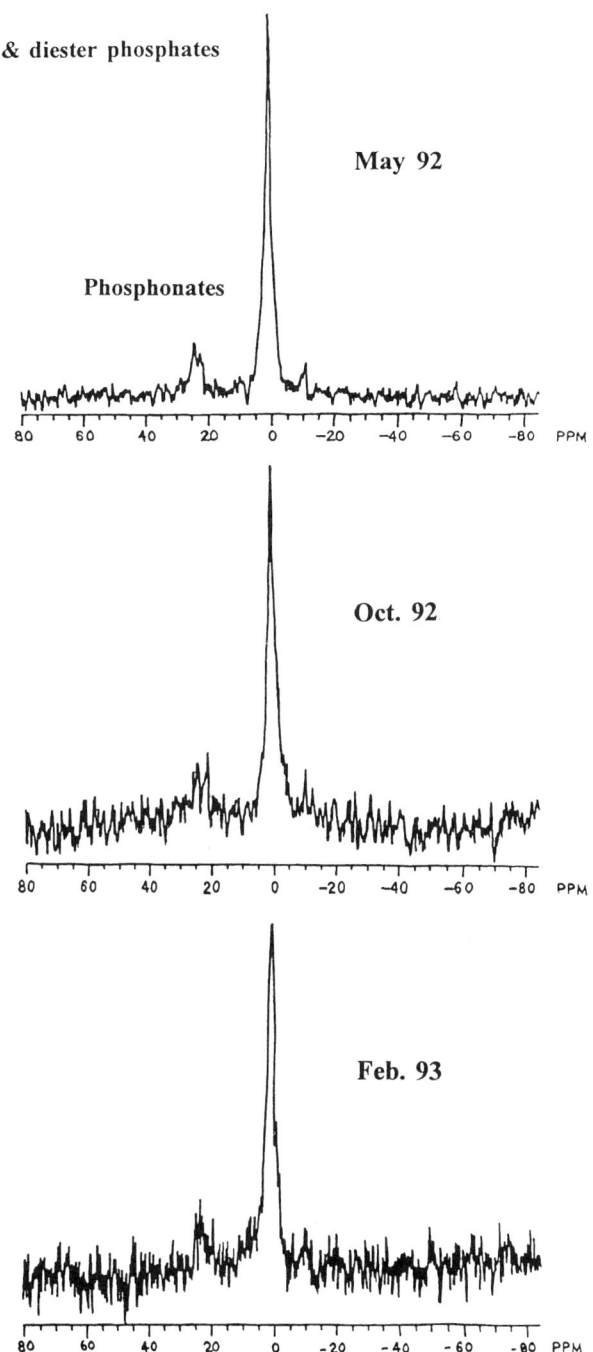

Figure 13.2 Seasonal ^{31}P FT-NMR spectra of high-molecular-weight samples, illustrating signals in the phosphonate region.

229

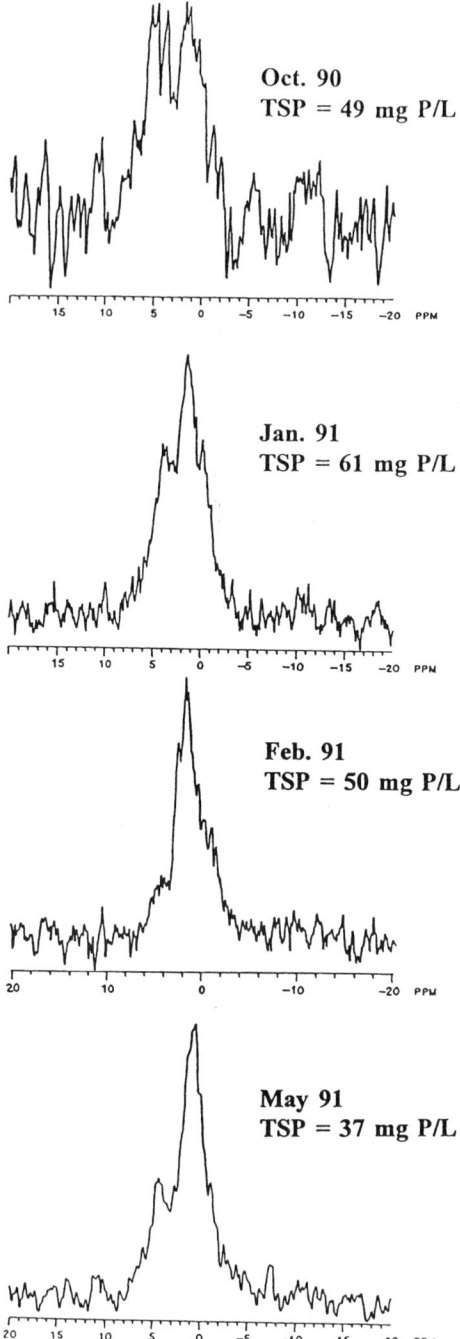

Figure 13.3 Seasonal ^{31}P FT-NMR spectra of intermediate-molecular-weight fractions (30 kDa < x < 1 kDa).

230

spiked into a sample, a signal in the region of 2.0 to 4.0 ppm appears. Also, the position of the orthophosphate signal depends upon the sample pH, and so it is easy to identify the orthophosphate signal once the sample pH is known.[1] A second, smaller signal is present in the chemical shift region of 4.0 to 5.0 ppm, which is the monoester phosphate region. The height of this smaller signal relative to the ubiquitous peak at 1.0 to 1.5 ppm changes with season. For samples taken from early September to mid-January, the 4.00 to 5.00 ppm signal is approximately three-quarters the height of the large broad signal. In mid-February, the 4.00 to 5.00 ppm signal disappeared and only the large broad signal at 1.00 to 1.50 ppm was present (Figure 13.4). At this time, the heavy snow and ice cover over the lake had broken up and the water color changed from a drab olive–brown to a bright green, indicating new algal growth. As the season progressed into spring, the 4.00 to 5.00 ppm signal reappeared and gradually grew in intensity until it was again approximately three-quarters of the height of the large broad signal. Figure 13.4 shows the relative height of the two peaks for 11 intermediate-molecular-weight samples as indicated by the solid line. The height of the signals in the diester phosphate region compared to the height of the large ubiquitous signal at 1.0 to 1.5 ppm is indicated by the broken line in Figure 13.4. The changes in the relative heights were not as dramatic, but a seasonal change was detected at the same time that the change in the 4.00 to 5.00 ppm signal occurred.

If the signals in the region of 4.00 to 5.00 ppm and in the diester phosphate region arose from truly dissolved DOP and the large broad signal at 1.00 to 1.50 ppm was due to DOP that was incorporated into an aggregate or micelle structure, then a simple hypothesis can be formulated about the variations in the relative heights of the signals. In the late winter and early spring when new algal blooms occur, there is a high demand for phosphorus. Orthophosphate will be rapidly consumed and algal cells will need to hydrolyze organic phosphorus in order to obtain additional phosphorus. DOP that becomes incorporated with micelles or aggregates could be more difficult to degrade. Hence, the ^{31}P FT-NMR spectrum during this season only displayed the large broad signal at 1.00 to 1.50 ppm. During the algal growing season, the 4.00 to 5.00 ppm signal appeared, but was still small. The DOP compounds responsible for this signal were probably from DOP that is biologically produced, i.e., DOP that was released during algal and plant cell lysis. In the fall, the 4.00 to 5.00 ppm signal began to increase until it reached the same intensity as the large broad signal. This increase was probably due to the large input of DOP from fallen leaves into the lake. During the winter, the ratio decreased slightly, which was presumably due to the decreased input of DOP from algal cells and leaves and to the loss of DOP absorption to suspended particles and settling out. The behavior of the ratio of the diester phosphate signal to the large broad signal probably had the same causes.

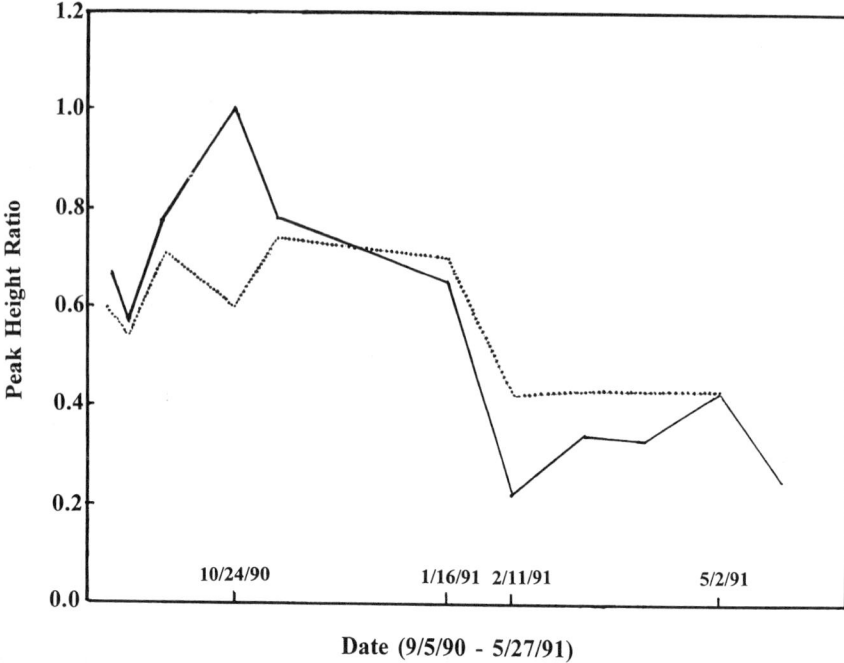

Figure 13.4   Seasonal changes in the relative heights of signals in the interemdiate-molecular-weight fractions. Solid line, 4.0 to 5.0 ppm signals/1.0 to 1.5 ppm signals, broken line, 0.0 to −1.0 ppm signals/1.0 to 1.5 ppm signals.

Low-molecular-weight Samples

It is difficult to obtain ^{31}P FT-NMR spectra of low-molecular-weight samples because TSP concentrations are often near or at the ^{31}P FT-NMR detection limit. Even when the TSP concentrations are adequate, sensitivity is very low. This is believed to be caused by the high salt concentrations that result from the sample being a retentate of an RO membrane. Figure 13.5 presents four ^{31}P FT-NMR spectra of low-molecular-weight samples collected at different seasons. These samples have signals present primarily in the monoester phosphate and orthophosphate regions. Orthophosphate, as expected, is usually a very prominent signal in these spectra. Very little signal is seen in the diester phosphate regions.

Extraction, Isolation, and Degradation Techniques

It is obvious from the ^{31}P FT-NMR spectra presented above that to obtain any further information regarding identity and characteristics of DOP present in lake water, the concentrated samples need to be modified so that the DOP can be freed from matrix interferences. This will allow greater sensitivity and

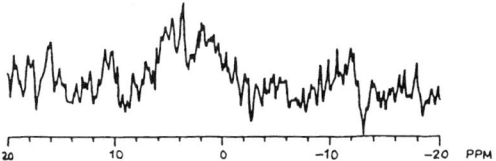

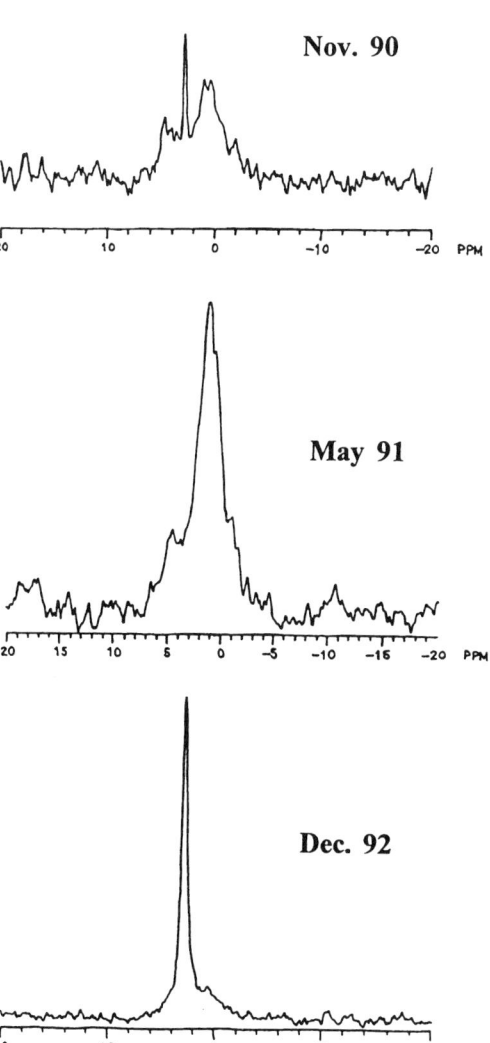

Figure 13.5 Seasonal ^{31}P FT-NMR spectra of low-molecular-weight fractions (1 kDa $< x <$ 300 Da).

better resolution, in addition to enhancing the understanding of the chemistry occurring between the DOP and the highly concentrated lake water matrix. The chemistry that can occur in these concentrated samples is varied and complex because of the multitude of components present. The formation of aggregates from either humic and fulvic compounds, silicate polymers, or clay colloids, with the concurrent surface adsorption and incorporation of DOP into the aggregate interior, could cause loss of sensitivity and line broadening. Figure 13.6 presents the possible mechanisms of interactions between DOP and the aggregates. Hydrogen bonding, metal bridging complexation, and hydrophobic interactions are all possible mechanisms for surface interaction of DOP with aggregates. DOP could become incorporated into the interior of the aggregates by attaching to the aggregate surface during its growth phase and becoming encapsulated as the aggregate increases in size. Micelles are another type of structure that could interact with DOP to reduce spectral quality. Phospholipids could form bilayer micelles which would have phosphate groups in their interior (Figure 13.7). Dissolved organic carbon with surfactant qualities could also be involved in micelle formation with phospholipids. In addition to micelles shielding phosphate groups from the solution environment, they could contribute to line broadening by rigidly holding phospholipids and preventing rotational motion. Another mechanism that could account for poor spectral quality is the formation of very small inorganic phosphate particles that remain suspended in solution. These particles result from the binding of calcium and ferric ions with the phosphate group of DOP. It is thought that the possibility of this mechanism occurring is relatively small because the sample is passed through a cation exchange resin before concentration, replacing soluble cations with Na^+. The low-molecular-weight sample, i.e., the RO retentate, is passed through a second cation exchange after concentration, to further insure that this kind of precipitation will not occur. Finally, the viscosity of the concentrated sample could contribute to the line broadening and sensitivity loss by hindering DOP rotational motion. Thus, to obtain better ^{31}P FT-NMR spectra of concentrated lake samples and to understand the chemistry that is occurring between the DOP and the concentrated sample matrix, a technique or procedure is required to isolate the DOP from the interfering components. The following techniques were examined for this purpose: addition of chelating compounds; XAD-8 extraction; liquid–liquid extraction; freeze-drying followed by solvent extraction; diafiltration; alkaline bromination; and lanthanide shift reagents.

Chelation Compounds

Tetrasodium ethylenediaminetetraacetate (EDTA) and *trans*-1,2-diaminocyclohexane-N, N, N', N'-tetraacetic acid (CDTA) were added to the sample to complex any soluble iron and magnesium ions. Addition of CDTA to adenosine triphosphate solutions has been found to complex magnesium so that

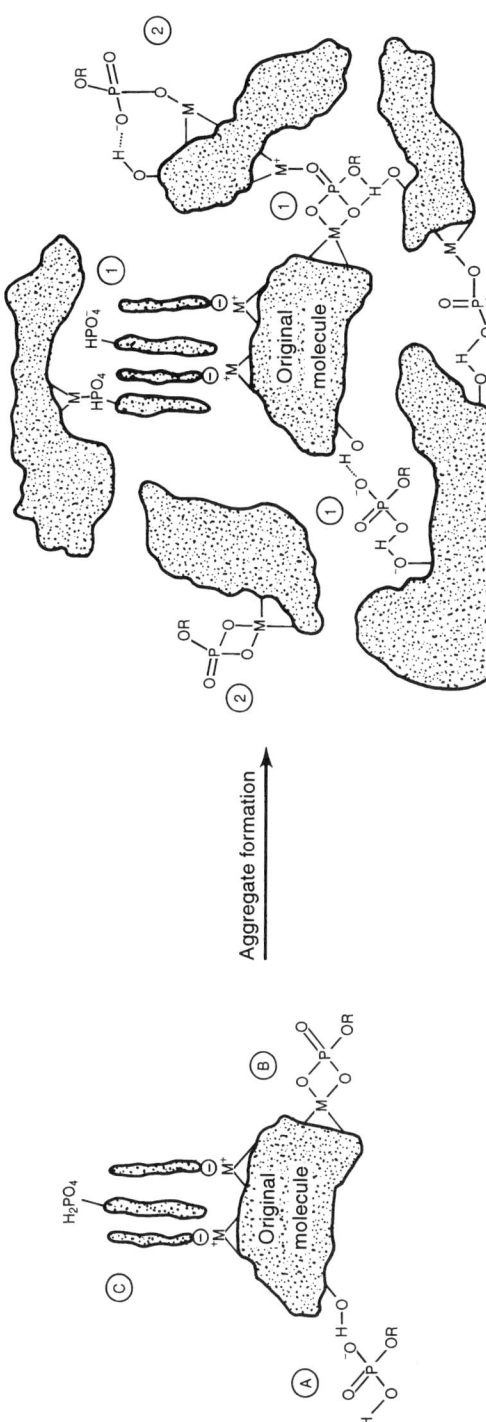

Figure 13.6 Proposed model for the incorporation of organic phosphorus during aggregate growth. Organic phosphorus can interact with humic and fulvic molecules or condensed silicate polymers by three mechanisms: (A) hydrogen bonding; (b) metal ion bridging; and (c) hydrophilic bonding with an amphiphile molecule that is attached to the molecule. After aggregation, interior organic phosphorus molecules (1) are present as well as surface organic phosphorus molecules (2).

235

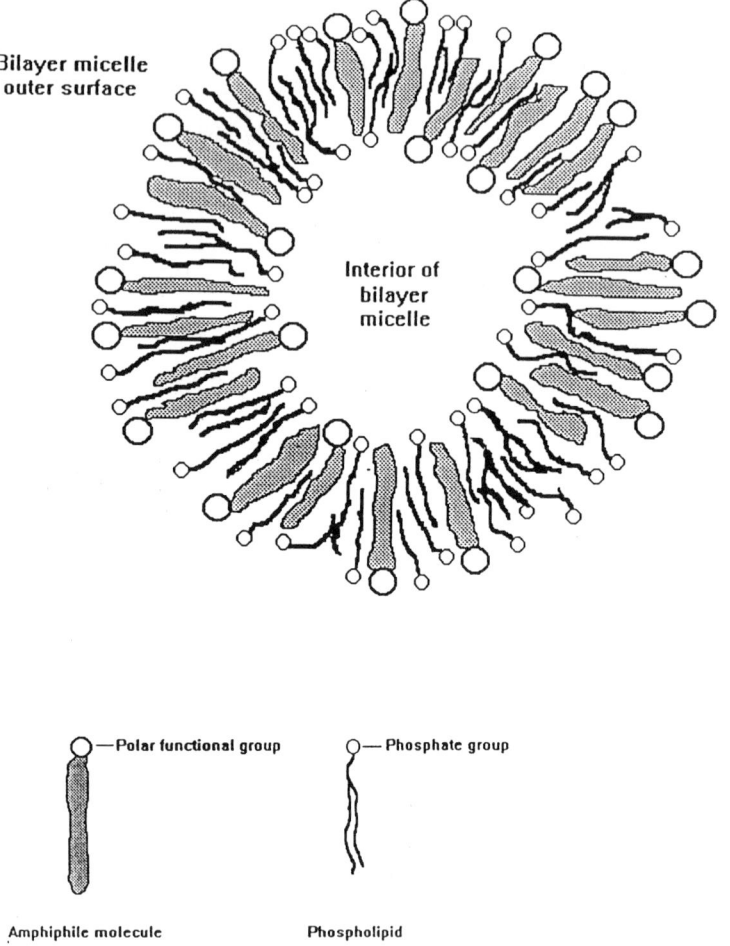

Figure 13.7   Model of bilayer micelle formed from amphiphilic molecules and phosopholipids.

spectral sensitivity is enhanced and the β and γ phosphate groups become visible in the spectrum.[16] No improvement in spectral quality or the appearance of new signals was detected upon the addition of either EDTA or CDTA. This would support the idea that any paramagnetic ions that are interacting with DOP are inaccessible to the complexing agents. An example could be the interaction of a phosphate group with a paramagnetic ion occurring inside an aggregate or micelle. It is also possible that the ferric ions are tightly complexed with the DOP, and EDTA or CDTA is unable to chelate the paramagnetic ion. An example is the interaction of ferric ions with inositol phosphates, which strongly chelate ferric ions.

XAD-8 Resin

XAD-8 resin was used to remove hydrophobic humic and fulvic acids,[9] which could interact with DOP by forming aggregates and micelles and incorporating the DOP into its structure, either by surface adsorption or encapsulation. A major disadvantage of this technique is the potential for acid-induced hydrolysis of the DOP upon decreasing the pH to 2.0. Fortunately, none of the DOP in the concentrated samples from Crystal Lake was hydrolyzed. Even though approximately 70 to 80% of the dissolved organic carbon was removed with XAD-8 and approximately 90 to 95% of the DOP for the high- and intermediate-molecular-weight samples passed through the XAD-8 resin, there was no effect upon the ^{31}P FT-NMR spectrum. This indicates that the DOP does not interact with a majority of the dissolved organic carbon in the sample or that the humic aggregates or micelles that contain DOP are not hydrophobic enough, even at a pH of 2.0, to be absorbed on the XAD-8 resin. The fact that very little DOP in the high- and intermediate-molecular-weight fractions is retained by the XAD-8 resin indicates that phospholipids are not present in these samples.

Organic Solvent Extraction

Organic solvent extractions of the samples with methanol–chloroform mixtures were performed to remove any hydrophobic DOP such as phospholipids from the sample. It was also hoped that the presence of organic solvents would cause the humic/fulvic aggregates to break apart. Even though most of the DOP remained in the aqueous phase, ^{31}P FT-NMR spectra could not be observed. Thin layer chromatography did detect lipids in the organic fraction, but no phospholipids were seen. Extraction with organic solvents illustrates that the DOP present is not composed of phospholipids which could form micelles.

Freeze-drying

Freeze-drying the samples, followed by extraction with various solvents, was used in the hope of isolating the DOP from interfering components in the concentrated sample. The first extraction solvent used was 0.1 M NH$_4$OH. All of the DOP in the lake water samples was extracted into the liquid phase; in fact, all of the freeze-dried material redissolved. The ^{31}P FT-NMR spectrum shows very little difference from the ^{31}P FT-NMR spectrum of the original sample before freeze-drying (Figure 13.8). The next extractant used was a 75% acetonitrile solution with tetrabutylammonium bromide (TBA) present. It was hoped that the phosphate groups of the DOP would form neutral complexes with the tetrabutylammonium cation and then be soluble in the acetonitrile solution. The ^{31}P FT-NMR spectrum of the sample only had a weak signal from orthophosphate (Figure 13.8). It is unclear whether this operation only extracted orthophosphate in the sample, or whether the DOP that was extracted was hydrolyzed to orthophosphate.

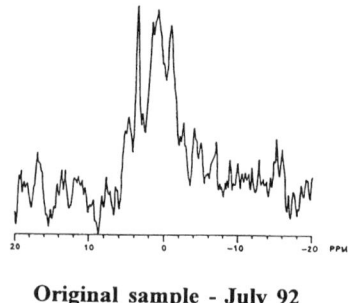

Original sample - July 92

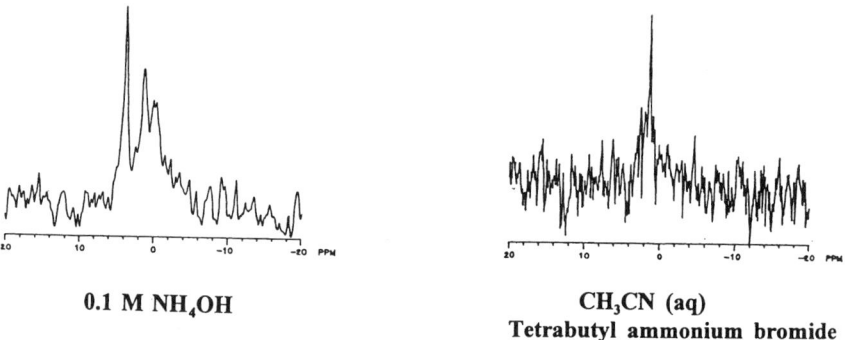

0.1 M NH₄OH

CH₃CN (aq)
Tetrabutyl ammonium bromide

Figure 13.8 ^{31}P FT-NMR spectra of a high-molecular-weight sample after freeze-drying, followed with various extractions (see text).

Diafiltration

Diafiltration of the samples was used to remove dissolved salts and paramagnetic ions from solution. The removal of salts is especially crucial for the low-molecular-weight samples, i.e., the RO membrane retentate. It was also hoped that reducing the ionic strength would disturb or break up aggregates and release bound DOP. The conductivity, total organic carbon concentration, and the TSP and SRP concentrations of the filtrate were measured after each diafiltration. There was no change in the ^{31}P FT-NMR spectrum of the high-molecular-weight sample after diafiltration, indicating that there is no release of DOP from disturbed or disrupted aggregates. The ^{31}P FT-NMR spectra of the two low-molecular-weight samples were different; one had poor sensitivity while the second spectrum consisted of a very sharp, single signal from ortho-phosphate, with a small shoulder in the monoester phosphate region. Unfortunately, removal of the salts from the sample with low sensitivity did not improve the spectrum. The orthophosphate was removed from the sample with the high orthophosphate concentration, leaving the organic phosphorus (Figure 13.9). The ^{31}P FT-NMR spectrum of this sample after diafiltration shows the signals due to DOP clearly. Even though the sensitivity of this spectrum was not improved, it is much easier to analyze the DOP with the signal from orthophosphate greatly decreased in size. Hence, diafiltration can be a valuable tool to remove unwanted orthophosphate from a sample without having to use chemical precipitation methods.

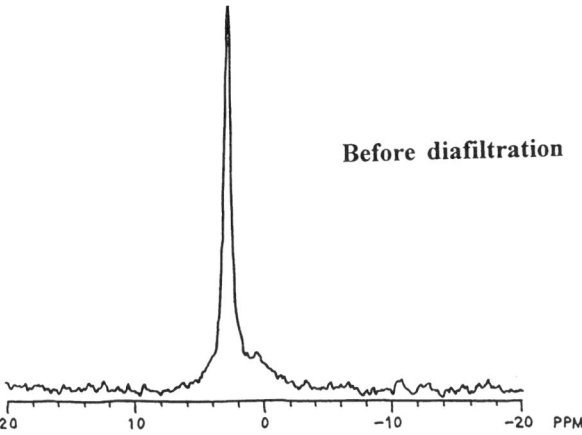

Before diafiltration

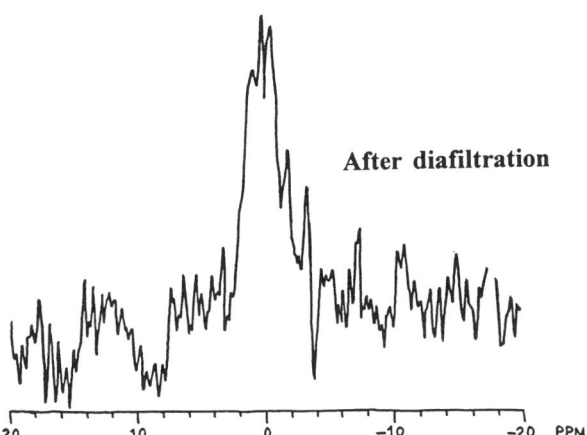

After diafiltration

Figure 13.9 ^{31}P FT-NMR spectra of a low-molecular-weight sample, before and after diafiltration.

Alkaline Bromination

Alkaline bromination is an extreme oxidative degradation technique that oxidizes all DOP to orthophosphate except for DNA[17] and inositol hexaphosphate (IHP).[11] The idea behind its use was to break apart any aggregation and release the incorporated DOP. Using alkaline bromination we hydrolyze a major portion of the DOP, but even so, because IHP and DNA are resistant to hydrolysis, alkaline bromination provides a method to test for their presence. ^{31}P FT-NMR has the ability to distinguish easily between the diester phosphate (DNA) and the monoester phosphate (IHP). A ^{31}P FT-NMR spectrum of a high-molecular-weight sample, collected in the winter of 1992, has a large broad signal with a maximum at 1.42 ppm and two smaller signals at 3.38 and 4.45 ppm (Figure 13.10). The signal at 3.38 ppm is probably due to orthophosphate, based upon the sample pH and the signal position. Alkaline bromination changes this spectrum to two distinct signals, a small broad signal at 0.46 ppm and a signal at 4.43 ppm. Since alkaline bromination degrades all DOP except DNA and IHP, it is presumed that the signal at 0.46 ppm arises from DNA, even though this is a little high for the chemical shift position of a diester phosphate. The signal at 4.43 ppm is probably due to orthophosphate, based upon the sample pH. Examining an intermediate-molecular-weight sample after alkaline bromination reveals a peak in the diester phosphate region, indicative of DNA (Figure 13.10). The middle signal is probably a remnant of whatever gives rise to the original signal, and the largest peak is presumably orthophosphate.

Lanthanide Shift Reagents

Lanthanide shift reagents were used in an attempt to separate overlapping signals. The broad signal in the high- and intermediate-molecular-weight samples could be caused by the overlap of several monoester phosphates whose ^{31}P FT-NMR signals appear close to each other and have become broadened by factors from the sample matrix. Another possibility is that the broad signals are instead due to DOP incorporated into the interior and surface of aggregates. If this is true, the use of lanthanide shift reagents will help distinguish between the surface and the interior DOP by causing the signal position of the surface DOP to change as a function of lanthanide shift concentration, while not affecting the signal position of the interior DOP.

The best results so far are from the use of the lanthanide shift reagent, praseodymium ethylenediaminetetraacetic acid (PrEDTA) with high-molecular-weight samples. Figure 13.11 illustrates two high-molecular-weight samples, one collected in May 1992 and the other collected in October 1992. In the May sample the PrEDTA separates the broad envelope into two signals, and the October sample is split into three signals. Of these signals, one always remains in the original position while the others change position as a function of PrEDTA concentration. This would indicate that a portion of the DOP is isolated from the PrEDTA, i.e., DOP that is incorporated into the interior of

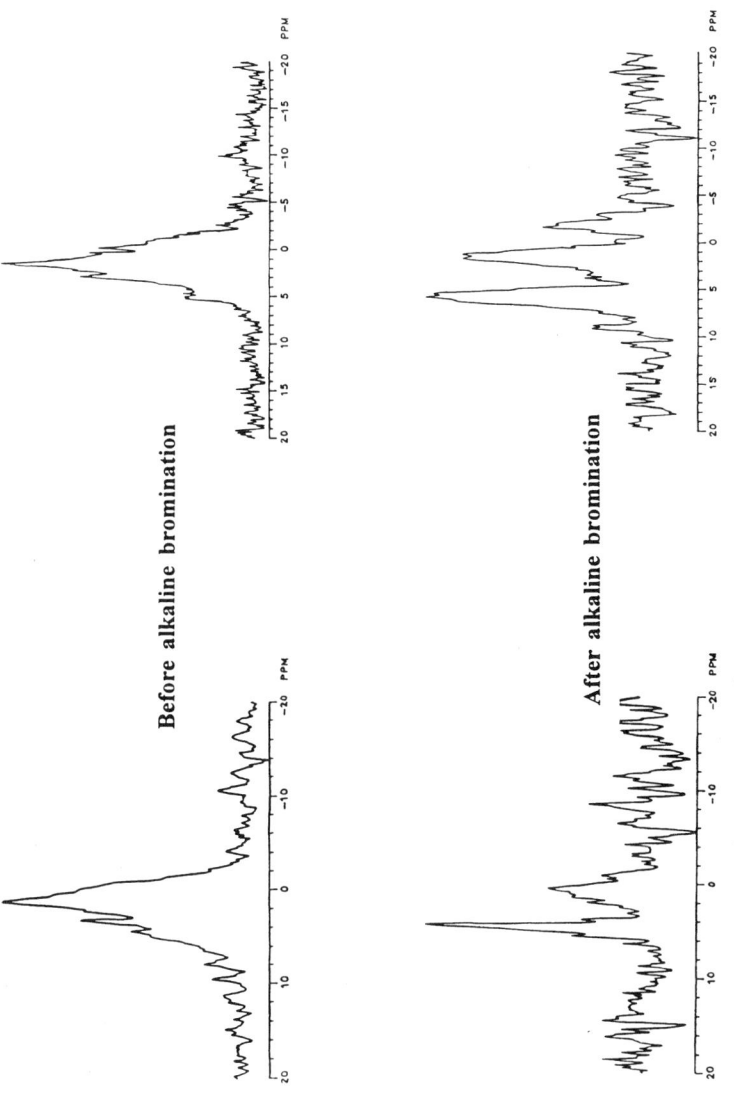

Before alkaline bromination

After alkaline bromination

**High molecular
weight sample**

**Intermediate molecular
weight sample**

Figure 13.10 ³¹P FT-NMR spectra of a high-molecular-weight and an intermediate-molecular-weight sample, before and after alkaline bromination.

241

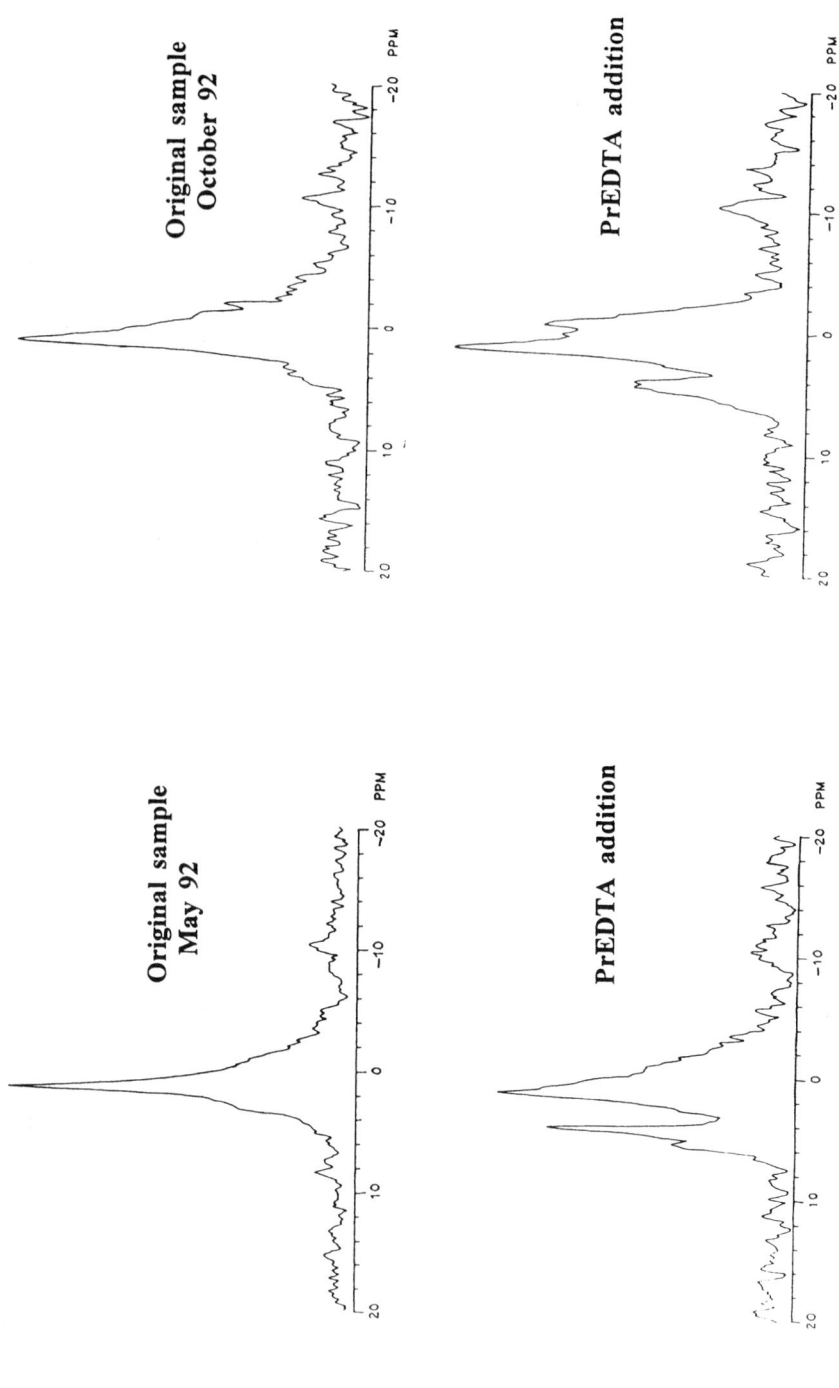

Figure 13.11 ^{31}P FT-NMR spectra of high-molecular-weight samples, before and after the addition of lanthanide shift reagent, praseody-mium ethylenediamine tetraacetate (PrEDTA).

an aggregate. The DOP that gives rise to the signals that change position could be truly dissolved DOP or DOP that is adsorbed to the aggregate surface but can still interact with PrEDTA. Typical Pr/P ratios required to induce signal shifting ranged from 8 to 10.

Conclusion

Even though ^{31}P FT-NMR spectroscopy does not yet provide identification of individual DOP compounds, it still provides a wealth of information about DOP characteristics and its interactions with the concentrated sample matrix. This study has shown that ^{31}P FT-NMR spectra can be obtained for high-, intermediate-, and low-molecular-weight DOP compounds present in lake water and that there are changes in these fractions which are probably due to seasonal changes in the lake. The types of DOP compounds detected in each of the molecular size fractions are as listed:

> High-molecular-weight fraction: Mono- and diester phosphates, phosphonates, and possibly pyrophosphate. Alkaline bromination indicates that some of the diester phosphate DOP could be DNA.
> Intermediate-molecular-weight fraction: Mono- and diester phosphates. Alkaline bromination indicates that some of the diester phosphate DOP could be DNA.
> Low-molecular-weight fraction: Monoester phosphates and orthophosphate.

The detection of phosphonates in the high-molecular-weight fraction is of special interest because this is the first report of phosphonates detected in fresh water.

The major difficulty in the ^{31}P FT-NMR spectral analysis of DOP compounds is the line broadening and low sensitivity that occur. Numerous mechanisms exist which could be responsible for this behavior. To improve spectral quality and also to gain further understanding of the chemistry that occurs between the DOP compounds and the concentrated sample matrix, the samples were treated in several ways to reduce the factors that could possibility be degrading the spectral sensitivity. The addition of chelation compounds to complex paramagnetic ions, use of XAD-8 resin to remove humic and fulvic hydrophobic acids, organic solvent extractions to isolate hydrophobic DOP, freeze-drying of samples followed by solvent extraction, diafiltration to remove salts, degradation of the sample by alkaline bromination, and the use of lanthanide shift reagents to separate overlapping signals were all used on various samples. Based upon the results, it appears that the most likely scenario for the interaction of the DOP with the concentrated matrix is that the DOP interacts with aggregates from humic and fulvic compounds or from condensed silicates, which form polymeric silicate polymers. The DOP can become adsorbed or bonded to these aggregates early in their growth stage so that, as the aggregate grows, the DOP is encapsulated and isolated from the solution. Hydrogen bonding or bonding through a metal bridging atom is likely to be the most probable mechanism between DOP and the aggregate. The fact that the spectra are not greatly improved by adding chelation agents indicates that the paramagnetic ions with which the DOP interacts are probably already complexed, presumably by humic and fulvic complexes or by the surface oxygen atoms on the silicate polymers and clay colloids. Since very little DOP was isolated with the XAD-8 resins and by the organic solvent extractions, it is unlikely that any of the DOP is hydrophobic. Thus hydrophobic bonding with the humic and fulvic compounds is unlikely to be a prevalent mode of interaction. The results from adding lanthanide shift reagents to the sample indicate that some of the DOP is isolated from the solvent. This reinforces the idea that DOP becomes incorporated into aggregates and the remaining DOP is either adsorbed to the aggregate surface or freely dissolved in solution. This hypothesis agrees with the previous results, mentioned at the beginning of this chapter, obtained when the sample pH was varied. Also mentioned in the introductory paragraphs are the results from spiking the samples with known DOP compounds. These results also fit the hypothesis of DOP binding to aggregate surfaces. The idea of DOP compounds and orthophosphate forming complexes with humic and fulvic compounds in the presence of paramagnetic bridging ions such as ferric ions has been proposed and examined extensively by Francko and Heath[18] and Jones et al.[19] using size exclusion chromatography. It has been found that orthophosphate, in the presence of ferric ions, will form phosphate–ferric–humic complexes in the presence of high-molecular-weight humic materials. These phosphate–ferrous–humic complexes break apart in the presence of UV light, producing orthophosphate, ferrous ions, and the humic material.

Future Research

One goal of future ^{31}P FT-NMR research into environmental samples is to increase the sensitivity and resolution so that phosphorus compounds can be easily identified and characterized. Another goal is to use ^{31}P FT-NMR to study the chemical and physical interactions of orthophosphate and DOP compounds with other species present in the aqueous environment. As these goals are approached, a much better understanding of phosphorus cycling in the aquatic ecosystem will be achieved. This in turn will provide better control of eutrophication in lakes and reservoirs, better insight into the use of wetlands for the removal of nutrients, and improved treatment and removal of phosphorus from municipal wastewaters.

   To continue current ^{31}P FT-NMR work with concentrated lake water samples, or any concentrated aqueous sample for that matter, a better concept of the behavior of phosphorus with the concentrated sample matrix is required, in addition to the development of techniques to release bound phosphorus from aggregate structures. Besides alkaline bromination, there are various chemical oxidation methods that could be easily applied to the sample as alternative degradation techniques. One interesting oxidation method might be ozonation of the sample. Photo-oxidation is another technique that may be effective in releasing bound DOP, especially since previous research has found that UV light breaks down phosphate–ferric ion–humic complexes.

However, to avoid interferences from the concentrated sample matrix, isolation or extraction of the DOP before or during the concentration procedure is preferred. This is undoubtedly a very difficult task, considering the low concentration of DOP in most aqueous samples and also that aqueous samples are very complex mixtures composed of ionic compounds similar in chemical nature to the phosphate group, i.e., carboxylic acids, phenolic groups, etc. One potential method for the isolation of DOP is the use of ion-pairing reagents which can form neutral complexes with DOP. Extraction could be achieved by removal with a resin such as XAD-8. A major advantage is that pH values can be kept at neutral values, thus eliminating the risk of acid- and base-induced hydrolysis. Another advantage is that, as neutral complex, the DOP–ion pairing reagent could be dissolved in an organic solvent, which would allow the use of lanthanide shift reagents and T_1 relaxation agents that are insoluble in aqueous solvents. Also, the use of different organic solvents can further influence the interaction behavior of phosphorus with dissolved organic carbon present in the sample.

References

1. Nanny, M. A., and Minear, R. A., Organic phosphorus in the hydrosphere: characterization via ^{31}P FT-NMR. In *Environmental Chemistry of Lakes and Reservoirs*, Baker, L. (ed.), American Chemical Society, Washington DC, 1993, pp. 161–191.

2. Lean, D. R. S., Movements of phosphorus between its biologically important forms in lake water. *J. Fish. Res. Board Can.* 30, 1525–1536, 1973; Francko, D. A., and Heath, R. T., Functionally distinct classes of complex phosphorus compounds in lake water. *Limnol. Oceanogr.* 24(3), 463–473, 1979.

3. Stevens, R. J., and Stewart, B. M., Concentration, fractionation and characterization of soluble organic phosphorus in river water entering Lough Neagh, *Water Res.* 16, 1507–1519, 1982.

4. Minear, R. A., Characterization of naturally occurring dissolved organophosphorus compounds. *Environ. Sci. Technol.* 4, 431–437, 1972.

5. Chrzeszczyck, A., Wishnia, A., and Springer, Jr., C. S., Hyperfine induced splitting of free solute nuclear magnetic resonances in small phospholipid vesicle preparations. In *Mangetic Resonance in Colloid and Interface Science*, Resing, H. (ed.), American Chemical Society, Washington DC, 1976.

6. Henderson, T. O., Glonek, T., and Myers, T. C., Phosphorus-31 nuclear magnetic resonance spectroscopy of phospholipids. *Biochemistry* 13, 623–628, 1974.

7. Nanny, M. A., Kim, S., Gadomski, J. E., and Minear, R. A., Aquatic soluble unreactive phosphorus: concentration by ultrafiltration and reverse osmosis membranes. *Water Res.* 28(6), 1355–1365, 1994.

8. APHA et al., *Standard Methods for the Examination of Water and Wastewater*, 17th edn, American Public Health Association, American Water Works Association, Water Pollution Control Federation, Washington DC, 1989, pp. 437–452.

9. Leenheer, J. A., Comprehensive approach to preparative isolation and fractionation of dissolved organic carbon from natural waters and wastewaters. *Environ. Sci. Technol.* 15(5), 578–587, 1981.

10. Bligh, E. G., and Dyer, W. J., A rapid method of total extraction and purification. *Can. J. Biochem. Phys.* 37(8), 911–917, 1959.

11. Irving, G. C. L., and Cosgrove, D. J., The use of hypobromite oxidation to evaluate two current methods for the estimation of inositol polyphosphates in alkaline extracts of soils, *Commun. Soil Sci. Plant Anal.* 12, 495–509, 1981.

12. DeFlaun, M. F., Paul, J. H., and Davis, D., Simplified method for dissolved DNA determination in aquatic environments. *Appl. Environ. Microbiol.* 52(4), 654–659, 1986.

13. Karl, D. M., and Bailiff, M. D., The measurement and distribution of dissolved nucleic acids in aquatic environments. *Limnol. Oceanogr.* 34, 543–558, 1989.

14. Kittredge, J. S., and Roberts, E., A carbon–phosphorus bond in nature. *Science* 164, 37–42, 1969.

15. Gorenstein, D. G., *Phosphorus-31 NMR, Principles and Applications*, Academic Press, New York, NY, 1984, p. 563.

16. Bass, M. B., and Fromm, H. J., *trans*-1,2-Diaminocyclohexame-N,N,N',N'-tetraacetic acid is superior to ethylenediamine-N,N,N',N'-tetraacetic acid for sequestering Mg^{2+} in ^{31}P NMR experiments involving ATP spectra at neutral and acidic pH. *Anal. Biochem.* 145, 292–301, 1985.

17. Clarkin, C. M., and Minear, R. A., University of Illinois at Urbana—Champaign, unpublished results.

18. Francko, D. A., and Heath, R. T., Abiotic uptake and photodependent release of phosphate from high-molecular-weight humic–iron complexes in bog lakes. In *Aquatic and Terrestrial Humic Materials*, Christman, R. F., and Gjessing, E. T. (eds), Ann Arbor Science, Ann Arbor, MI, 1983.

19. Jones, R. I., Shaw, P. J., and De Haan, H., Effects of dissolved humic substances on the specification of iron and phosphate at different pH and ionic strength. *Environ. Sci. Technol.* 27, 1052–1059, 1993.

14.

Use of ^{31}P NMR in the Study of Soils and the Environment

LEO M. CONDRON, EMMANUEL FROSSARD, ROGER. H. NEWMAN,
PIOTR TEKELY, & JEAN-LOUIS MOREL

Phosphorus (P) is an essential nutrient for plants and animals because of its vital role in energy transformation processes such as photosynthesis and glycolysis. Soil is the primary source of P for plants (and animals), and while native soil P is mainly derived from the mineral apatite ($Ca_{10}(PO_4)_6(OH,F)_2$) present in soil parent material, supplementary P is added in fertilizers. In addition, large quantities of P are present in household and industrial chemicals such as detergents, and as a result organic wastes such as municipal sewage sludge contain significant amounts of various chemical forms of inorganic and organic P.

The biogeochemical cycling of P in soil is determined by a complex interaction of chemical, biochemical and biological processes, which in turn are influenced by a variety of environmental and anthropogenic factors in natural and agro-ecosystems.[1-4] It is clear that P is a key element in the environment; it is indispensable for plant growth, but its release into water bodies such as rivers and lakes can cause significant environmental damage as a result of eutrophication.[5] The detailed chemical nature and associated transformations of P in the soil–plant system, and the fate of native and applied P in particular, must be fully understood in order to maximize the agronomic benefits of P while minimizing any adverse environmental impacts. The latter is particularly important in view of the growing importance of land application as a disposal option for organic wastes such as animal manures and municipal sewage sludge.

The large gyromagnetic ratio of the ^{31}P nucleus and its 100% natural abundance make ^{31}P easy to detect by nuclear magnetic resonance (NMR) spectroscopy,[6] and accordingly NMR has been used to examine P in a wide variety of environments. In particular, NMR spectroscopy has been shown to be a valuable tool for investigating the chemical nature and transformations of P in the soil environment and the associated fate of fertilizers, pesticides and organic wastes.

This chapter includes a brief summary of the use of liquid-state ^{31}P NMR to investigate the chemical nature and cycling of P in the soil–plant system and the fate of selected pesticides. Recent advances in the application of solid-state ^{31}P NMR to study the forms and bioavailability of soil P and selected organic P wastes–fertilizers are presented, and future directions in the use of ^{31}P NMR in soil and environmental research are discussed.

Use of ^{31}P NMR in Soil and Environmental Studies

Liquid-state ^{31}P NMR

The use of liquid- and solid-state ^{31}P NMR in the study of P cycling in the soil environment has been included in recent reviews by Pierzynski,[7] Sanyal and DeDatta,[2] and Magid et al.[4] The main purpose of this section is to review briefly the major aspects of NMR analysis of soil and pesticide P species, and to highlight some recent developments.

Soils

Initial studies conducted by Newman and Tate[8] showed that ^{31}P NMR could be used to identify different P species in concentrated alkali (NaOH) extracts of soils; thus separate signals for orthophosphate inorganic P, orthophosphate monoester organic P, orthophosphate diester organic P, polyphosphate, and pyrophosphate were distinguished in extracts from a range of native grassland soils in New Zealand. Subsequent studies showed that ^{31}P NMR could be used to establish and define quantitative relationships between the chemical forms of P in soils and various management and environmental factors.[11–19] ^{31}P NMR is a more convenient and less tedious method of determining different forms of organic P in soil extracts than the traditional "wet chemistry" techniques such as those employed in the separation and analysis of inositol phosphates (orthophosphate monoester organic P).[9] On the other hand, while NMR can distinguish broad categories of P compounds, the different P species which make up the different categories such as orthophosphate monoester and diester organic P cannot be easily distinguished in most cases.

Numerous workers have used ^{31}P NMR analysis of NaOH extracts to study the chemical nature and cycling of organic P in various natural and disturbed environments. These studies, which have included the impacts of soil type (e.g., soil water regime), climate, vegetation, cultivation, and long-term fertilizer inputs, have clearly shown that orthophosphate diester forms of organic P are mineralized more readily in the soil than orthophosphate monoesters. This general principle has been demonstrated by:

(1) their predominance in soils under cold, wet climatic conditions;[10]

(2) a marked decrease in diester organic P as a result of long-term cultivation;[11,12]

(3) the observed predominance and persistence of monoester organic P in
a variety of soil environments[12–15] and the fact that monoesters made
up most of the organic P which accumulated in pasture soils in
response to long-term P fertilizer additions.[11,16]

These findings indicate that diester forms of organic P are an important
source of P for plants.[15] Furthermore, these studies revealed the presence of
hitherto unidentified soil organic P species, namely phosphonates (C–P) and
teichoic acids (polyribitol phosphates) in acid temperate native grassland soils
and forest litter, respectively.[8,12,17]

Some recent studies have attempted to use milder reagents than sodium
hydroxide to extract "labile" organic P components from soil and to study
their nature and dynamics under different conditions. Thus, Adams and
Byrne,[18] and Adams,[19] used a buffered cation exchange resin (Chelex 20:
pH 7–10) to examine organic P cycling in soils under eucalypt forest in
Australia. These studies found that although the concentrations of P in
resin extracts were considerably lower than those obtained using NaOH,
the extracts could be successfully concentrated by freeze-drying. Results
reported by Adams[19] clearly demonstrated that during forest regeneration
following fire, the inorganic P in the soil was rapidly converted to organic
P, which in turn was dominated by diester P forms. These observations
tentatively confirm that the mild resin treatment extracts a "labile" fraction
of the total organic P from the soil which may be useful in studying short-
term organic P dynamics. Condron and Newman (unpublished data, 1993)
used aqueous extracts with Chelex 20 and freeze-drying to study the effect of
radiata pine *(Pinus radiata)* on organic P in native grassland soils in New
Zealand. Using this technique it was possible to obtain very high total P
concentrations for liquid-state ^{31}P NMR (2500–4500 μg P mL^{-1}). Preliminary
results indicate that despite a dramatic reduction in total soil organic P under
the radiata pine, the relative decreases in orthophosphate monoesters and
diesters were similar in three different soils studied (these investigations are
continuing).

Pesticides

Phosphorus–sulphur (P–S) and phosphorus–nitrogen (P–N) bonds are present in many pesticides (mainly insecticides and herbicides). In liquid-state ^{31}P
NMR, these forms of P resonate at chemical shifts between $+30$ and
$+100$ ppm,[20–22] which means that they can be easily distinguished from
other P species (inorganic orthophosphate, orthophosphate esters, pyrophosphate, polyphosphates, and phosphonates) which generally resonate between
-20 and $+20$ ppm.[8] Several studies have demonstrated that ^{31}P NMR can
been used quantitatively to identify P-containing pesticides and pesticide
residues in various environments and to assess their biological transformations

and associated toxicology.[22-24] For example, Mortimer and Dawson[22] found that ^{31}P NMR could be used to detect very low concentrations (0.5 µg g^{-1}) of organophosphorus pesticides such as diazinon (*O,O*-diethyl *O*-(2-isopropyl-6-methylpyrimidin-4-yl) phosphorothionate), dimethoate (*O,O*-dimethyl *N*-methylcarbamoylmethyl phosphorothiolothionate) and parathion (*O,O*-diethyl *O*-(4-nitrophenyl) phosphorothioate) in commercial food crops. Krolski et al.[24] were able to measure metabolites of sulprofos (*O*-ethyl *O*-[4-(methylthiophenyl)] *S*-propyl phosphorodithioate) in methanol extracts of soil treated with very small amounts of pesticide (7.5 µg g^{-1} soil). The latter study also showed that ^{31}P NMR compared favorably with high-performance liquid chromatography (HPLC) as a quantitative analytical technique for studying the fate of sulprofos in soil; in general, however, further work is required on soil extraction procedures and improving accuracy by the use of appropriate internal standards.

Solid-state ^{31}P NMR

Introduction to CP/MAS

Magic Angle Spinning NMR frequencies depend on the orientation of chemical bonds relative to the static magnetic field of the spectrometer. Solid powders packed with random orientations show statistical distributions of signal strength as in Figure 14.1, trace a. Molecules in solution tumble so fast that only an averaged chemical shift is observed. The effects of random tumbling can be approximated by spinning the sample around an axis inclined at 54.7° relative to the static magnetic field, as in Figure 14.2. A typical magic angle spinning (MAS) frequency is 5 kHz, i.e., 300 000 rpm. This can be achieved with high-performance materials, e.g., sapphire or zirconia, machined to great precision. Samples should be ground to powder for even packing and a well-balanced rotor. The MAS NMR spectrum (Figure 14.1, trace b) shows a peak at a chemical shift which is the mean of the three principal values of a "chemical shift tensor".

MAS NMR spectra show "spinning sideband" (SSB) signals spaced at intervals of the MAS frequency on either side of the centerband (CB) signal. Patterns of intensity resemble the shape of the spectrum of a static sample. The SSB strength can be suppressed to negligible levels if the first-order SSB signals fall outside the limits of the powder pattern.

Cross Polarization Solution NMR spectra are normally obtained with a powerful radiofrequency (rf) pulse followed by a period of data acquisition and a recovery delay before the sequence is repeated. This "single-pulse excitation" (SPE) procedure can be used on solids, but the recovery delay can sometimes be prohibitively long because only limited spin relaxation pathways are available for ^{31}P nuclei in solids.

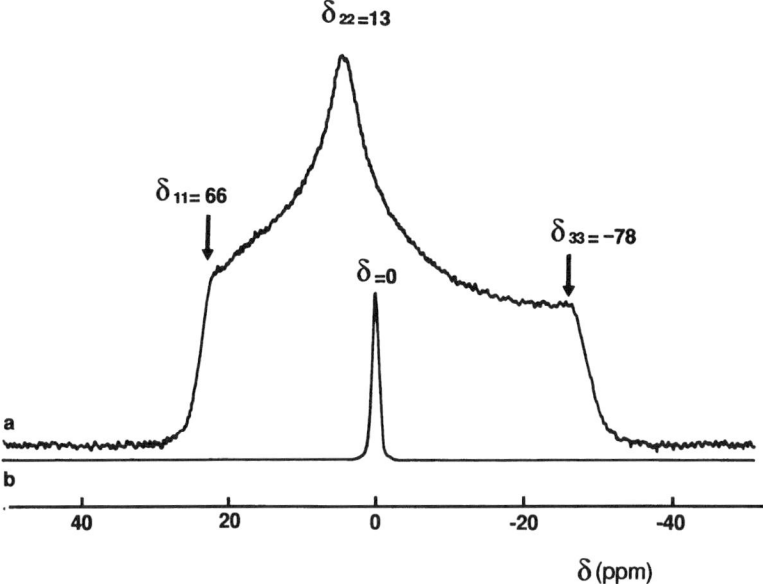

Figure 14.1 Phosphorus-31 CP/MAS NMR spectra of 3-phosphoglyceric acid, barium salt. Trace a, static sample; trace b, with magic angle spinning.

Proton (^{1}H) spin polarization recovers relatively rapidly because of spin diffusion to and from sites at which spin relaxation is particularly efficient. The protons themselves do not move. Spin diffusion involves exchange of spin information between adjacent nuclei. A relaxation mechanism available at just one site in a solid particle could, in principle, result in complete recovery of ^{1}H polarization for the entire particle. The dominant relaxation process for humic substances probably involves paramagnetic centers, e.g., ferric ions.

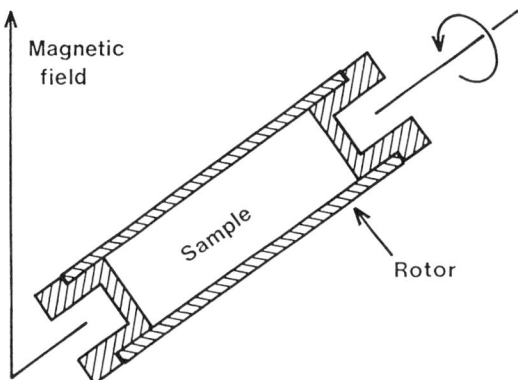

Figure 14.2 Representational cross section of a rotor spinning at the "magic angle."

The "cross-polarization" (CP) technique utilizes the relatively rapid recovery of ^{1}H polarization by transferring that polarization to the ^{31}P spins via dipole–dipole interactions. Figure 14.3 illustrates the process in terms of a hydraulic model. The "valve" between the two upper reservoirs is opened for a "contact time" of the order of 1 ms by subjecting the sample to rf energy generated simultaneously at both the ^{1}H and ^{31}P NMR frequencies. Relative output powers must be precisely adjusted to meet the "Hartmann–Hahn" condition, i.e., the magnetic field strength oscillating at the ^{31}P NMR frequency must be 1.25 times that oscillating at the ^{1}H NMR frequency. Only then will the "valve" open.

While the "valve" is open, polarization is transferred between "reservoirs" with time constant T_{PH}, but polarization is also lost from each "reservoir" with decay time constants $T_{1\rho H}$ and $T_{1\rho P}$. The latter is generally assumed to be so long that loss of P magnetization can be ignored. The overall result is a NMR signal that increases over the time scale of T_{PH} and then decays over the time scale of $T_{1\rho H}$.

Efficient transfer obviously requires $T_{PH} \ll T_{1\rho H}$, and this in turn requires the presence of protons close enough to ^{31}P nuclei for efficient cross polarization. Preliminary experiments on orthophosphate monoesters in a humic acid indicated values of $T_{PH} = 0.2$ ms and $T_{1\rho H} = 1.8$ ms. If these values are typical, then the condition is satisfied. A contact time of 1 ms ensures a reasonable degree of cross polarization without allowing excessive loss of signal strength through rotating-frame relaxation.

The mathematical description of CP for ^{31}P differs from that for ^{13}C because the ^{31}P isotopic abundance is 100% and proton polarization can become depleted in any P-rich domains during the contact time. Consequential effects on spectra of soil have not been fully explored. It might mean, for example, that signals from P-rich minerals such as apatite could be suppressed relative to P distributed at rare sites through domains rich in organic matter.

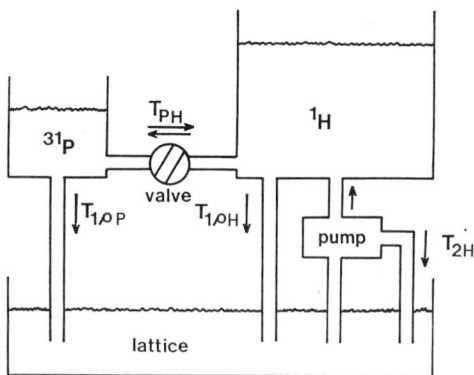

Figure 14.3 Hydraulic model illustrating the principles of cross polarization (depths of liquid in reservoirs represent nuclear spin polarization).

Table 14.1 Chemical Shifts (ppm) for Orthophosphates of Formula M_nH_{3-n} $PO_4.xH_2O^{26}$

Cation	$n = 3$	$n = 2$	$n = 1$
Li^+	10.8	—[a]	—
Na^+	7.8	6.6	2.3
K^+	—	4.3	2.1
Ammonium	—	1.5	0.9

[a] —, Data not available.

Chemical Shifts   Liquid-state NMR frequencies reflect averaging of chemical shifts over the numerous cation–anion encounters that occur during the time scale of an NMR experiment. Solid-state NMR frequencies reflect discrete values for each combination of cations around each type of anion. This means that NMR signals from inorganic orthophosphate P can be spread over a band about 10 ppm wide (Table 14.1). Two trends emerge:

1. The chemical shift decreases with increasing cation size.
2. The chemical shift generally increases as the ionic charge on the orthophosphate anion increases.

Signals can be incorrectly assigned if the pH of the material is not taken into account. Most solution NMR spectra have been run on solutions with high pH, but humic matter can be neutral or even acidic, especially if it has been pretreated by washing with HCl and/or HF. Signals at −2 and −3 ppm in spectra of marine sediments have been assigned to phosphate diesters, e.g, phospholipids,[25] but the signals could just as easily be assigned to organic monoesters, as in Figure 14.1, trace b.

Humic Acids and Soils

The examples shown here were obtained by packing samples in cylindrical rotors of 7 mm external diameter and spinning them at about 5 kHz for ^{31}P NMR at 81 MHz in a Varian XL-200 NMR spectrometer. Each 4 µs proton preparation pulse was followed by a 1 ms contact time, 10 or 20 ms of data acquisition and a recovery delay of 300 or 500 ms. Chemical shifts were referenced relative to the signal from ammonium dihydrogen phosphate, placed at 0.9 ppm.[26]

The spectrum of a humic acid extracted from a Summit Hill soil (Dystrochrept from New Zealand) is shown in Figure 14.4 after averaging 9457 transient signals, requiring an hour of data accumulation. The dominant signal at −0.6 ppm is assigned to orthophosphate monoesters. A weak and broad band centered on 23 ppm is assigned to phosphonates.

Figure 14.5 shows the spectrum of a Pomare soil after averaging 129 728 transient signals, requiring 11 h of data accumulation. The dominant signal at −1.7 ppm is assigned to orthophosphate monoesters. A poorly resolved

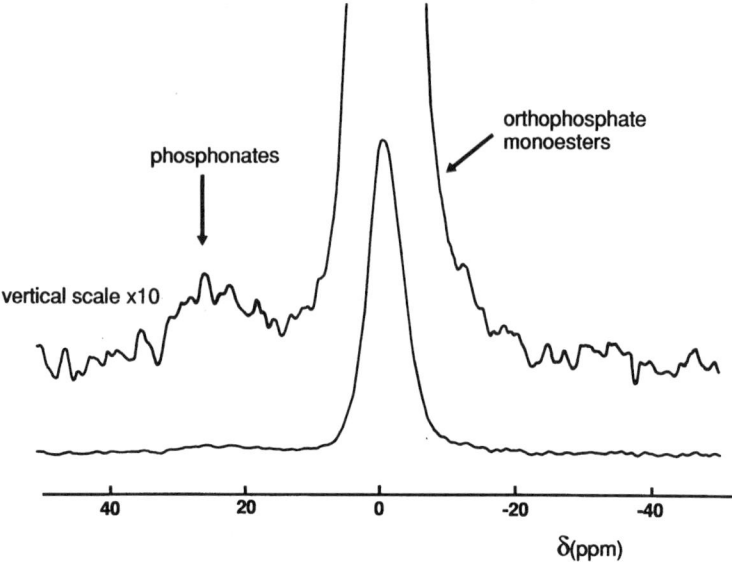

Figure 14.4 Phosphorus-31 CP/MAS NMR spectrum of humic acid extracted from a Summit Hill soil.

shoulder at −11 ppm is assigned to diesters, end groups on polyphosphate chains, plus pyrophosphates or related esters. A weaker band centered on −22 ppm is assigned to chain units in polyphosphates.

The overlapping bands in Figures 14.4 and 14.5 hinder any attempt at a detailed interpretation. One possible approach to simplification is proton spin

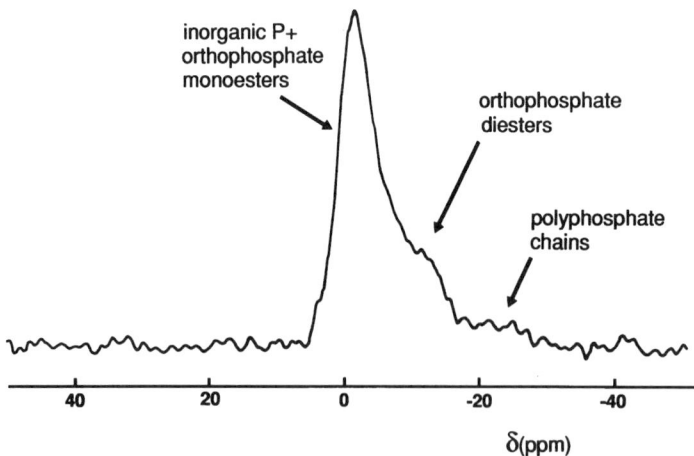

Figure 14.5 Phosphorus-31 CP/MAS NMR spectrum of a Pomare soil under pasture, air-dried and ground to powder but otherwise not pretreated.

relaxation editing (PSRE), exploiting differences in spin relaxation properties between distinct components of a mixture of two or more types of materials. The earliest combination of PSRE and CP techniques required inversion and partial recovery of the proton magnetization.[27] The recovery interval was adjusted until the proton magnetization in one component was just passing through null, thus eliminating the subspectrum of that component from the ^{13}C NMR spectrum. Such a trial-and-error adjustment is not practical when the NMR signals are very weak. A modified version of PSRE requires two spectra to be run and combined in different proportions to achieve separation of subspectra.[28] In the present case, two spectra designated S and S' were run with the normal CP pulse sequence, and with prior inversion and partial recovery, respectively. Linear combinations of S and S' were then adjusted to maximize mutual exclusion of signals from the two subspectra without causing any signal to become inverted. This method has been applied to ^{13}C NMR of soils and humins with some success.[29]

The use of PSRE NMR is illustrated in Figure 14.6. The normal ^{31}P CP/ MAS NMR spectrum of a Pukaki soil (Ustochrept) from New Zealand shows a broad and unresolved band (Figure 14.6(a)). This spectrum was separated into subspectra by generating linear combinations with a second spectrum (not shown) obtained with inversion of proton magnetization and a 10 ms interval for partial recovery before the CP pulse sequence. The linear combinations in Figure 14.6(b) and (c) correspond to selection of signals from domains with proton spin–lattice relaxation time constants of 17 ms and 8 ms, respectively. The band in Figure 14.6(a) has been separated into three distinct signals. The signal at -2 ppm (Figure 14.6(b)) is assigned to inorganic orthophosphate and/or monoesters, the signal at -9 ppm (Figure 14.6(c)) to monoesters and/or diesters, and the weak signal at -24 ppm (Figure 14.6(b)) to polyphosphates.

Ambiguities in signal assignments are inevitable at this stage, because so little is known about which specific cations are associated with each P species. Clues can however be gleaned from comparisons between ^{31}P and ^{13}C PSRE NMR subspectra. The ^{13}C NMR spectrum (not shown) can be separated into subspectra associated with ^{1}H spin–lattice relaxation time constants of 24 ms (38% of C) and 12 ms (62% of C). The former subspectrum shows signals associated with partly degraded organic matter; the latter shows signals associated with more humified matter.[29] The ^{31}P NMR signal at -9 ppm is therefore likely to be associated with relatively inert P species that survive through humification processes. Relatively rapid proton spin relaxation in these domains may be the result of interactions between humic matter and minerals containing paramagnetic ions.

If signal areas in Figure 14.6 are assumed representative of relative amounts of P in each type of domain, then the 935 µg P g^{-1} in the soil can be divided into 355 and 580 µg P g^{-1} associated with domains in which ^{1}H spin relaxation is slow or fast, respectively. This soil was sampled directly under a 15 year-old Douglas fir tree (*Pseudotsuga menziesii*). Soil sampled 2.0

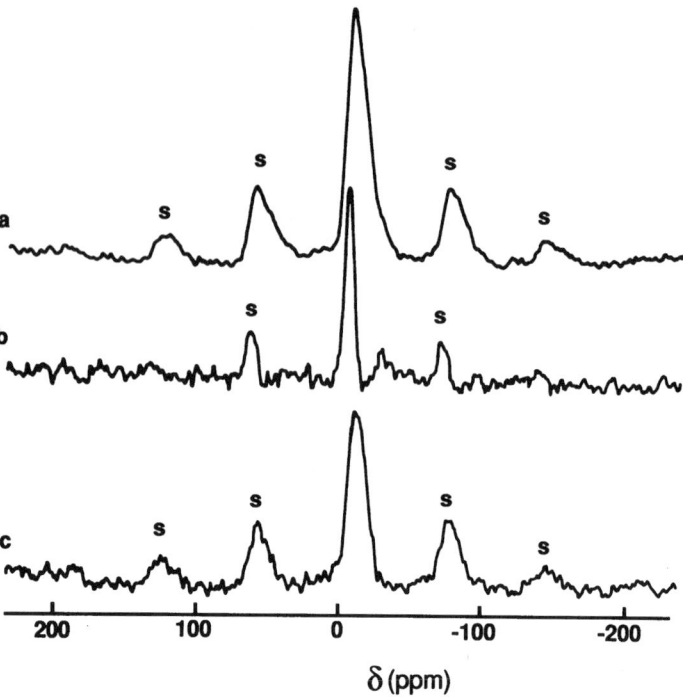

Figure 14.6 PSRE ^{31}P NMR spectra of a Pukaki soil sampled under a Douglas fir tree: (a) total spectrum; (b) subspectrum of domains with $T_{1\rho^H} = 17\,\text{ms}$; (c) subspectrum of domains with $T_{1\rho^H} = 8\,\text{ms}$ (s = spinning sideband peaks).

m away, under a grass/legume pasture, had a similar P content. Relative areas of PSRE subspectra (Figure 14.7) were used to divide this into portions of 147 and 788 µg P g^{-1} associated with domains with ^{1}H spin–lattice relaxation time constants of 19 and 8 ms respectively. Differences between the two soils can be explained in terms of uptake of P into fine roots and associated fungal hyphae (ecto-mycorrhizae) under the tree. Subsequent senescence of the roots and hyphae leaves the P content in the category of organic matter labeled "partly degraded". This interpretation depends on the assumption that relative NMR signal areas are representative of relative amounts of P, an assumption that will require further investigation.

The experiment illustrated in Figure 14.6 required 22 h of spectrometer time. Poor signal response might help to explain why the early ^{31}P NMR studies of soils were confined to fertilized sites.[30] Data averaging times might be shortened by the recent introduction of commercial CP/MAS probes using rotors that can hold larger sample volumes. Ambiguities in signal assignments and uncertainties regarding quantization seem to present more formidable obstacles to routine use of ^{31}P NMR in characterization of soils.

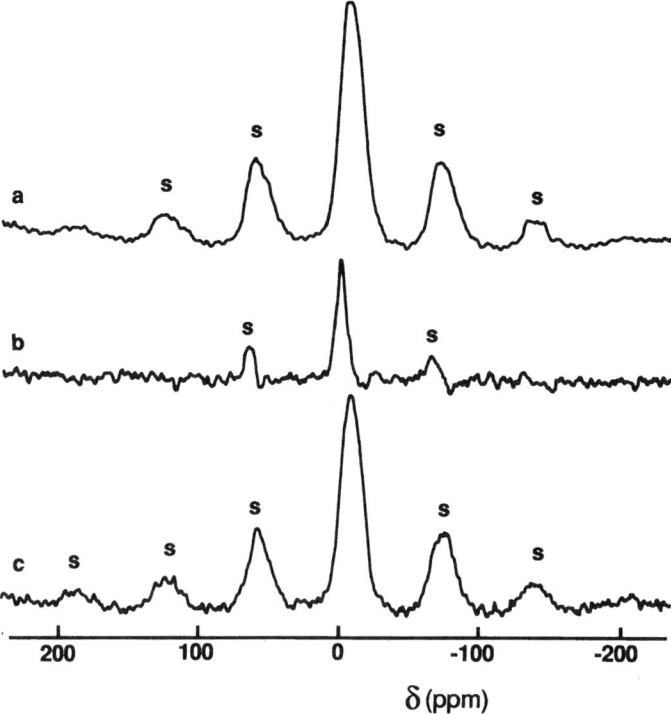

Figure 14.7 PSRE ^{31}P NMR subspectra of a Pukaki soil sampled under grass/legume pasture: (a) total spectrum; (b) subspectrum of domains with $T_{1\rho^H} = 17$ ms; (c) subspectrum of domains with $t_{1\rho^H} = 8$ ms (s = spinning sideband peaks).

Organic P Fertilizers

Introduction Phosphorus is a key element for plant growth and development, and the application of P fertilizers to food and fiber crops is a vital component of sustainable agro-ecosystems. However, when fertilization is provided with water-soluble P, orthophosphate ions react strongly with soil colloids; this reduces the immediate bioavailability of the applied P. An alternative is to apply P with organic amendments. This may increase the recovery of the added P by crops, because some organic and orthophosphate ions may compete for similar sorption sites on soil colloids. The application of P and organic matter can be made either with organic wastes, such as urban (e.g., sludges, composts) or agricultural wastes (e.g., animal manure), or with synthetic organophosphorus fertilizers. However, the forms of P in these fertilizers are not well known and solid-state ^{31}P NMR can be a useful tool to characterize them. The use of solid-state ^{31}P NMR will be shown: (1) for the characterization of the P species present in three urban sewage sludges, and

(2) for the characterization of an organo-P fertilizer ("chitosan–polyphosphate complex") and for studying its fate in the soil environment.

Sewage Sludge Due to their high P content, sewage sludges are potentially valuable P fertilizers.[31] However, very large differences in their efficiency as fertilizers have been reported,[32] which may be related to the formation of different P species during the sludge treatments.[31,33,34] Sequential extraction[35,36] and solubility curves[37–39] have been used to characterize P species but these methods are indirect and their use is hampered by a number of limitations, including an inability to describe properly the phosphate minerals in soil.[7] The use of classical methods such as X-ray diffraction or scanning electron microscopy is precluded by the low concentration of P, the poor crystallinity of the P minerals, and the small size of crystallites encountered.[40]

Solid-state ^{31}P NMR allows for the identification of P minerals in complex mixtures by: (1) comparing the spectrum characteristics (i.e., the δ_{iso} chemical shift and shape of the spectrum) to the spectra of known P species, and (2) comparing spectra obtained in single-pulse (SP) or cross-polarization (CP) sequences. The SP sequence allows for a nonselective observation of all ^{31}P nuclei while the CP sequence allows only for the observation of ^{31}P nuclei located in the close vicinity of ^{1}H nuclei.[41,42] Preston et al.[43] obtained ^{31}P CP/MAS NMR spectra of fish composts. However their spectra were poorly resolved, probably because of a low P content (between 1.6 and 2.3%) and of the large number of P species present. Using ^{31}P MAS NMR with single-pulse sequences, Hinedi et al.[40] detected calcium phosphates, aluminum phosphates, and pyrophosphates in an anaerobically digested and alum-flocculated sludge, and calcium phosphates in an anaerobically digested and windrow-composted sludge.

The objective of this section is to show that the use of single-pulse and cross-polarization sequences allows for a better interpretation of the solid-state NMR spectra of selected sewage sludges. The P species thus identified are discussed in the light of the sludge treatments and of their agronomic efficiency as P fertilizers.

Three urban sewage sludges from different regions in France were studied: an anaerobically digested and heat-treated (at 200 °C and 18 bar for 45 min) sludge dewatered on drying beds (Paris–Achères); an activated sludge (Briare); and an anaerobically digested sludge flocculated with an organo-cationic polymer and dewatered by suction under vacuum (Nancy). The sludges were air-dried and ground ($< 250 \mu m$) before analysis (Table 14.2).

High-resolution solid-state ^{31}P nuclear magnetic resonance spectra were recorded on a Bruker MSL 300 spectrometer ($B_0 = 7.04$ T). Samples were introduced in a zirconium oxide (ZrO_2) rotor and spun at the magic angle (54.74°) at macroscopic speeds between 6.0 and 7.0 kHz (resonance frequency, 121.494 MHz). Two pulses sequences were used; the first was a single-pulse (SP) sequence which excited all the nuclei nonselectively; while the second, a standard cross-polarization (CP) sequence,[41] consisted in the

Table 14.2 Selected Chemical Characteristics of Sewage Sludges

	Paris–Achères	Briare	Nancy
pH	7.1	6.1	7.2
Organic matter $(mg\,g^{-1})$	331	308	427
Total N $(mg\,g^{-1})$	13.4	22.2	26.2
Total P $(mg\,g^{-1})$	33.7	9.28	18.4
Olsen P[a] $(mg\,g^{-1})$	0.38	0.87	0.46
Total Ca $(mg\,g^{-1})$	10.55	5.65	9.86
Total Fe $(mg\,g^{-1})$	18.02	9.76	21.96

[a] Olsen P: mineral P extracted with 0.5 M $NaHCO_3$ at pH 8.5.

transfer of magnetization from the ^{1}H system to the ^{31}P system with a contact time of 1.0 ms. The CP experiment allowed for the observation of ^{31}P located within a few ångströms from ^{1}H.[41] Chemical shifts (δ ppm) were measured relative to an external signal for aqueous H_3PO_4 (14 M). In all cases a high-power proton decoupling was applied during the acquisition time. Chemical shifts were reproducible within 0.1 ppm. Linewidths were measured at half-height of the isotropic peaks.

For the Paris–Achères sludge, a large signal was obtained with the single-pulse sequence yielding a chemical shift (δ_{iso}) at $+3.0$ ppm and a small shoulder located between 0 and -5 ppm (Figure 14.8); the dominant signal was also observed with the cross-polarization sequence. This indicated that one family of P species dominated in this sludge and ^{1}H nuclei were present within a few ångströms from the ^{31}P. This chemical shift was ascribed to carbonate apatite $[Ca_{10}(PO_4, CO_3)6_2^F]$, hydroxyapatite $[Ca_{10}(PO_4)_6(OH)_2]$, octacalcium phosphate $[OCP, Ca_8H_2(PO_4)_6 \cdot 5H_2O]$ or to amorphous calcium phosphates.[42,44] The shoulder located between 0 and -5 ppm on the SP spectrum showed the presence of a small proportion of ^{31}P nuclei far from the ^{1}H nuclei which may be dehydrogenated pyrophosphates.[45] The half-height linewidth of the isotropic peak obtained in the SP sequence was 0.8 kHz and almost reached 1 kHz in the CP sequence (Table 14.3). This linewidth is much greater than those reported in the literature for pure calcium phosphate (CaP) minerals.[42] This should be related to: (1) the structural inhomogeneity of these ^{31}P nuclei (i.e., a mixture of OCP and apatites is present), (2) the presence of poorly ordered minerals, and (3) the presence of paramagnetic impurities (Fe^{3+}—Table 14.2).

In the SP sequence (Figure 14.9) of the Briare sludge a single large signal was obtained with a δ_{iso} of $+1.5$ ppm which was attributed to brushite $[CaHPO_4.2H_2O]$.[44] A narrower resonance signal was observed using the CP sequence. This suggested that a significant part of the ^{31}P nuclei giving a large resonance signal between 1 and 10 ppm was located far from ^{1}H nuclei. This resonance region was attributed to dehydrogenated CaP such as fluorapatite $[Ca_{10}(PO_4)6_2^F]$ or tricalcium phosphate $[Ca_3(PO_4)_2]$.[42,46]

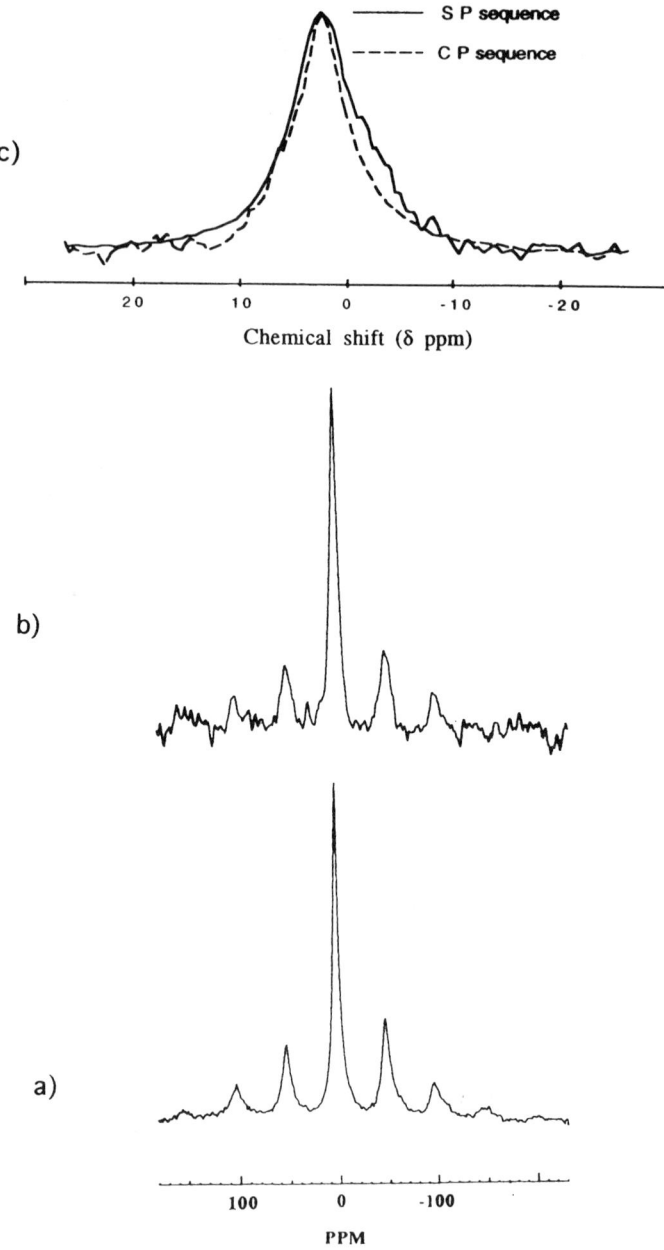

Figure 14.8 High-resolution solid-state ^{31}P NMR spectra of the Paris–Achères sludge obtained by: (a) a single-pulse (SP) experiment (repetition time DO = 20 s; number of scans NS = 624; macroscopic rotation speed RS = 6.0 kHz); (b) a cross-polarization (CP) sequence (DO = 10 s; contact time CT = 1 ms; NS = 5416; RS = 6.0 kHz); (c) enlargement of the central part of the SP and CP spectra.

260

Table 14.3 ^{31}P NMR Chemical Shift and Half-height Linewidth Data for Sewage Sludges

Sludges	Sequence	Chemical Shift, δ_{ISO} (ppm)	Half-height Linewidth (kHz)
Paris–Achères	SP	3.0	0.8
		0 to −5 (shoulder)	
	CP	3.0	1.0
Briare	SP	1.5	1.5
		10 to 1 (shoulder)	
	CP	1.5	0.8
Nancy	SP	2.4 to −4.0	2.3
	CP	−2.0	1.7

For the Nancy sludge, two broad peaks were obtained with the SP sequence with chemical shifts at $+2.4$ ppm and -4.0 ppm (Figure 14.10). With the CP sequence, a relatively narrow peak at -2.0 ppm dominated. As in the previous cases, the very large linewidths observed for this sludge in both experiments (CP and SP) resulted from the presence of different P species. The ^{31}P nuclei resonating at -2.0 ppm in CP experiment were attributed to monetite $[Ca(H_2PO_4)_2]$.[42] The peak at δ_{iso} $+2.4$ ppm should belong to dehydrogenated CaP such as fluorapatite or tricalcium phosphates. The peak at -4.0 ppm was interpreted tentatively as belonging to dehydrogenated pyrophosphates.

The results described above show that ^{31}P NMR allowed for the observation of calcium phosphates and pyrophosphates. These phosphate species may have formed during the sludge treatment. Lucas[47] showed that, in the presence of Ca, the microbial breakdown of organic P resulted in the precipitation of apatite. However, the abundance of organic compounds in the sludges probably limited the precipitation and growth of apatite crystals.[48,49] This is consistent with the presence of OCP and apatites observed in the three sludges. The acidic conditions prevailing in Briare hampered most of the transformations of brushite to OCP or apatites.[50,51] Most of the pyrophosphates present in biologically digested sludges may be related to the activity of microorganisms which synthesize them when in the presence of high orthophosphate concentrations.[52]

Zhang et al.[33] recently published data on the P fertilizing value of these three sludges (i.e., Paris–Achères, Briare, and Nancy). A test crop (perennial ryegrass, *Lolium perenne*) was grown for four months in a clayey soil (clay 46.5%; pH 7.7) and a loamy soil (clay 26.3%; pH 7.1) amended with the sludges. The quantity of P derived from the sludge and taken up by the crop (i.e., the percentage of sludge-P recovered by the crop) was measured. This experiment showed that Achères was the poorest source of P for *Lolium perenne* while Briare was the best (Table 14.4).

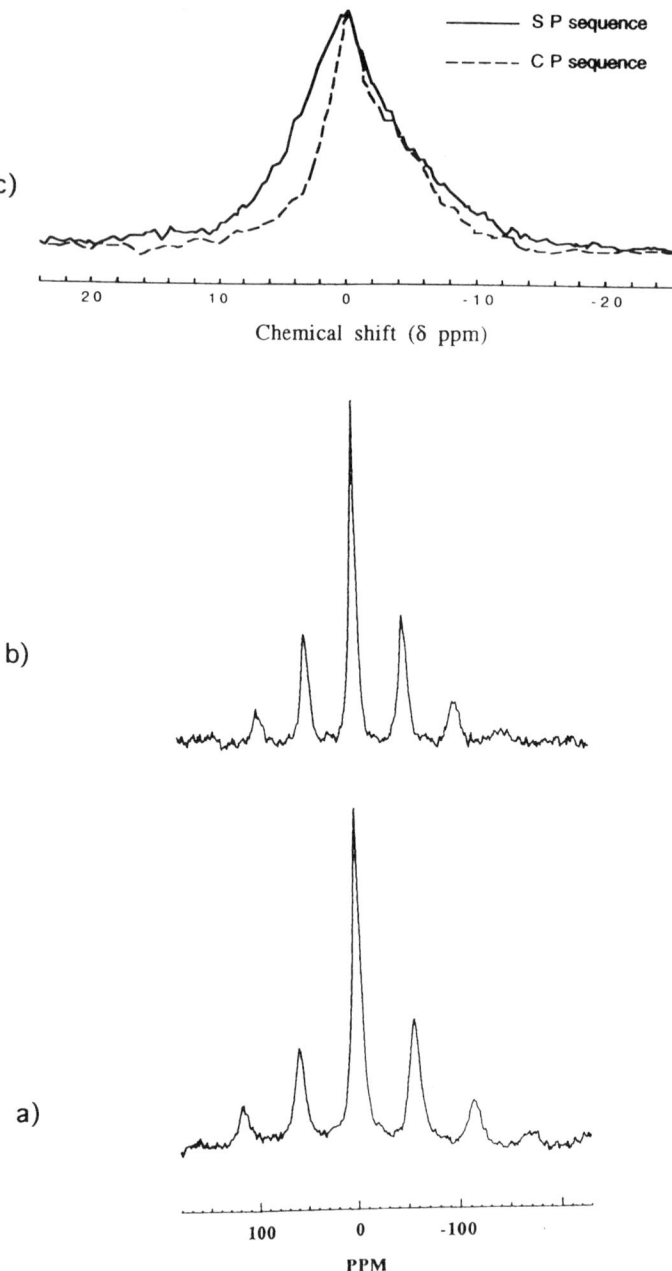

Figure 14.9 High-resolution solid-state ^{31}P NMR spectra of the Briare sludge obtained by: (a) a single-pulse (SP) experiment (DO = 30 s; NS = 2140; RS = 7.0 kHz); (b) a cross-polarization (CP) sequence (DO = 10 s; CT = 1 ms; NS = 6380; RS = 5.9 kHz); (c) enlargement of the central part of the SP and CP spectra.

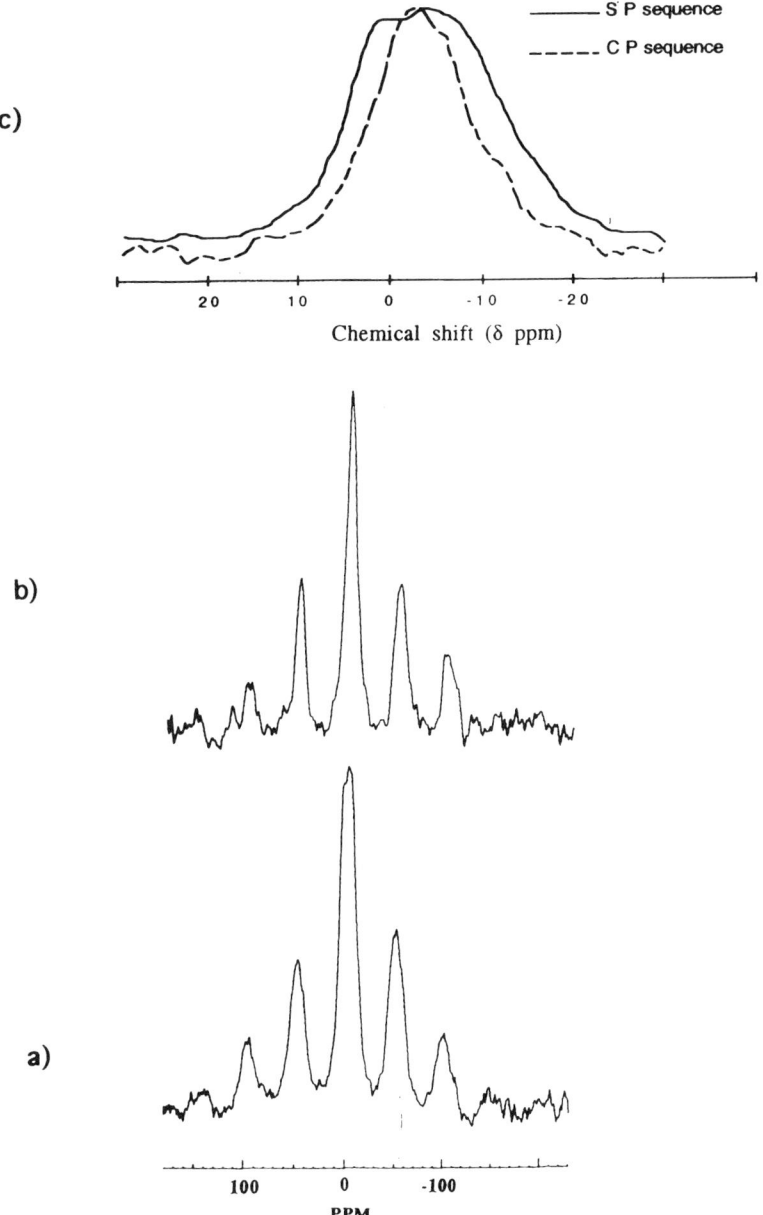

Figure 14.10 High-resolution solid-state ^{31}P NMR spectra of the Nancy sludge obtained by: (a) a single-pulse (SP) experiment (DO = 30 s; NS = 1834; RS = 6.0 kHz); (b) a cross-polarization (CP) sequence (DO = 10 s; CT = 1 ms; NS = 8492; RS = 6.0 kHz); (c) enlargement of the central part of the SP and CP spectra.

263

Table 14.4 Percentage of Sludge-P Recovered by *Lolium perenne* in a Clayey and a Loamy Soil after Four Months of Growth[a]

	Loamy Soil (% of added P)	Clayey Soil (% of added P)
$Ca(H_2PO_4)_2$	6.6a	11.6a
Paris–Archères	2.1b	2.9c
Nancy	3.2b	7.4b
Briare	2.8b	8.2b

[a] Sludges were mixed with soils at the rate of $87\,mg\,P\,kg^{-1}$ soil.
Data followed by different letters are statistically different at the 5% level (for vertical comparison only).

These results are consistent with the forms of P identified with solid-state ^{31}P NMR spectroscopy. Octocalcium phosphate and apatites, the prevalent P species in Achères, are not water-soluble and need an acidic environment to be dissolved. This explains the very low efficiency of Achères in neutral soils. On the other hand, brushite is much more soluble than OCP or apatite.[50] The large proportion of brushite in Briare may account for the relatively high P fertilizing value of this sludge. The presence of monetite and pyrophosphate in Nancy also account for its high P fertilizing value since monetite is soluble[50] and polyphosphates have been shown to be as good a source of P as water-soluble fertilizers.[53]

Chitosan–Polyphosphate Complex Frossard et al.[54] synthesized an organo-P fertilizer, the "chitosan–polyphosphate complex" (CH–PP). This fertilizer resulted from the precipitation of chitosan, a deacetylated derivative of chitin, with a mixture of Na-pyro- and Na-metaphosphates. Chitosan-polyphosphate had a P fertilizing value equivalent to that of monocalcium phosphate. Solid-state ^{31}P NMR was used to characterize this complex and study its fate during a soil incubation.

The CH–PP complex contained $273\,g\,C\,kg^{-1}$ and $147\,g\,P\,kg^{-1}$. Its ^{31}P spectra are presented in Figure 14.11 and the attribution of the chemical shifts is given in Table 14.5 according to the references published by Griffiths et al.[45] Phosphate was present in CH–PP as a mixture of pyro- and metaphosphate physically entrapped within sheets of chitosan.[54] The fate of CH-PP was studied during an 8-week incubation in which it was added to a loamy soil at the rate of $20\,g\,P\,kg^{-1}$ soil. The incubation of CH–PP in the loamy soil produced changes in P species as revealed with solid-state ^{31}P NMR (Figure 14.12). At the beginning of the incubation, resonance signals were observed at δ_{iso} of 0, −7 to −10 and −23 ppm. They were attributed respectively to pyrophosphates (0 to −10 ppm) and to metaphosphates (−23 ppm). Their relative intensities were different than those observed with the initial complex (Figure 14.11). In the soil, P_2O_7 gave a dominant signal while P_6O_{18} remained a minor compound. This was probably related to the presence of

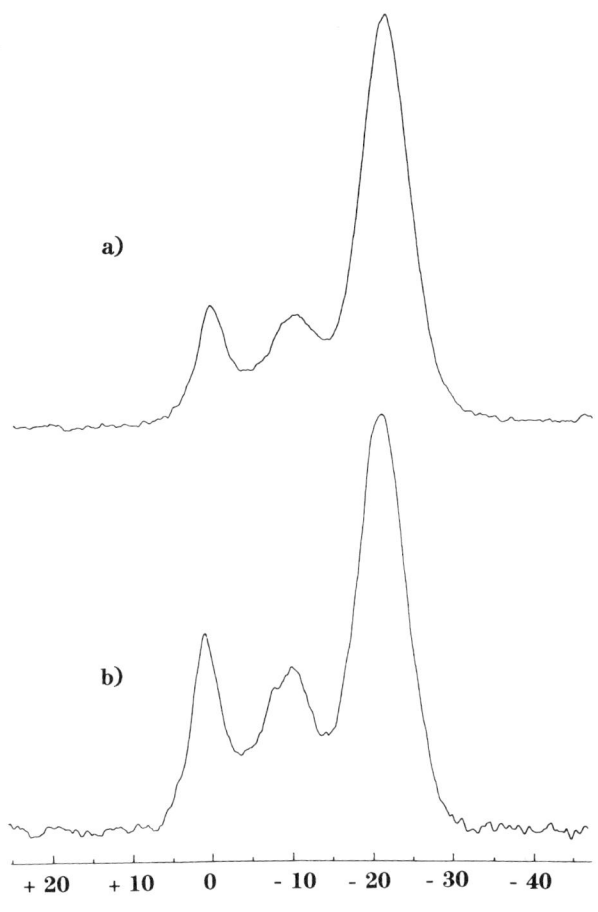

chemical shift ppm

Figure 14.11 Enlargement of the central part of high-resolution solid-state ^{31}P NMR spectra of the chitosan–polyphosphate (CH–PP) obtained (a) with a single-pulse experiment (DO = 30 s) and (b) with a cross-polarization sequence (CT = 1 ms). The macroscopic rotation frequency was 8.0 kHz. Number of scans = 20 (redrawn from Frossard et al.[54]).

Table 14.5 ^{31}P NMR Isotropic Chemical Shifts (δ_{iso}) Obtained in the Single-Pulse Experiment on Chitosan–Polyphosphate (CH–PP)

	P Species			
	$Na_4P_2O_7$	$Na_3HP_2O_7$	$Na_2H_2P_2O_7$	$Na_6P_6O_{18}$
CH–PP	−0.4	−8.0 (shoulder)	−9.9	−21.7

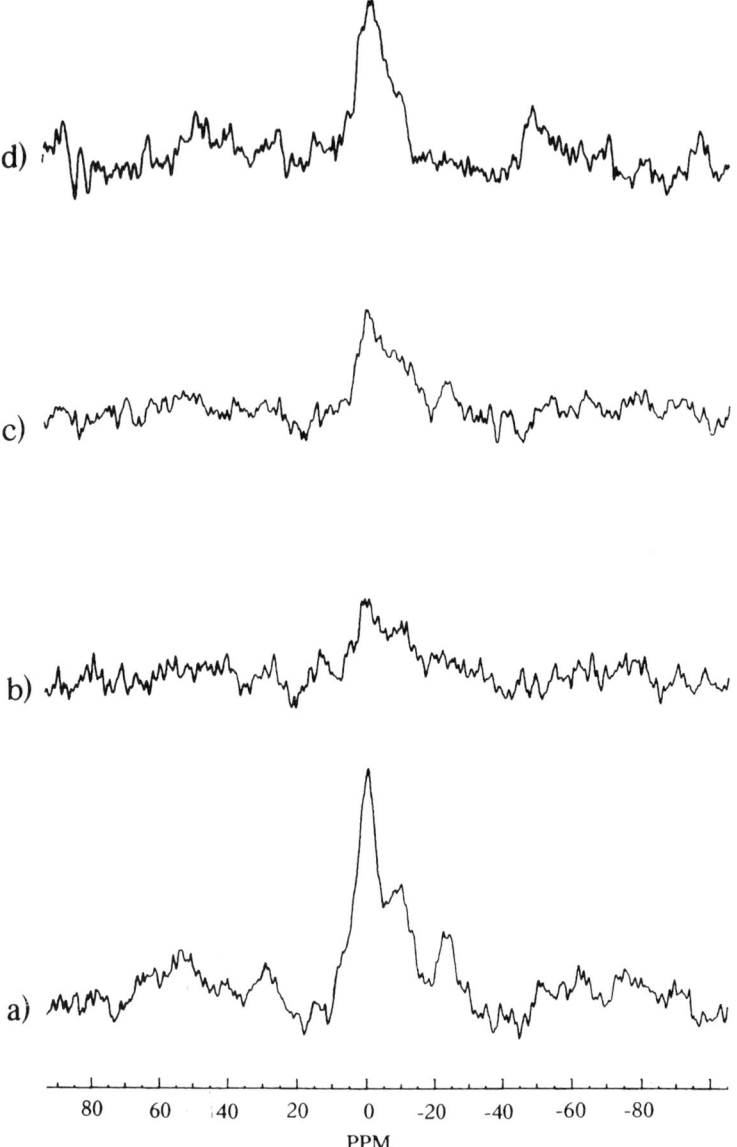

Figure 14.12 Changes in the structure of the chitosan–polyphosphate (CH–PP) after various incubation periods in the loamy soil monitored by ^{31}P solid-state NMR (SP sequence; DO = 60 s, spinning frequency = 6.1 kHz). (a) 0 week, 1700 scans; (b) 2 weeks, 1720 scans; (c) 4 weeks, 1700 scans; (d) 8 weeks, 2200 scans (redrawn from Frossard et al.[54]).

266

paramagnetic impurities (Fe^{3+}) in the vicinity of P_6O_{18} groups while some of the P_2O_7 would have remained protected within the chitosan sheets. Polyphosphate ions are indeed known to sorb strongly on iron oxides.[55] After two, four, and eight weeks of incubation the relative intensities of these three peaks decreased strongly. At the end of the incubation period, only the peak at 0 ppm (P_2O_7) was still clearly present while peaks at -10 and -23 ppm had almost disappeared. These results were attributed to structure modifications of the CH–PP due to its biodegradation in the soil. Changes were probably due to enzymic activities and may explain the release of P as orthophosphate in the soil solution, thus accounting for the high P fertilizing value of this compound.[54]

Future Prospects

In recent years, high-resolution solid-state ^{31}P NMR has been made possible by the development of more powerful (higher-frequency) NMR spectrometers. Preliminary studies on the application of solid-state ^{31}P NMR to soils has so far shown only broad peaks. The use of ^{31}P NMR on solid materials such as soils is restricted by low P concentrations (a minimum concentration of 1 mg P g^{-1} is necessary). Furthermore, the presence of paramagnetic impurities (e.g., Fe^{3+}, Mn^{2+}) limit the useful application of this technique since they decrease the signal to noise ratio and increase the intensity of spinning sidebands. Signal broadening in solid-state ^{31}P NMR may be attributed mainly to variability in the numbers and types of cation associated with each P-containing anion.

Solid-state ^{31}P NMR has been shown to be a valuable technique for the study of P in organic wastes and fertilizers, and should be considered as a unique tool for validating data obtained from routine analysis, allowing for the characterization of P species such as chemical sequential extractions of soil P.[54] However, this technique is not by itself sufficient to describe the bioavailability of P applied in organic wastes and fertilizers, but should be used together with other approaches such as ^{32}P isotopic exchange which allow for the measurement of the mobility of PO_4 ions in soil– or sludge–solution environments.[34,55]

Despite the restrictions described above, ^{31}P NMR has considerable potential for application in the analysis, identification, and associated chemical and biological transformations of P species encountered in soils and wastes because it is a noninvasive method which allows for almost unbiased observations. Further methodological research is needed to develop the most suitable techniques for obtaining meaningful ^{31}P NMR spectra from soils and wastes. Bleam[56] has noted that "the power of NMR is its capacity to probe interactions. The extent to which we can translate our concept of the chemical environments existent in soil organic polymers, at mineral surfaces, in poorly crystalline solids or other natural materials into this language of interactions will largely determine the utility of modern pulse NMR spectroscopy in soil science." Further development of PSRE methods will provide the tools for probing these interactions. This in turn will greatly assist in improving our understanding of P cycling in the soil environment in relation to primary production and the disposal and re-use of organic wastes.

References

1. Stewart, J. W. B., and Tiessen, H., Dynamics of soil organic phosphorus. *Biogeochemistry* 4, 41, 1987.
2. Sanyal, S.K., and DeDatta, S.K., Chemistry of phosphorus transformations in soil. *Adv. Soil Sci.* 16, 1, 1992.
3. Frossard, E., Brossard, M., Hedley, M. J., and Metherell, A., Reactions controlling the cycling of P in soils. In *Phosphorus in the Global Envrionment*, Tiessen, H. (ed.), Wiley, Chichester, 1995, pp. 107–137.
4. Magid, J., Tiessen, H., and Condron, L., Dynamics of organic phosphorus in soils under natural and agricultural ecosystems. In *Humic Substances in Terrestrial Ecosystems*, Piccolo, A. (ed.), Elsevier, Amsterdam, 1996, pp. 429–466.
5. Sequi, P., Ciavatta, C., and Vittori Antisari, L., Phosphate fertilizers and phosphorus loadings to rivers and seawater. *Agrochimica* 35, 200, 1991.
6. Wilson, M. A., *NMR Techniques and Applications in Geochemistry and Soil Chemistry*, Pergamon Press, Oxford, 1987.
7. Pierzynski, G. M., The chemistry and mineralogy of phosphorus in excessively fertilized soils. *Crit. Rev. Environ. Control* 21, 265, 1991.
8. Newman, R. H., and Tate, K. R., Soil characterised by ^{31}P nuclear magnetic resonance. *Commun. Soil Sc. Plant Anal.* 11, 835, 1980.
9. Anderson, G., Assessing organic phosphorus in soil. In *Role of Phosphorus in Agriculture*, Khasawaneh, F. E., Sample, E. C., and Kamprath, E. J. (eds), American Society of Agronomy, Madison, WI, 1980, pp. 411–431.
10. Tate, K. R., and Newman, R. H., Phosphorus fractions of a climosequence of soils in New Zealand tussock grassland. *Soil Biol. Biochem.* 4, 191, 1982.
11. Hawkes, G. E., Powlson, D. S., Randall, E. W., and Tate, K. R., A ^{31}P nuclear magnetic resonance study of the phosphorus species in soils from long-continued field experiments. *J. Soil Sci.* 35, 35, 1984.
12. Condron, L. M., Frossard, E., Tiessen, H., Newman, R. H., and Stewart, J. W. B., Chemical nature of organic phosphorus in cultivated and uncultivated soils under different environmental conditions. *J. Soil Sci.* 41, 41, 1990.
13. Zech, W., Alt, H. G., Haumaier, L., and Blasek, R., Characterisation of phosphorus fractions in mountain soils of the Bavarian Alps by ^{31}P NMR. *Z. Planzenernahrung Bodenkunde*, 150, 119, 1987.
14. Trasar-Cepeda, M.C., Gil-Sotres, F., Zech, W., and Alt, H. G., Chemical and spectral analysis of organic P forms in acid high organic matter soils in Galicia (NW Spain). *Sci. Tot. Environ.* 81/82, 429, 1989.
15. Forster, J. C., and Zech, W., Phosphorus status of a soil catena under Liberian evergreen rain forest: results of ^{31}P NMR spectroscopy and phosphorus adsorption experiments. *Z. Planzenernahrung Bodenkunde*, 156, 61, 1993.
16. Condron, L. M., Goh, K. M., and Newman, R. H., Nature and distribution of soil phosphorus as revealed by a sequential extraction method followed by ^{31}P nuclear magnetic resonance analysis. *J. Soil Sci.* 36, 199, 1985.

17. Ogner, G., ^{31}P NMR spectra of humic acids: a comparison of four different raw humus types in Norway. *Geoderma* 29, 215, 1983.
18. Adams, M. A., and Byrne, L. T., ^{31}P NMR analysis of phosphorus compounds in extracts of surface soils from selected Karri (*Eucalyptus diversicolor F. Muell.*) forests. *Soil Biol. Biochem.* 21, 523, 1989.
19. Adams, M. A., ^{31}P NMR identification of phosphorus compounds in neutral extracts of mountain ash (*Eucalyptus regnans F. Muell.*) soils. *Soil Biol. Biochem.* 22, 419, 1990.
20. Emsley, J., and Niazi, S., The analysis of soil phosphorus by ICP and P NMR spectroscopy. *Phosphorus Sulfur*, 16, 303, 1983.
21. Greenhalgh, R., Blackwell, B. A., Preston, C. M., and Murray, W. J., Phosphorus-31 nuclear magnetic resonance analysis of technical organophosphorus insecticides for toxic contaminants. *J. Agric. Food Chem.* 31, 710, 1983.
22. Mortimer, R. D., and Dawson, B. A., A study to determine the feasibility of using ^{31}P NMR for the analysis of organophosphorus insecticide residues in cole crops. *J. Agric. Food Chem.* 39, 911, 1991.
23. Nihira, M., Hirakawa, K., Hayashida, M., Watanabe, T., Susaki, S., and Yama-moto, Y., Rapid analysis of organophosphorus pesticides using ^{31}P fourier transform nuclear magnetic resonance spectroscopy (FT-NMR). *Jap. J. Toxicol.* 3, 57, 1990.
24. Krolski, M. E., Bosnak, L. L., and Murphy, J. J., Application of nuclear magnetic resonance spectroscopy to the identification and quantisation of pesticide residues in soil. *J. Agric. Food Chem.* 40, 458, 1992.
25. Ingall, E. D., Schroeder, P. A., and Berner, R. A., The nature of organic phosphorus in marine sediments: new insights from ^{31}P NMR. *Geochim. Cosmochim. Acta* 54, 2617, 1990.
26. Turner, G. L., Smith, K. A., Kirkpatrick, R. J., and Oldfield, E., Structure and cation effects on phosphorus-31 NMR chemical shifts and chemical shift aniso-tropies of orthophosphates. *J. Magn. Reson.* 70, 408, 1986.
27. Zumbulyadis, N., Selective carbon excitation and the detection of spatial hetero-geneity in cross-polarization magic-angle-spinning NMR. *J. Magn. Reson.* 53, 486, 1983.
28. Van der Hart, D. L., and Perez, E., A ^{13}C NMR method for determining the partitioning of end groups and side branches between the crystalline and noncrys-talline regions of polyethylene. *Macromolecules* 19, 1902, 1986.
29. Preston, C. M., and Newman, R. H., Demonstration of spatial heterogeneity in the organic matter of de-ashed humin samples by solid-state ^{13}C CPMAS NMR. *Can. J. Soil Sci.* 72, 13, 1992.
30. Williams, R. J. P., Giles, R. G. F., and Posner, A. M., Solid state phosphorus NMR spectroscopy of minerals and soils. *J. Chem. Soc., Chem. Commun.* 20, 1051, 1981.
31. Kirkham, M. B., Agricultural use of phosphorus in sewage sludge. *Adv. Agron.*, 129, 1982.
32. Coker, E. G., and Carlton-Smith, C. H., Phosphorus in sewage sludges as a fertilizer. *Waste Management Res.* 4, 303, 1986.
33. Zhang L. M., Morel, J. L., and Frossard, E., Phosphorus availability in sewage sludge. In *Proc. 1st Congress of the Europ. Soc. Agronomy*, Scaife, A. (ed.), European Society ofAgronomy, Paris, 1990.
34. Frossard, E., and Morel, J. L., Assessment of the phosphate fertilizing value of urban sewage sludges. In *Soil Management in Sustainable Agriculture*, Cook, H., and Lee. H. (eds),Wye College, University of London, 1995, pp. 226–230.

35. Häni, H., Gupta, S. K., and Furrer, O. J., Availability of phosphorus fractions in sewage sludge. In *Phosphorus in Sewage Sludge and Animal Waste Slurries*, Hucker, T. W. G., and Catroux, G. (eds), Reidel, Dordrecht, 1981, p. 177.

36. Hanotiaum, G., Heck, J. P., Rocher, M., Barideau, L., and Marlier-Geets, O., The content of phosphorus in the sludge of wallonian water purification plants and the form of phosphorus in these materials. In *Phosphorus in Sewage Sludge and Animal Waste Slurries*, Hucker, T. W. G., and Catroux, G. (eds), Reidel, Dordrecht, 1981, p. 99.

37. Soon, Y. K., and Bates, T. E., Extractability and solubility of phosphate in soils amended with chemically treated sewage sludges. *Soil Sci.* 134, 89, 1982.

38. O'Connor, G. A., Knudtsen, K. L., and Connell, G. A., Phosphorus solubility in sludge amended calcareous soils. *J. Environ. Qual.* 15, 308, 1986.

39. Hinedi, Z. R., and Chang, A. C., Solubility and phosphorus 31 magic angle spinning nuclear magnetic resonance of phosphorus in sludge amended soils. *Soil Sci. Soc. Am. J.* 53, 1057, 1989.

40. Hinedi, Z. R., Chang, A. C., and Yesinowski, J. P., Phosphorus 31 magic angle spinning nuclear magnetic resonance of wastewater sludges and sludge amended soil. *Soil Sci. Soc. Am. J.*, 53, 1053, 1989.

41. Hartmann, S. R., and Hahn, E. L., Nuclear double resonance in the rotating frame. *Phys. Rev.* 128, 2042, 1962.

42. Belton, P. S., Harris, R. K., and Wilkes, P. J., Solid-state phosphorus-31 NMR studies of synthetic inorganic calcium phosphates. *J. Chem. Solids* 49, 21, 1988.

43. Preston, C. M., Ripmeester, J. A., Mathur, S. P., and Lévesque, M., Application of solution and solid-state multinuclear NMR to a peat-based composting system for fish and crab-scrap. *Can. J. Spectrosc.* 31, 63, 1986.

44. Aue, W. P., Roufosse, A. H., Glimcher, M. J., and Griffin, R. G., Solid-state phosphorus 31 nuclear magnetic resonance studies of synthetic solid phases of calcium phosphate: potential models of bone mineral. *Biochemistry* 23, 6110, 1984.

45. Griffiths, L., Roots, A., Harris, R. K., Packer, K. J., Chippendale, A. M., and Tromans, F. R., Magic-angle spinning phosphorus-31 nuclear magnetic resonance of polycrystalline sodium phosphates. *J. Chem. Soc., Dalton Trans.* 14, 2247, 1986.

46. Rothwell, W. P., Waugh, J. S., and Yesinowski, J. P., High-resolution variable temperature ^{31}P NMR of solid calcium phosphates. *J. Am Chem. Soc.* 102, 2637, 1980.

47. Lucas, J., Les dépôts de phosphates sur le continent africain. In *Phosphorus Cycles in Terrestrial and Aquatic Systems; Regional Workshop 4: Africa*, Tiessen, H., and Frossard, E. (eds), SCOPE/UNEP, University of Saskatchewan, Saskatoon, 1992, pp. 157–168.

48. Amoros, B., André, L., and Lacout, J. L., Influence de la matière organique sur l'évolution en solution du phosphore monocalcique. *C.R. Séances Acad. Sci. Paris Sér. 2* 303(14), 1355, 1986.

49. Inskeep, W. P., and Silvertooth, J. C., Inhibition of hydroxyapatite precipitation in the presence of fulvic, humic and tannic acids. *Soil Sci. Soc. Am. J.* 52, 941, 1988.

50. Lindsay, W. L., Vlek, P. L. G., and Chien, S. H., Phosphate minerals. In *Minerals in Soil Environment*, 2nd edn., Dixon, J. B., and Weed, S. B. (eds), Soil Science Society of America, Madison, WI, 1989, p. 1089.

51. Grossl, P. R., and Inskeep, W. P., Precipitation of dicalcium phosphate dihydrate in the presence of organic acids. *Soil Sci. Soc. Am. J.* 55, 670, 1991.

52. Florentz, M., Granger, P., and Harteman, P., Use of ^{31}P nuclear magnetic resonance spectroscopy and electron microscopy to study phosphorus metabolism of microorganisms from wastewaters. *Appl. Environ. Microbiol.* 47, 519, 1984.
53. Dick, R. P., and Tabatabai, M. A., Polyphosphates as a source of phosphorus for plants. *Fertil. Res.* 12, 107, 1987.
54. Frossard, E., Tekely, P., and Morel, J. L., Chemical characterization and agronomic effectiveness of phosphorus applied as a polyphosphate–chitosan complex. *Fertil. Res.* 37, 151, 1993.
55. Al Kanani, T., and MacKenzie, A. F., Sorption and desorption of orthophosphate and pyrophosphate by mineral fractions of soils, goethite and kaolinite. *Can. J. Soil Sci.* 71, 327, 1991.
56. Bleam, W., Soil science applications of nuclear magnetic resonance spectroscopy. *Adv. Agron.* 46, 91, 1991.

15.

Characterization of Nitrogen in Plant Composts and Native Humic Material by Natural-Abundance ^{15}N CPMAS and Solution NMR Spectra

HEIKE KNICKER, RÜDIGER FRÜND, & HANS-DIETRICH LÜDEMANN

Soil organic matter (SOM) provides one of the major deposits for carbon and nitrogen on the surface of the Earth.[1-3] It is continuously produced, mainly from dead plant material, by composting and humification processes. During these processes microorganisms thoroughly convert the starting material, which consists mostly of insoluble lignocelluloses. The end products of these processes in average middle-European soils, that contain typically 1 to 5% w/w of organic material, are clay–SOM complexes which are insoluble in all the usual organic and inorganic solvents. The standard aqueous sodium hydroxide extraction procedure[4] dissolves at most 40% of the total organic carbon in all the soils tested by our group.[5] The insoluble majority, the humin fraction, remains as poorly defined aluminosilicate–SOM complexes.

During the decomposition and conversion processes the carbon to nitrogen ratio decreases. Compared to the starting material, SOM is enriched in nitrogen. Under natural conditions, i.e., without the artificial addition of nitrogen in the form of manure or fertilizer, SOM provides the major part of the nitrogen available for plant growth.[1]

The chemical characterization of this ubiquitous but ill-defined material has only been partly successful until now. For characterization of the organic carbon in complete soils and extracts, nuclear magnetic resonance (NMR) methods appear to be most promising,[5,6] especially since the application of high-resolution solid-state methods has become almost a laboratory routine. The combination of proton–carbon cross polarization with high-speed magic angle rotation (the CPMAS technique) permits the study of complete native soils, and thus provides detailed information about the gross chemical structure of the total SOM, without introducing any of the chemical modifications that could result from aggressive chemical extraction procedures.

It has been shown by ^{13}C CPMAS and high-resolution (HR) solution ^{13}C NMR studies of a series of typical European soils, in which the concentration of paramagnetic metal ions was fairly low and which contained humic material with an aromatic carbon content $\leq 20\%$, that the carbon could be quantitatively assigned.[7] The measurement of the ^{13}C CPMAS spectra of complete native soils with a carbon content in the region of 1% w/w is rather instrument-time consuming, and appeared to be at the limit of sensitivity.

The nucleus of ^{14}N, the most abundant nitrogen isotope, does not yield any HR spectra in chemical compounds because of its large quadrupole moment. The ^{15}N isotope occurs at a natural abundance of only 0.37% and has a low and negative magnetogyric ratio γ. The receptivity of a ^{15}N NMR study of material that is not ^{15}N-enriched is thus approximately 50-fold lower than that of a ^{13}C NMR study, and it was probably assumed that natural-abundance ^{15}N CPMAS NMR spectra could not be obtained in feasible spectrometer times. The first ^{15}N spectra in humic material were consequently obtained with ^{15}N-enriched composts and melanoidins.[8–11]

Figure 15.1 gives the ^{15}N CPMAS NMR spectrum of a humic acid extracted from an orthic humic Gleysol after seven months of incubation with $Na^{15}NO_3$ and sodium acetate at 25 °C and 67% moisture by Benzing-Purdie et al.[8] This is, to our knowledge, the first ^{15}N CPMAS NMR spectrum obtained from humic material, and it shows that amides and especially secondary amides, i.e., most probably peptide nitrogen in proteinaceous material, contribute approximately 90% of the total signal intensity. Unfortunately this paper has often been overlooked[6,12] or not given the proper credit[13] by the scientists writing the most recent reviews about nitrogen in soils and humic material. In ref. 13, p. 236, a fairly pessimistic statement, characteristic of the reception of this spectroscopic result by the soil scientists, is given: "The application of modern NMR techniques has provided some tantalizing evidence, in the form of ^{15}N spectra, comparing the chemical environments of the N atoms in typical Maillard products with those of a humic acid (Benzing-Purdie et al., 1983). As has happened previously with the application of other advanced spectroscopic techniques, the evidence is vague and inconclusive, but it holds some promise, especially if the method were to be applied to isotopically labeled humic substances."

Wet chemical analysis of dissolved SOM leaves approximately 50% of the total nitrogen unidentified and assigns a major part of this unidentified fraction to heteroaromatic structures[6,12,13] and even nitrile groups,[14] while the identified fraction consists mostly of amino acids. The spectrum given in Figure 15.1, which is representative of all ^{15}N-SOM spectra published hitherto, assigns about 90% of the total intensity to amide structures, and gives weak signals in the chemical shift range of free amino groups, and no intensity in the region of typical heteroaromatic phenazine or pyrrole structures.

It is attempted to show that the few spectra given by Benzing-Purdie et al.[8] are really characteristic for soil organic matter and contain all the spectral

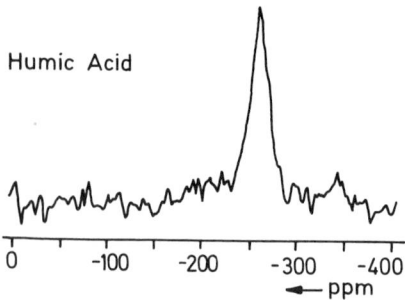

Figure 15.1 CPMAS ^{15}N spectrum of a humic acid from an orthic humic gleysol (18.3 MHz) (after Benzing-Purdie et al).[8]

features observed with the more advanced technology available now. In this chapter ^{13}C and ^{15}N NMR spectra of composts and their extracts obtained from plants grown on potassium nitrate with 90% ^{15}N enrichment are discussed. These studies have two main objectives:

(1) To study systematically the spectral changes observed as a function of the composting period and to correlate the ^{13}C and ^{15}N NMR spectra.

(2) To determine all relevant relaxation times for a strategically chosen set of ^{15}N-enriched composts and their aqueous sodium hydroxide extracts, in order to adjust all the parameters of the CPMAS pulse sequence for the optimization of the signal to noise ratio.

After this optimization process it became possible to measure the ^{15}N NMR spectra of humic extracts from native soils at the natural ^{15}N-abundance. All laboratory-produced material has been fermented at most for a couple of years, and it could be argued that the chemical processes that form the characteristic nitrogen compounds occur on a much longer time scale, since a major fraction of the SOM has been in the soil for several hundred to several thousand years.

Experimental

Sample Preparation

The composts were prepared from *Lolium perenne, Lolium rigidum,* and *Tritium sativum* plant material grown on a nutrient solution,[15] containing potassium nitrate (90% ^{15}N-enriched) as the sole nitrogen source. A 10 g portion of freeze-dried and milled plant material was mixed with 100 g of quartz sand and kept at 25 °C either at 60% or 100% water holding capacity (water saturation). The samples were inoculated with 1 mL of an aqueous extract from a natural compost. On most samples fungi started growing after an incubation time of some weeks. For several samples these were separated from the starting mixture, freeze-dried and used as a ^{15}N-rich reference material. The soil samples used for the study of the natural-abundance ^{15}N NMR spectra have been characterized previously[5] by ^{13}C CPMAS NMR and elemental analysis. Table 15.1 contains the relevant results. In Figure 15.2 the aqueous sodium hydroxide extraction procedure for the composts and soils is described. No attempts were made to characterize the organic products that diffused out of the dialysis bags by (Servapor 16, separation size 10—15 kDa; Serva, Heidelberg, Germany) by ^{15}N NMR.

NMR Methods

The CPMAS ^{13}C NMR spectra were obtained at a Bruker MSL 100 spectrometer operating at 2.3 T (^{13}C resonance frequency 25.2 MHz). The spinning rate was 4 kHz. A commercial Bruker double-bearing probe with 7 mm o.d. and phase-stabilized zirconium dioxide rotors were used. A conventional cross-polarization pulse program with a contact time of 1 ms was used. To improve the signal to noise ratio, a line-broadening of 10 Hz was applied. Details of the determination of the relevant nuclear relaxation times are given by Dev et al.[16] The CPMAS and solution ^{15}N NMR spectra were measured on a Bruker MSL 300 spectrometer operating at 7 T (^{15}N resonance frequency 30.4 MHz); for the CPMAS NMR spectra a probe identical to the one described above was used. The rotation frequency was set between 4 and 4.5 kHz. The proton spin–lattice relaxation time T_{1H} was determined indirectly by detecting the ^{13}C or ^{15}N magnetization.[17]

By variation of the contact time in a series of CPMAS NMR spectra, the time relevant for the polarization transfer T_{XH} and the proton spin–lattice relaxation time in the rotating frame $T_{1\rho H}$ were determined. The ^{13}C chemical shifts are given on the tetramethylsilane (TMS = 0 ppm) scale. ^{15}N NMR spectra are referenced to external nitromethane (= 0 ppm). For the solution ^{15}N NMR spectra the standard pulse program used was a Ridegate sequence described by Gerothanassis[18] and modified by Fründ.[18] The protons were decoupled by the inverse-gated broadband decoupling scheme. In order to obtain preliminary estimates of the spin–lattice relaxation times T_1, a series of spectra was collected in which the delay time between two pulses was increased from 3 s to 90 s. (The normal inversion recovery techniques for the exact determination of T_1 would, for T_1 values larger than 100 s, consume weeks of spectrometer time and most probably not lead to quantitative results because of spectrometer drifts.) The preliminary results showed that for all ^{15}N signals observable $T_{1N} \leq 2$ s. The individual T_{1N} values were then determined exactly with the standard inversion recovery pulse sequence. The distortionless enhanced polarization transfer (DEPT) pulse sequence has been described in detail.[20]

Table 15.1 Origin and Some Characteristics of the Soil Samples (Top 10 cm, Late Winter)

	Geographical Origin	Order	Composition (%)	
			C	N
Field	Pfaffenhofen, Bavaria	Cambisol	1.2	0.1
	Harthausen, Bavaria	Luvic Cambisol	3.4	0.3
Grassland	Oberwarngau, Bavaria	Rendzina	4.6	0.4
	Ismaning, Bavaria	"Black" calcaric Regosol	11.5	0.9
Forest	Solling D1, Lower Saxony	Spodoc distric Cambisol	3.4	0.2
	Göttingen, Lower Saxony	Chromo-calcic Cambisol	4.4	0.4

Theoretical

The emphasis in the NMR spectroscopy of heteronuclei in SOM is to determine via chemical shift assignments the gross chemical structure of this material, and to attempt a quantitative correlation of the different signal intensities to the chemical composition.

In single-resonance HR solution NMR the relative intensity of a signal is directly proportional to the relative concentration of the chemical species, if

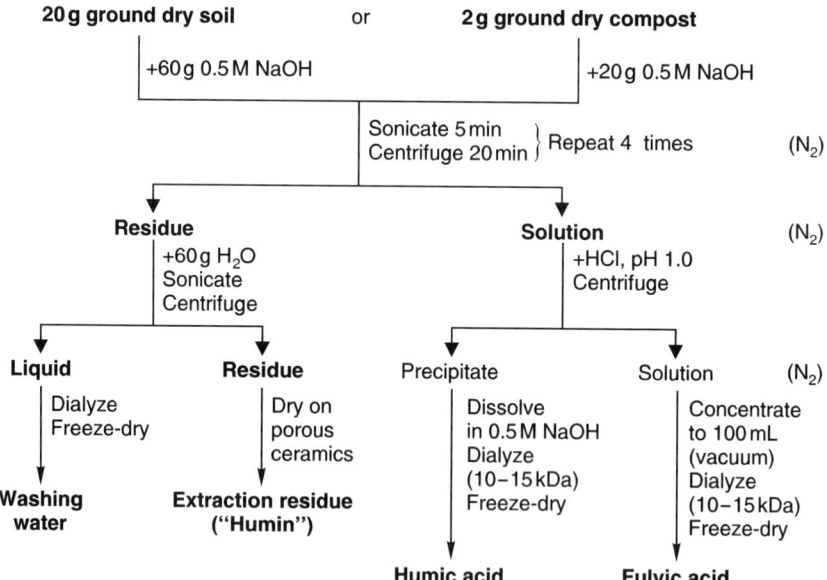

Figure 15.2 Extraction procedure for soils and composts. The organic material in the compost was separated mechanically from the quartz sand, milled, and freeze-dried.

saturation effects are avoided, i.e., if the time span between two single-pulse experiments is approximately five times greater than the longest spin–lattice relaxation time T_1^{max} observed for the nucleus under study.[21] ^{13}C and ^{15}N NMR spectra are usually obtained as double resonance spectra under broad-band proton decoupling. In general, double–resonance experiments can lead to significant intensity changes by transfer of magnetization between the two spin systems. This transfer is effectively suppressed by the inverse-gated decoupling technique.[21]

Solution NMR spectra obtained with these precautions can thus be integrated over the various chemical shift ranges and interpreted quantitatively. A comparison of these integrals with the results obtained from CPMAS NMR spectra should provide a thorough check for the limits of quantitative analysis obtainable from the solid-state spectra.

For CPMAS NMR spectra the intensity I of a signal is a fairly complex function of three relaxation or polarization transfer times:

$$I_t = I_0(T_{1H}, T_{XH}, T_{1\rho H}, t) \qquad (1)$$

Here I_t is the intensity observed after the Hartmann–Hahn contact time t. I_0 is directly proportional to the number of nuclei in the probe. I_t is a function of the proton spin–lattice relaxation time T_{1H}, the cross-polarization time T_{XH} (in our case X = [15]N, ^{13}C), the proton spin–lattice relaxation time in the rotating frame $T_{1\rho H}$, and the contact time t. In the experiments described below, the saturation dynamics of the X nuclei are determined by T_{1H} entirely.[21] Setting the repetition time $D \geq 5T_{1H}$ thus eliminates saturation effects. However, the efficiency of magnetization transfer to a group of X nuclei depends on the local proton concentration, and on the rotational correlation time of the X nucleus. Also, the X nucleus must relax entirely by the dipole–dipole interaction with the protons to permit the maximal cross-polarization effect. Under these rather restrictive boundary conditions and if the inequality $T_{XH} \ll t \ll T_{1\rho H}$ is applicable, equation (1) simplifies to $I_t = I_0 \exp(-t/T_{1\rho H})$, and one can hope to obtain quantitative results from the CPMAS spectra. In addition, it must be kept in mind that SOM has a fairly high concentration of organic free radicals, providing efficient relaxation sinks. Only in the case where their contribution to X nuclei relaxation is negligible or approximately the same for all X nuclei observable in the spectrum, can one hope for quantitative results.[6,22,23]

Results and Discussion

In Figure 15.3 a first series of CPMAS ^{13}C and ^{15}N NMR spectra obtained from a complete *Lolium rigidum* compost, kept at 60% water holding capacity (WHC), is given.[11] During the 70 days of fermentation about 80% of the carbon was lost from the starting biomass. The nitrogen losses were negligible. Rather pronounced changes are seen in the ^{13}C series of NMR spectra. Around 165 ppm in the carboxyl/carbonyl region of chemical shift a new peak

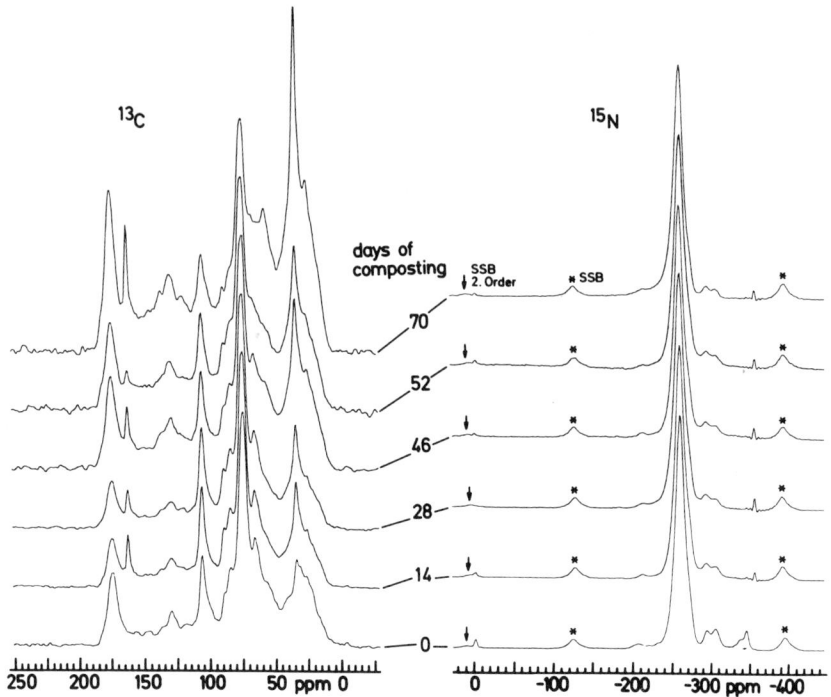

Figure 15.3 CPMAS ^{13}C (25.2 MHz) and ^{15}N (30.4 MHz) spectra from the complete freeze-dried composts of *Lolium rigidum* plant material (Almendros et al.[11]). *SSB, spinning sideband (first order).

is developing that is most probably to be assigned to aromatic carboxylic acids, formed during the degradation of lignin, and to the amide carbons in proteinaceous material. In Table 15.2 the results of the integration of these spectra are compiled.[11] Overall, the chemical shift range between 110 and 60 ppm loses intensity. The range between 60 and 45 ppm remains constant and all other ranges increase in intensity. Qualitatively these results are explained by a preferential decomposition of the carbohydrate fraction of the plant material. Considering the large carbon loss (80%), proteinaceous and lignin structures must also be destroyed, albeit at a lower rate. It is generally accepted that amino acids are removed from the plant material and reincorporated into the proteinaceous structures of the microbial SOM without major chemical modification, a process that must leave the ^{13}C and ^{15}N NMR spectra of the nitrogen-carrying structures unchanged.

Another remarkable feature of these spectra is that the intensity distribution within the aromatic region, i.e., the ratio of the intensity I (160–140 ppm)/I (140–110 ppm) remains quite constant, although the overall intensity in the range between 160 and 110 ppm increases from 10 to 16%. This feature, which has also been observed in all other composts studied hitherto,

Table 15.2 Carbon Composition (%) of the Composted *Lolium rigidum* Material Featured in Figure 15.2 as Determined by Integration of the ^{13}C CPMAS Spectra

Plausible assignment of the ranges
220–160 ppm: Carboxyl/carbonyl
160–110 ppm: Most aromatic structures
110–60 ppm: Carbohydrate-derived structures, α-C of peptides
60–45 ppm: Methoxy groups, α-C of peptides
45–0 ppm: Aliphatic structures

Composting Time (days)	Integration Range (ppm)				
	220–160	160–110	110–60	60–45	45–0
0	10	10	48	10	22
14	9	12	50	8	21
28	9	12	51	8	20
42	11	14	40	10	26
56	11	14	40	10	26
70	13	16	31	10	30

can most probably be explained by the assumption that the substitution of the aromatic rings is not significantly changed during the composting process. This experimental fact leaves very little space for any model of humification chemistry that involves formation of larger fractions of polyaromatics or heteroaromatics during these stages of composting.

Compared to the rather drastic changes observed in the ^{13}C NMR spectra of Figure 15.3, the ^{15}N NMR spectra are remarkably uninfluenced. No new signals, indicative of the formation of new chemical compounds (heteroaromatics), can be observed in the spectra. In Table 15.3 are compiled the signals and their most probable assignments together with the trends in the intensity changes. More than 80% of the total intensity is found in all spectra in the chemical shift range between -190 and -285 ppm and is most probably to be assigned to peptide nitrogen in proteinaceous material. Free amino groups decrease in intensity while the weak signal around -300 ppm remains almost constant. Assuming that these spectra detect all forms of nitrogen present in the samples, a very simple explanation for the nitrogen metabolism can be given. Practically all the metabolic and chemical transformations of the nitrogen-containing molecules that occurred during the composting process described here converted the plant proteins into microbial and fungal proteins and proteinaceous structures.

This explanation is fully corroborated by the analysis of the ^{13}C NMR spectra but contradicts most accepted models for the chemical structure of nitrogen-containing compounds in the soil[12–14,23,24] Thus it has to be checked very carefully whether these short-term composts are representative of the structure of aged SOM that has been exposed to the environment for several hundred to several thousand years,[25] and it should also be tested thoroughly

Table 15.3 Possible Assignments for the ^{15}N NMR Spectra

Chemical shift range (ppm)	Assignment
−145 to −220	Indole, imidazole, uric acid, purine/pyrimidine, pyrrole
−220 to −285	Amide/peptide, lactam
−285 to −300	NH in guanidine
−300 to −325	NH_2 and NR_2 groups
−325 to −370	NH_3^+ and NR_3^+ groups, NH_4^+
Missing signals:	
25 to −25	Nitrate, nitrite, nitro groups[a]
−25 to −90	Imine, phenazine, pyridine
−90 to −145	Purine (N7), nitrile groups

[a] A signal around −4 ppm is visible in the very early stages of composting experiments under addition of $K^{15}NO_3$. Also, in the final stages of composting, after $\sim$400 days, the nitrate signal is observed (cf. Figure 15.12).

whether the CPMAS method is able to pick up all the nitrogen structures present. As stated in the Theoretical section, only ^{15}N nuclei cross-polarized by protons are monitored by this technique and it could be claimed that quaternary and tertiary nitrogen structures cannot be observed; also, relaxation interaction with the organic free radicals, present in SOM, can suppress signals in the CPMAS spectra. For the ^{13}C CPMAS NMR spectra of SOM it could be shown that for all normal SOM samples the spectra can be evaluated quantitatively.[5,6] The main argument for this conclusion was derived from a comparison of solution and CPMAS NMR spectra and from a determination of all relevant relaxation times for both types of spectra. This comparison is of course confined to the soluble fraction of SOM.

In Figure 15.4 ^{15}N CPMAS NMR spectra of the fractions (complete composts, insoluble residue, and NaOH extracts) obtained from the sodium hydroxide extraction of series of composts are given. Comparison of the spectra shows that some minor fractionation has taken place during this extraction procedure. The soluble fraction is enriched in free amino groups; the signals between −325 and −375 ppm are more pronounced. In the insoluble residues the high- and low-field "feet" of the main peptide signal are clearly visible, which are almost completely suppressed in the extracts. Figure 15.5 presents the inverse-gated proton-decoupled solution ^{15}N NMR spectra of the sodium hydroxide extracts. The solution and the solid-state spectra of the extracts are surprisingly similar.

In order to avoid saturation effects, which would distorted the intensities in the solution spectra, a preliminary estimate of the ^{15}N spin–lattice relaxation times was obtained by varying the delay time between two pulses. It was considered technically unfeasible and imprecise to attempt a raw determination of the time range of the ^{15}N T_1 inversion recovery experiments with delay times longer than 100 s. Such experiments would have consumed weeks of

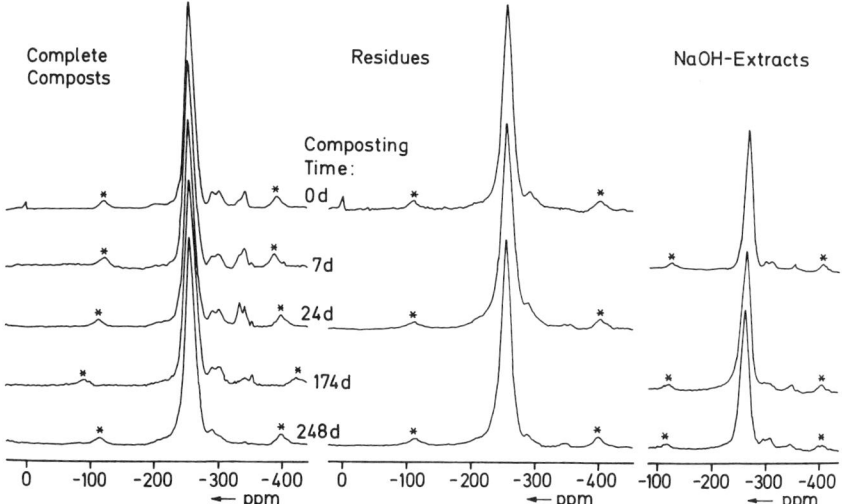

Figure 15.4 ^{15}N CPMAS NMR spectra (30.4 MHz) of composts from *Lolium perenne* plant material grown on 90% ^{15}N KNO$_3$. The complete composts are compared to the NaOH extracts and to the residues of this extraction process. Asterisks (*) indicate spinning sidebands.

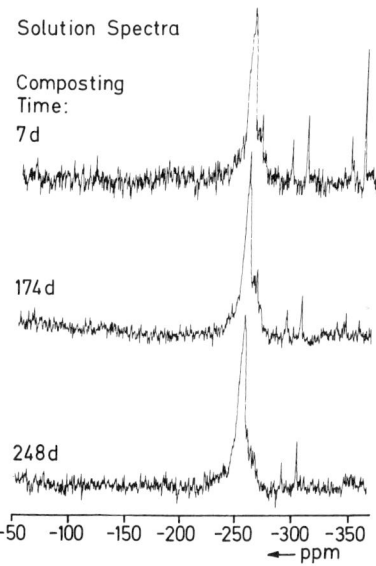

Figure 15.5 ^{15}N inverse-gated decoupled solution NMR spectra (30.4 MHz) of the sodium hydroxide extracts from the composts characterized in Figure 15.4.

spectrometer time and probably would not have yielded very meaningful results.

In Figure 15.6 inverse-gated proton-decoupled spectra of a fungal extract are given as a function of the delay time. Minor quantitative differences in the spectra are seen when the first two obtained with 3 s and 10 s delay are compared. Increasing the delay to 90 s does not lead to any further changes and no additional signals appear.

From this result it becomes clear that all observed signals have $T_{1N} \leq 2$ s. In a second set of experiments T_{1N} for all visible signals was determined with the inversion recovery technique.

Table 15.4 compiles some representative results. The standard protocol for the solution ^{15}N NMR spectra operated with a delay time of 15 s, which was also applied to the samples shown in Figure 15.5. The intensity distribution in the solid state ^{15}N CPMAS NMR spectra is determined by T_{1H}, T_{NH} and $T_{1\rho H}$, as stated in the Theory section. T_{1H} has been determined for starting material, composts and extracts by the pulse sequence described previously.[16] Table 15.5 contains some of the results obtained. In a heterogeneous and complex probe the proton spin–lattice relaxation is most certainly not described by a single exponential. The T_{1H} values given in Table 15.5 thus

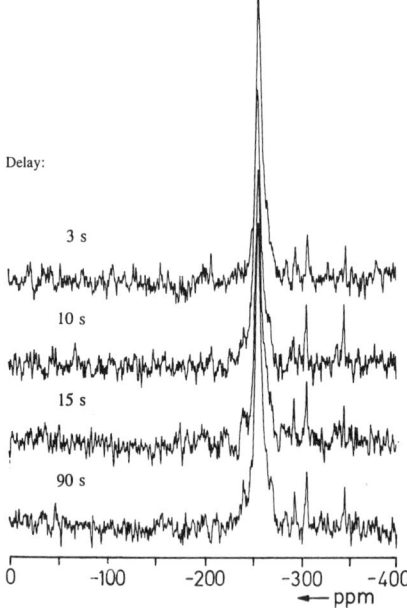

Delay:

3 s

10 s

15 s

90 s

0   -100   -200   -300   -400
                                    ◀— ppm

Figure 15.6 Inverse-gated decoupled ^{15}N NMR solution spectra (30.4 MHz) of a sodium hydroxide extract from a fungus as a function of delay time t. Note that the S/N ratio and the relative signal intensities remain constant at $t \geq 10$ s. ^{15}N enrichment: 80%. Number of transients: 630.

Table 15.4 ^{15}N-T_1 (s) in Aqueous Solutions of Sodium Hydroxide Extracts from *Lolium perenne* Composts (100% WHC)

	Signal (ppm)						
	−245	−257	−269	−293	−306	−346	−358
Sample:							
Fungus	2.5	0.9	0.5	0.5	0.8	0.7	2.0
Lolium perenne							
Start	—	0.9	0.5	0.9	0.9	1.1	—
7 days	—	0.6	0.4	0.6	0.7	1.2	3.0
174 days	—	0.5	0.7	0.6	0.7	—	—
541 days	—	0.6	0.6	—	—	0.4	0.6

most probably are weighted averages of the different relaxation rates and are only to be taken as a coarse characterization for the time scale of the proton spin–lattice relaxation processes. The repetition time in the standard ^{15}N CPMAS NMR experiments was set to 300 ms. The natural-abundance ^{15}N CPMAS NMR spectra were normally obtained with a repetition time of 100 ms. For a few fractions of the natural ^{15}N abundance material the spectra were repeated with the longer standard repetition time of 300 ms. No differences outside experimental error were detected between the spectra obtained by these two procedures.

Table 15.5 T_{1H} (ms) for the Different ^{15}N Signals as Determined from the CPMAS ^{15}N Spectra of Plant and Compost Materials

	Signal (ppm)			
	−210	−257	−306	−346
Starting material:				
^{15}N-*Lolium perenne*[a]	—	220	240	290
^{15}N-Wheat[a]	1300	450	420	760
Composts:				
^{15}N-Wheat[a]	30	40	—	30
Wheat[b]	40	50	—	30
Beechwood[b]	30	40	—	30
Fractions from a				
Lolium perenne compost				
Fulvic acid	10	20	20	20
NaOH-Extract	—	40	40	30
Residue	—	40	30	—

[a] Plants grown on ^{15}N-enriched potassium nitrate (99% enrichment).
[b] Nitrogen natural-abundance plant material composted after the addition of ^{15}N-enriched ammonium sulfate (99% enrichment).

T_{NH} and $T_{1\rho H}$ were estimated from a determination of the spectral intensities I as a function of the contact time t. Figure 15.7 compiles some typical results. The steep initial increase of I at the application of fairly short contact times and the rather slow decrease of I at longer t show that the inequality $T_{NH} \ll T_{1\rho H}$ is fulfilled. From these curves standard computer programs calculate T_{NH} and $T_{1\rho H}$ routinely. The lines drawn through the experimental points result from these fits. Within the precision of the intensities measured, each curve can be described by one T_{NH} and one $T_{1\rho H}$. However, as already stated for the T_{1H} above, these numbers should not be taken to mean that the magnetic relaxation of all fractions of these complex mixtures hidden under one signal can really be characterized by a single exponential. A distribution of relaxation times would almost certainly be found, if the curves had to be fitted to a greater number of more precise $I(t)$ points. The times determined here should rather be taken as practical hints for the optimization of the experimental parameter in the CPMAS NMR experiments. Table 15.6 presents typical results.

In order to assess the possibility of evaluating quantitatively the ^{15}N CPMAS NMR spectra, several extracts were studied by ^{15}N solution NMR and by the standard CPMAS protocol presented above. Some of these comparisons are given in Figures 15.5 and 15.6. In Figure 15.8 the two spectra of the extract from a seven-day-old ^{15}N-enriched *Lolium perenne* compost are directly compared. The result of the integration process is given in the figure. Much to our surprise, the two sets of intensities are identical within the precision of the data. This in our opinion indicates definitely that we can, at least for the enriched composts and their extracts, obtain quantitative data.

Verification and Comparison of NMR Measurements
for Nitrogen in Soil and Extracts

For the spectra from the enriched compost 20 000 free induction decays were accumulated, while approx. 1 000 000 transients had to be collected for the compounds with natural ^{15}N abundance. In Figure 15.9 the ^{15}N CPMAS NMR spectra of a *Lolium perenne* compost and a black calcaric regosol from Ismaning, and their fractions as obtained from the NaOH extraction, are compared.

The line broadening applied to the spectra from the soil and its NaOH-extracted fraction was 50–100 Hz higher than the 10–50 Hz used for the *Lolium* compost. However, the overall features of these two series are surprisingly similar. In both sets more than 80% of the intensity is covered by the intense amide/peptide peak. All signals assigned and well characterized for the ^{15}N-enriched composts are also identifiable in the soil fractions. No indications for any "new" nitrogen-containing structures, not present in the composts, can be found. The Ismaning soil residue was the only residue that after 2 000 000 transients yielded something similar to a spectrum, but it has a very poor signal to noise (S/N) ratio and, together with the spectra of the complete

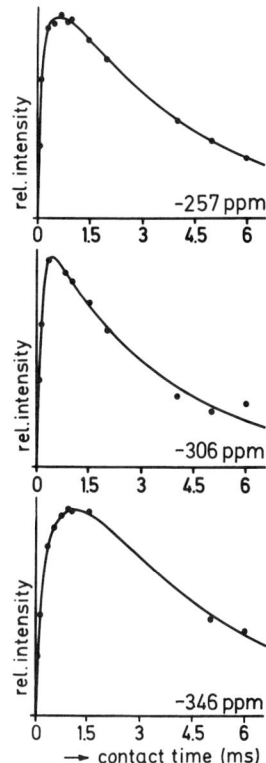

Figure 15.7 Relative intensity of three ^{15}N signals in a compost from ^{15}N-labeled wheat straw (631 days of incubation) as a function of the contact time t. The line drawn through the experimental points results from fitting $I(t)$ with one T_{NH} and one $T_{1\rho H}$.

Table 15.6 Typical Results for the Distribution of Relaxation Times of Plant Material, Composts and Humic Substances in a CPMAS ^{15}N NMR Experiment

Signal (ppm)	$T_{1\rho H}$ (ms)	T_{NH} (µs)	T_{1H} (ms)
Starting material			
−256	6	230	220
−306	6	90	240
−345	7	300	290
Compost			
−256	6	100	70
−306	6	60	70
NaOH extract			
−256	5	110	40
−306	4	60	25
Fulvic acid			
−257	3	250	15
−340	6	290	15

285

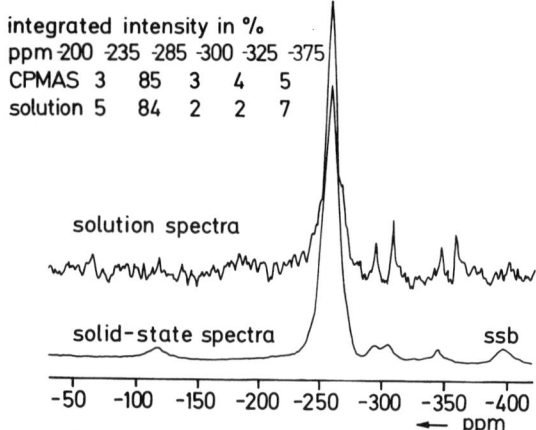

Figure 15.8   Comparison of the solution and ^{15}N CPMAS NMR spectra (30.4 MHz) of a sodium hydroxide extract from a *Lolium perenne* compost. In the table insert, the relative integrated intensities for the two spectra are compiled. The chemical shift (ppm) values indicate the range of integration (-200 to -375 ppm).

soils from Oberwarngau and Göttinger Wald (Figures 15.10 and 15.11), shows the absolute limits that our group can reach with the present equipment. The residues obtained after NaOH extraction for the last two soils did not yield any assignable spectrum after two million transients and half a week of spectrometer time for each attempt. Overall the four series of spectra compiled in Figures 15.9 to 15.11 are surprisingly similar. The main peak of the complete soil spectra has definitely a larger half-width or rather a more pronounced unsymmetrical foot than any of the extracts. This observation is corroborated by a comparison with the Ismaning residue and also with the *Lolium* compost. It cannot be decided at the moment whether it results from a fraction of insoluble proteinaceous material with very short ^{15}N spin–spin relaxation times or whether differences in chemical composition exist between the insoluble and the soluble fraction.

Nitrogen Functional Groups in Composts, Soil, and Extracts

No signals clearly assignable to heteroaromatics were found in any of the compost or soil spectra. However, under its broad foot on the low-field side (-190 to -230 ppm) the broad and intense peptide signal could cover resonances originating from indoles, pyrimidine and purine bases, quinone, iminium salts, and pyrroles. The occurrence of a larger concentration of pyrimidine and purine derivatives can be excluded because for these heterocycles about 30% of the nitrogen would show up in the chemical shift range between -140 and -180 ppm,[26,27] a region that is completely free of signals in all the spectra studied hitherto. Indoles and quinone iminium salts could be

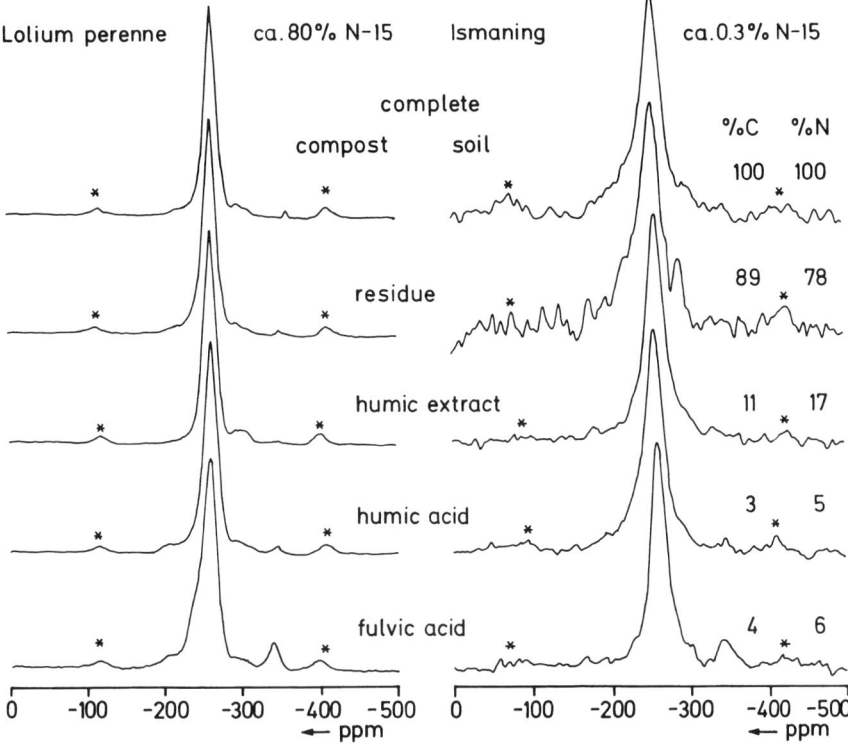

Figure 15.9 Comparison of the ^{15}N CPMAS NMR spectra (30.4 MHz) of a *Lolium perenne* compost and a black calcaric Regosol from Ismaning, and their fractions as obtained from the sodium hydroxide extraction. For the soil and its extracts, %C and %N give the percentages of the total organic C and N, respectively, contained in the various fractions.

hidden in the chemical shift range given above, but even for the very improbable case that this whole area should be assigned to these structures, they cannot contribute more than 10% to the total signal intensity. In Figure 15.12 the spectra of the sodium hydroxide extracts, the humic acids, and the fulvic acids are collected for the soils. The NaOH-extract and humic acid spectra are very similar. This is to be expected, because in most extracts the humic acids account for the biggest fraction. In Table 15.7 the results of the integration of the spectra are presented. The poor S/N ratio makes it unreasonable to give the integrals for the individual spectra. The reproducibility of the integration procedure depends upon the S/N ratio, and upon the baseline corrections necessary. In order to determine the practical limits of the reproducibility, the free induction decays of several spectra of the ^{15}N-enriched materials were three times independently Fourier-Transformed, phase- and baseline corrected, and integrated. The integrals could in most cases be reproduced

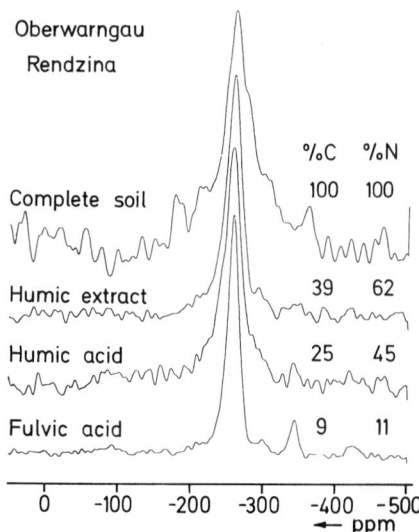

Figure 15.10 Natural-abundance ^{15}N CPMAS NMR spectra (30.4 MHz) of a Rendzina and its main fractions as obtained from the standard sodium hydroxide extraction. %C and %N: percentages of the C and N, respectively, bound in each fraction. Humic extract = complete NaOH extract.

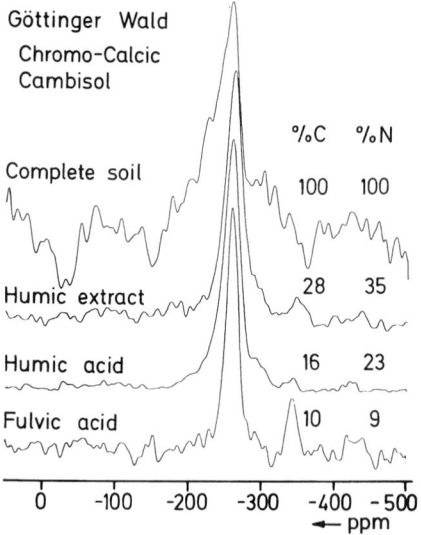

Figure 15.11 Natural-abundance ^{15}N CPMAS spectra (30.4 MHz) of a chromo-calcic Cambisol and its main fractions as obtained from the standard sodium hydroxide extraction. %C and %N: percentages of the C and N, respectively, bound in each fraction.

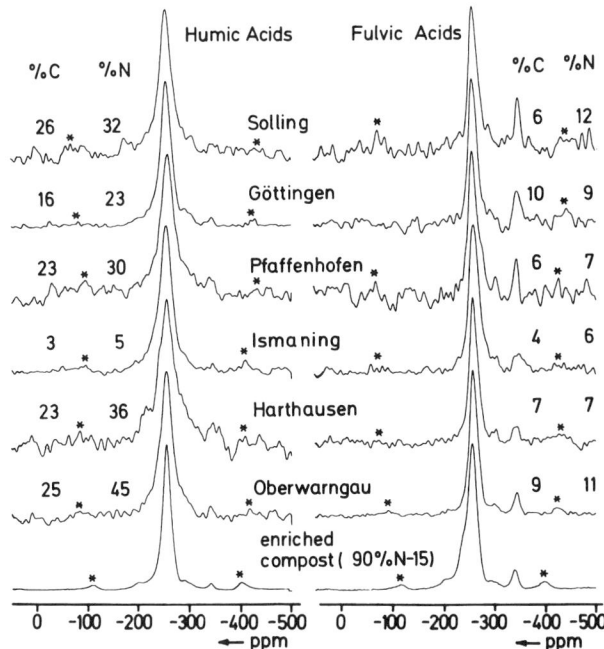

Figure 15.12 Comparison of the natural-abundance ^{15}N CPMAS NMR spectra (30.4 MHz) of humic acids and fulvic acids obtained from the six German soils given in Table 15.1 with spectra obtained from the ^{15}N-enriched humic and fulvic acids derived from a *Lolium perenne* compost after 200 days of fermentation (^{15}N abundance 90%). Asterisks (*) indicate spinning sidebands.

Table 15.7 Averages of the Relative Intensities in Spectra of Humic Matter Fractions of Soil with ^{15}N Natural Abundance[a]

Chemical Shift Range (ppm)	Assignment[b]	Average Intensity (%)		
		HE[c]	HA[d]	FA[e]
−190 to −285	"Amide"	87 (85–90)[f]	86 (84–90)	83 (81–87)
−285 to −325	N	8 (7–8)	10 (6–12)	6 (4–7)
−325 to −375	—NH$_2$	5 (3–7)	5 (3–6)	11 (7–13)

[a] The integration routine sets the total intensity at 100%. The reproducibility of the integration procedure depends on the signal to noise (S/N) ratio, which for the worst spectra is approximately 5%. Deviations from 100% stem from rounding-off errors.
[b] A more complete assignment is given in Table 15.3.
[c] HE, humic extract = NaOH/H$_2$O extract.
[d] HA = humic acid.
[e] FA = Fulvic acid.
[f] Numbers in parentheses: variation among different samples.

with a maximal variation of $\pm 2\%$ of the total intensity. Table 15.7 gives the average intensities obtained and the variations found for the three different fractions. It appears significant that the fulvic acid fractions have higher contents of free amino groups and lower contents of peptide nitrogen and NH derivatives. The larger concentrations of the free amino groups are consistent with the high solubility of fulvic acids in acidic aqueous solutions.

The general conclusion from all the spectra presented here is that hetero-aromatic nitrogen-containing structures or Schiff bases can contribute at most 10% to the total nitrogen in the SOM of these spectra. This is contrary to the 40 to 50% claimed from chemical analysis.[12,13] Also, the nitrile groups incorporated into a recent model for humic material[14] cannot be present in significant concentrations; they appear to be an artifact of the analytical pyrolysis process applied in order to characterize the fragments.

Conceptual Model for Nitrogen Incorporation in
Soils and Humic Substances

From a critical synopsis of the vast amount of data collected by ^{13}C NMR, and more recently also by ^{15}N NMR, a fairly simple and consistent model emerges for the structure and formation of the SOM in typical moderate-climate soils. The initial humification process preferentially removes the carbohydrates by microbial and fungal activity. Simultaneously lignin compounds and proteinaceous material are mineralized, although at a much lower rate. The residue of this normal biochemical activity is a complex mixture of macromolecules. Proteinaceous material, partially decomposed lignins and modified carbohydrates form the main constituents, in varying concentrations. Interaction of these mixtures with the mineral phase of the soil can form aluminosilicate–SOM complexes, which can be metabolized only very slowly because they are insoluble in aqueous solution or because the soluble fractions are covered by an insoluble, ill-defined, surface layer. Also, the application of the standard acid hydrolysis procedure to humic material for the determination of the "protein" content could be unreliable, and it may underestimate the peptide nitrogen content considerably, because the heterogeneous macromolecules are not very soluble in acidic water and during the initial stages of hydrolysis they may form acid-insoluble "coatings" that prevent further chemical attack.

This qualitative description, which does not involve any novel chemical reactions producing significant concentrations of structures not present in biological material, would also be in accord with the established metabolism of microbes and fungi.

From the types of soils analyzed here, it has been established for the ^{13}C NMR of SOM that most, if not all, of the carbon is detected by NMR and can be classified quantitatively, although with limited chemical resolution and quantitative precision.

The limited material available for ^{15}N NMR of composts and SOM[8–11,28–30] will certainly be criticized again with the claim that some nitrogen-containing structures are not cross-polarized by the protons and thus remain invisible in the spectra. Although it is impossible with the spectroscopic material available at present to completely rule out this possibility, we were able to find materials and substances related to SOM, showing ^{15}N NMR signals that had to be assigned definitely to nitrogen not bound directly to a proton. Figure 15.13 gives such an example. After about two years of fermentation of an enriched wheat straw at 60% WHC and more than 75% carbon loss from the sample, a rather intense signal developed around 0 ppm. This signal must most probably be assigned to nitrate ions (~0 ppm) formed in the late stage of mineralization. The only other assignment possible in this chemical shift range would be aliphatic and aromatic nitro compounds,[26,27] which in our opinion can be ruled out because no conceivable reaction sequence exists for their formation under the conditions of the composting experiments. At any rate, nitrogens in NO_2 groups or NO_3^- ions do not have any protons in their immediate atomic neighborhood, and are still visible in the CPMAS spectra. The "distance of closest approach" of a proton to a nitrogen nucleus in NO_3^- or NO_2 is certainly similar to the distances in heteroaromatics, and thus this spectrum gives reassuring evidence that one should be able to pick up nitrogen in tertiary and quaternary structures. Also, recently we obtained natural-abundance ^{15}N CPMAS NMR spectra of penguin guano and the soil under an antarctic algal ground layer. In Figure 15.14, some of these results are presented to show that significant variation can be observed in ^{15}N CPMAS spectra. The penguin guano shows well-resolved signals on the low-field side of the peptide peak, which most probably stem from uric acid, its derivatives, and its decomposition products. The antarctic "soil" spectrum shows a very well-resolved but until now unassigned peak at −200 ppm. For comparison, a ^{15}N CPMAS NMR spectrum from powdered and freeze-dried *Chlorella* material is included.

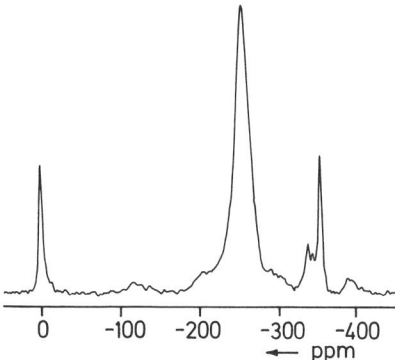

Figure 15.13 ^{15}N CPMAS NMR spectrum (30.4 MHz) from a complete freeze-dried wheat compost after 631 days of fermentation (final pH 5.8; carbon loss ~ 80%).

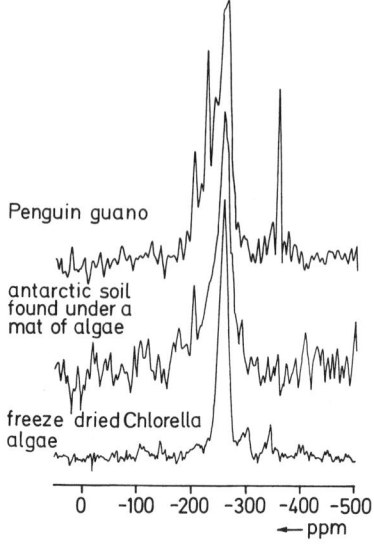

Penguin guano

antarctic soil
found under a
mat of algae

freeze dried Chlorella
algae

0   -100 -200 -300 -400 -500
←ppm

Figure 15.14 Natural-abundance ^{15}N CPMAS NMR spectra (30.4 MHz) of an antarctic penguin guano, an antarctic soil, and freeze-dried *Chlorella* algae.

Acknowledgments This work was supported by the BMFT (German Ministry of Science and Technology) by Grant No. BEO/51 0339137C. Professor K. Haider (Deisenhofen, Germany) and Professor F.J. González-Vila (Sevilla, Spain) gave invaluable advice and support. Those in the workshops of our faculty, especially Mr. R. Knott and Mr. G. Wührl, kept the instruments in working condition, and Mr. E. Treml helped with many of the analytical and preparative problems. All this support is gratefully acknowledged. The penguin guano and the antarctic soil samples were provided by Professor Blume and Dr. Beyer (University of Kiel, Germany).

References

1. Aiken, G. R., McKnight, D. M., Wershaw, R. L., and MacCarthy, P. (eds.), *Humic Substances in Soil, Sediment and Water*, John Wiley, New York, NY, 1985.
2. Frimmel, F. H., and Christman, R. F. (eds.), *Humic Substances and Their Role in the Environment*, John Wiley, Chichester, 1988.
3. Hayes, M. B., MacCarthy, P., Malcolm, R. L., and Swift, R. S. (eds.), *Humic Substances*, Vol. II, *In Search of Structure*, Wiley–Interscience, Chichester, 1989.
4. Parsons, J. W., Isolation of humic substances from soil and sediments. In *Humic Substances and Their Role in the Environment*, Frimmel, F. H., and Christman, R. F. (eds.), Wiley–Interscience, Chichester, 1988, pp. 3–14.
5. Fründ, R., and Lüdemann, H.-D., Quantitative characterization of soil organic matter and its fractionation products by solid state high resolution C-13 (CPMAS) spectroscopy. *Z. Naturforsch., C* 46, 982–988, 1991.
6. Wilson, M. A., *NMR Techniques and Applications in Geochemistry and Soil Chemistry*, Pergamon, Oxford, 1987.
7. Fründ, R., and Lüdemann, H.-D., The quantitative analysis of solution and CMAS C-13 NMR spectra of humic material. *Sci, Total Environ.* 81/82, 157–168, 1989.

8. Benzing-Purdie, L., Ripmeester, J. A., and Preston, C. M., Elucidation of the nitrogen forms in melanoidins and humic acid by nitrogen-15 cross polarization–magic angle spinning nuclear magnetic resonance spectroscopy. *J. Agric. Food Chem.* 31, 913–915, 1983.

9. Cheshire, M. V., Williams, B. L., Benzing-Purdie, L., Ratcliffe, C. J., and Ripmeester, J. A., Use of ^{15}N NMR spectroscopy to study transformations of nitrogenous substances during incubation of peat. *Soil Use Management* 6, 90–92, 1990.

10. Zhuo, S., Wen, Q., Du, L., and Wu, S., The nitrogen form of nonhydrolyzable residue of humic acid. *Chin. Sci. Bull.* 37, 508–511, 1992.

11. Almendros, G., Fründ, R., González-Vila, F. J., Haider, K. M., Knicker, H., and Lüdemann, H.-D., Analysis of ^{13}C and ^{15}N CPMAS NMR-spectra of soil organic matter and composts. *FEBS Lett.* 282, 119–121, 1991.

12. Schnitzer, M., Nature of nitrogen of humic substances from soil. In *Humic Substances in Soil, Sediment and Water*, Aiken, G. R., McKnight, D. M., Wershaw, R. L., and MacCarthy, P. (eds.), John Wiley, New York, 1985, pp. 303–325.

13. Anderson, H. A., Bick, W., Hepburn, A., and Stewart, M., Nitrogen in humic substances. In *Humic Substances*, Vol. II, *In Search of Structure*, Hayes, M. B., MacCarthy, P., Malcolm, R. L., and Swift, R. S. (eds.), Wiley–Interscience, Chichester, 1989, pp. 223–253.

14. Schulten, H.-R., and Schnitzer, M., A state of the art structural concept for humic substances. *Naturwissenschaften* 80, 29–30, 1993.

15. Hewitt, L. J., *Sand and Water Culture Methods Used in the Study of Plant Nutrition*, Commonwealth Agriculture Bureau, 1966, pp. 187–199.

16. Dev, S. B., Burum, D. P., and Rha, C. K., High resolution CPMAS and solution NMR studies of polysaccharides. Part 1: starch. *Spectrosc. Lett.* 20, 853–869, 1987.

17. Fründ, R., and Lüdemann, H.-D., ^{13}C-NMR-spectroscopy of lignins and lignocellulosic material. In *Physico-chemical Characterization of Plant Residues for Industrial and Feed Use*, Chesson, A., and Orskov, E. R. (eds.), Elsevier Applied Science, Brussels, 1989, pp. 110–117.

18. Gerothanassis, J. P., Simple reference baseline substraction −90° pulse sequence for acoustic ringing elimination in pulsed fourier transform NMR-spectroscopy. *Magn. Reson. Chem.* 24, 428–433, 1986.

19. Fründ, R., Quantitative Charakterisierung von organischen Bodeninhaltsstoffen mit Hilfe der C-13-NMR Spektroskopie. Dissertation, University of Regensburg, 1988.

20. Bendall, M. R., and Pegg, D. T. J., Editing of ^{13}C-NMR spectra; a pulse sequence for the generation of subspectra. *J. Am. Chem. Soc.* 103, 4603–4605, 1983.

21. Harris, R. K., *Nuclear Magnetic Resonance Spectroscopy*, Pitman, London, 1983.

22. Mehring, M., *High Resolution NMR in Solids*, 2nd edn., Springer, Berlin, 1983.

23. Flaig, W. J. A., Beutelspacher, H., and Rietz, E., Chemical composition and physical properties of humic substances. In *Soil Components*, Vol. 1, Gieseking, J. E. (ed.), Springer, New York, NY, 1975, pp. 1–211.

24. Flaig, W. J. A., Generation of model chemical precursors. In *Humic Substances and Their Role in the Environment*, Frimmel, F. H., and Christman, R. F. (eds.), John Wiley, Chichester, 1988, pp. 75–92.

25. Stevenson, F. J., Geochemistry of soil humic substances. In *Humic Substances in Soil, Sediment and Water*, Aiken, G. R., McKnight, D. M., Wershaw, R. L., and MacCarthy, P. (eds.), John Wiley, New York, NY, 1985, pp. 13–52.

26. Witanowski, M., Stefaniak, L., and Webb, G. A., Nitrogen NMR spectroscopy. *Annu. Rep. NMR Spectrosc.* Vol. 25, 1993.

27. Martin, G. J., Martin, M. L., and Gouesnard, J.-P., 15-N NMR spectroscopy. In *NMR Basic Principles and Progress*, Vol. 18, Diehl, P., Fluck, E., and Kosfeld, R. (eds.), Springer, Heidelberg, 1981.

28. Knicker, H., Fründ, R., Alemendros, G., González-Vila, F. J., Martin, F., and Lüdemann, H.-D., Characterisation of nitrogen on compost by N-15-NMR. *Humus Uutiset—Finnish Humus News* 3, 313–315, 1991.

29. Knicker, H., Fründ, R., and Lüdemann, H.-D., The chemical nature of nitrogen in native soil organic matter. *Naturwissenschaften* 8, 219–221, 1993.

30. Knicker, H., Fründ, R., and Lüdemann, H.-D., ^{15}N-NMR studies of humic substances in solution. In *Humic Substances in the Global Environment and Implications on Human Health*, Senesi, N., and Miano, T. M. (eds.), Elsevier, Amsterdam, 1994, pp. 501–506.

16.

Acquisition and Interpretation of Liquid-state ¹H NMR Spectra of Humic and Fulvic Acids

JERRY A. LEENHEER, T. I. NOYES, & ROBERT L. WERSHAW

Fourier Transform nuclear magnetic resonance (NMR) spectrometers have become available to many researchers studying humic substances over the last decade. As a result, liquid-state proton (¹H) NMR spectrometry has been commonly used to determine the nonexchangeable proton distribution in humic and fulvic acids.[1-4] The high sensitivity of the ¹H nucleus to NMR spectrometry allows spectra to be obtained on a relatively small quantity of sample (10–100 mg) in a short time (10–30 min).

¹H NMR spectrometric profiles of humic substances are useful to environmental scientists in determining the source, properties, and degree of transformation (humification) of organic matter that is operationally classified as humic substances. These ¹H NMR spectrometric profiles, which provide information about hydrogen distributions in humic substances, are more useful for structural and biogeochemical studies when combined with ¹³C NMR spectra, which provide information on carbon distributions, and infrared spectra, which provide information on oxygen distributions. These three spectra, in conjunction with elemental composition, molecular weight, and titrimetric data, can then be synthesized to provide average structural characteristics that can be related to source, properties, and degree of humification of the organic material being studied.

Special challenges, that are not met when obtaining the spectra of pure compounds, are encountered in obtaining ¹H NMR spectra of natural humic substances. These challenges include (1) lack of complete dissolution of macromolecular humic substances at the high concentrations required for NMR studies; (2) significant concentrations of exchangeable protons giving broad peaks that obscure portions of the spectrum; (3) broad peaks of non-exchangeable protons over the entire spectrum that cause difficulties in correct phasing; (4) unstable structures that oxidize, hydrolyze, and structurally

295

rearrange at the high pH conditions under which humic substances are the most soluble; and (5) the presence of unusual structures that complicate straightforward assignment of structure from handbook data.[5] The purposes of this chapter are to describe methods of sample preparation and to provide generally applicable structural assignments whereby ^{1}H NMR spectra suitable for quantitative studies of humic substance structure may be obtained and interpreted.

Acquisition of ^{1}H NMR Spectra

For this study, ^{1}H NMR spectra were obtained on Varian XL 300* and Varian FT-80A NMR spectrometers. On the 300 MHz spectrometer, acquisition parameters used to obtain quantitative spectra were: spectral window = 8000 Hz; tip angle = 25°; aquisition time = 1.0 s; and pulse delay = 5 s. On the 80 MHz spectrometer, acquisition parameters were: spectral window = 2000 Hz; tip angle = 45°; acquisition time = 2.0 s; and pulse delay = 5 s. These conditions were judged to give quantitative spectra because proton spin–lattice relaxation times for both humic and fulvic acids from the Suwannee River were determined to be less than or equal to 0.4 s by the progressive saturation method.[6]

One of the most important considerations in sample preparation is to avoid aggregation and precipitation of the humic or fulvic acid. To obtain quantitative spectra in liquid-state NMR, the molecules must be truly dissolved so that they are freely rotating in solution. Thus, the concentration of humic substances must be below concentrations where aggregation and precipitation occur.

Figure 16.1 illustrates differences in spectral integrations for high and low concentrations of fulvic acid isolated from the Suwannee River, Georgia. The spectral differences in Figure 16.1 are possibly caused by two factors. One factor is selective aggregation or precipitation of the lower-solubility components of fulvic acid at the higher concentration, which changes the composition of the dissolved fulvic acid molecules assayed in the ^{1}H NMR experiment. The other factor is the difficulty in phasing the rolling baseline obtained with the high-concentration sample. The dished baseline in Figure 16.1, spectrum A, is a result of the best compromise that could be made in the phasing of this spectrum. The spectra should be plotted from −5 to 15 ppm to assure correct phasing.

Humic acids, having larger molecular weights and more interactive functional groups than fulvic acids, aggregate to a greater degree than fulvic acids, and they have a greater concentration dependence for ^{1}H NMR spectra than fulvic acids. Therefore, to minimize spectral changes due to

---

*Any use of trade names in this manuscript is for descriptive purposes only and does not constitute endorsement by the US Geological Survey.

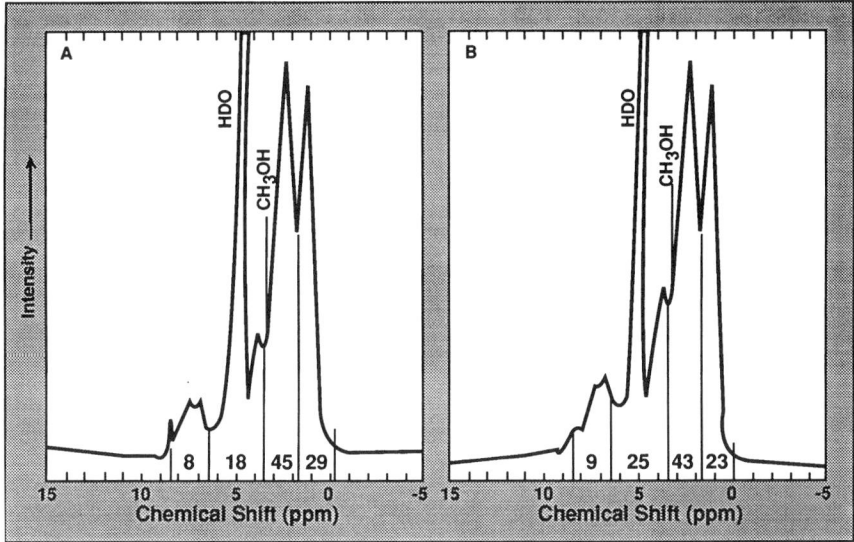

Figure 16.1 ^{1}H NMR spectra of fulvic acid from the Suwannee River. (A) 200 mg/mL D$_{2}$O, pH 6; (B) 50 mg/mL D$_{2}$O, pH 6.0. Spectra were determined on a Varian XL-300 NMR spectrometer. The numbers in the various spectral regions are percentages of the total spectral integral.

aggregation and phasing problems, ^{1}H NMR spectra of humic substances in aqueous solvent systems should be acquired at minimum concentrations relative to considerations of signal strength and the intensity of the exchangeable proton peak.

Humic substances are soluble in certain aqueous and nonaqueous solvents. For spectra obtained in an aqueous solvent, the important experimental parameters to be adjusted are pH, the extent of deuterium exchange, use of internal standards, and time. Fulvic acids have sufficient solubility at pH 6–8 to allow measurement of the spectrum; greater pH values should be avoided, to minimize ester hydrolysis.[7] Humic acids may have to be taken to pH 12–13 to ionize phenolic groups, to obtain sufficient solubility.

Humic substances contain groups that enolize (structurally rearrange from ketone to vinyl alcohol structures) at high pH values, as indicated by carbon methylation[6] where the carbon attached to the vinyl double bond adds a methyl group. Loss of proton intensity was observed for the peak at 2.3 ppm in Figure 16.1 when this fulvic acid remained at pH 13 for several days. Deuterium exchange with hydrogen on enolizable carbon structures causes this loss in intensity. Therefore, ^{1}H NMR spectra of humic acids in high-pH solutions should be measured immediately to minimize the extent of exchange reactions.

The spectra in Figure 16.1 were also optimized by minimizing the HDO solvent peak intensity by repeated evaporations of the neutralized sample from D$_2$O solution. Exchangeable protons that exchange with the deuterium are removed by this procedure. After this procedure is repeated two to three times, the sample is dissolved in "100%" D$_2$O to further minimize the HDO solvent peak.

Sometimes it is advantageous to use potassium salts of humic and fulvic acids, rather than sodium salts, to obtain ^{1}H-NMR spectra because of the greater solubility of the potassium salts of these organic acids. An internal reference standard, such as methanol at 3.30 ppm, should be used in aqueous solution spectra because the position of the HDO peak varies with pH.

Nonaqueous solvents for ^{1}H NMR studies of humic substances are advantageous in that they may solubilize hydrogen-saturated forms of the acids which are insoluble in water; they are solvents for derivatized (methylated and acetylated) humic substances which are insoluble in water; and they may provide observation of nonexchangeable protons in the 4.4–5.2 ppm region that is obscured by the exchangeable proton peak in aqueous solvents. ^{1}H NMR spectra of fulvic acid from the Suwannee River in various nonaqueous solvents are shown in Figure 16.2.

Spectra in aprotic nonaqueous solvents, such as N,N-dimethylformamide-d_7 and dioxane-d_8, are complicated by a broad peak consisting of exchangeable protons and associated water that usually obscures the aromatic proton region (6.5–8.5 ppm). This broad peak may be eliminated by homonuclear decoupling as was illustrated by Thorn[6] for fulvic acid from the Suwannee River dissolved in dimethyl sulfoxide-d_6. In the case of dioxane-d_8, repeated evaporations of this deuterated solvent also remove associated water from the fulvic acid so that the integral of the broad exchangeable proton peak from 6–8 ppm (Figure 16.2, spectrum D) is a quantitative measure of the exchangeable proton content after subtraction of the nonexchangeable aromatic protons.[3]

Methanol is an excellent solvent for fulvic acid, but it has two peaks that obscure portions of the spectra as shown in Figure 16.2, spectrum A. Aqueous–organic solvent mixtures, in which the water and fulvic or humic acid are proton donors and organic solvents are electron donors, are often excellent solvents for humic substances.[8,9] The acetone–water system, shown in Figure 16.2, spectrum C, is useful in that the position of the OH peak can be shifted upfield or downfield depending on the amount of acid added. Numerous solvent mixture systems have been used[8,9] to obtain complete solution of the humic fraction of interest for ^{1}H NMR studies. Minor differences between the position and integrals of the nonexchangeable proton peaks in the four spectra of Figure 16.2 are the result of uncertainties in phasing discussed previously, possible solubility limitations, and variable solvent effects on chemical shift positions of the various peaks.

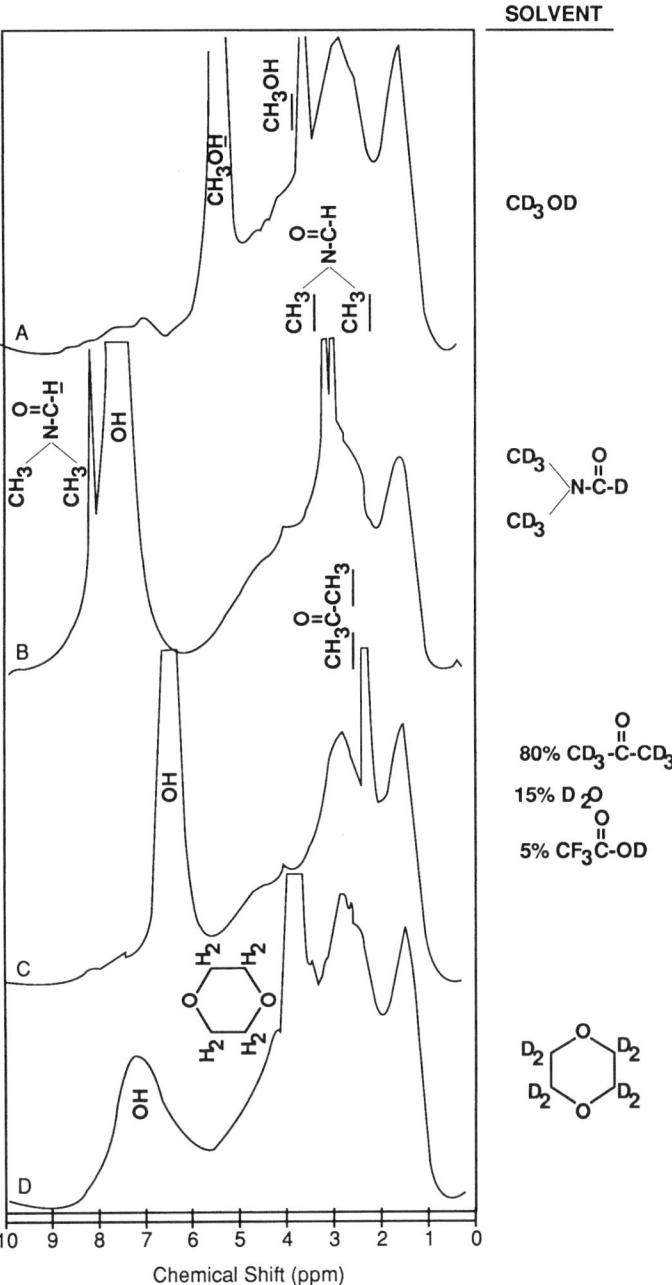

SOLVENT

Figure 16.2 ¹H NMR spectra of 100 mg/mL of fulvic acid from the Suwannee River dissolved in (A) methanol-d_4; (B) *N,N*-dimethylformamide-d_7; (C) 80% acetone-d_6, 15% D$_2$O, 5% trifluoroacetic acid; and (D) dioxane-d_8. Spectra were acquired on a Varian FT-80A NMR spectrometer.

299

Interpretation of ^{1}H NMR Spectra

A comprehensive characterization of fulvic acid from the Suwannee River[10] enabled detailed assignment of protons associated with various structural components. In this study, carboxyl groups were determined by potentiometric titration and by methylation followed by ^{13}C NMR and ^{1}H NMR measurement of methyl ester groups. Ester groups were determined by infrared spectrometry of the salt form of the acids and by ^{13}C NMR spectral measurement of the structural ester carbons that were not methylated by selective methylation of acid carboxyl groups. Ketone groups were measured by ^{13}C NMR spectrometry. Total hydroxyl groups were measured by acetylation with acetic anhydride and permethylation with sodium hydride and methyl iodide followed by NMR measurement of the added acetyl or methyl groups. Phenolic and alcoholic hydroxyls were measured by a combination of selective derivatization involving both acetylation and methylations followed by infrared and NMR measurements of the derivatives. In addition to this functional-group information, the spectrometric measurements provide quantitative information on structural features such as aromaticity, aliphatic structures such as alicylic rings, and ether linkages. After synthesizing the information of the multiple approaches and measurements, a summary of proton assignments for fulvic acid from the Suwannee River is given in Figure 16.3.

Much of the aliphatic 1 peak in Figure 16.3 is associated with branched methyl groups β to carbonyl groups (likely to be carboxylic acids). This assignment is supported by: (1) aliphatic methyl groups were detected by the attached-proton test by ^{13}C NMR;[6] (2) methyl groups give the largest peaks per carbon atom because the observed sensitivity ratio of detection of methyl/methylene/methine groups in ^{1}H NMR is 3:2:1; and (3) the chemical shift of the aliphatic 1 peak (1.2 ppm) corresponds to methyl groups β to carbonyl groups. Terminal aliphatic methyl groups are at lower frequency (near 0.9 ppm), and aliphatic straight-chain methylene groups are at higher frequency (near 1.3 ppm).[5]

The valley between the aliphatic 1 and 2 peaks is associated with aliphatic, alicyclic structures and with aliphatic hydrogens on carbons β to single-bond C–O linkages in alcohols, ethers, and esters. Methylene and methine groups predominate in this region; therefore, there is about the same number of structural carbons in this spectral region as carbons associated with the aliphatic 1 peak because of the H/C sensitivity ratio considerations.

The aliphatic 2 peak in Figure 16.3 is associated with protons on carbons attached to carbonyl groups or aromatic rings. Aliphatic ketones have the greatest potential to be observed in the aliphatic 2 peak because they may contain up to five protons on carbons on both sides of the carbonyl groups (acetone excepted). A maximum of three protons can be associated with esters and methyl-substituted aromatics, and a maximum of two protons can be associated with carboxylic acids (acetic acid excepted). When aliphatic

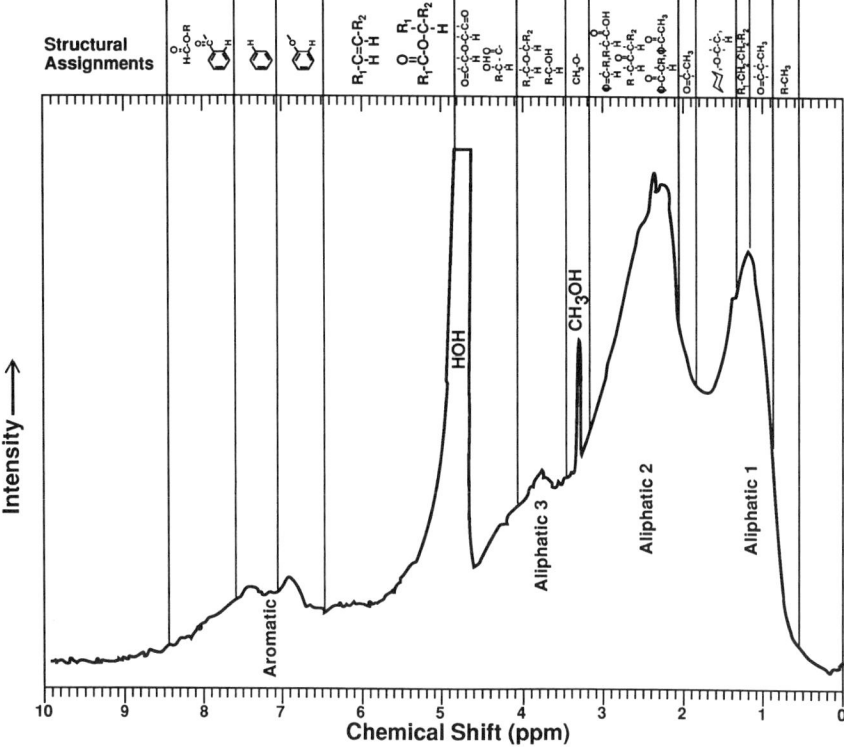

Figure 16.3 Structural assignments for interpretation of the ^{1}H NMR spectrum of fulvic acid from the Suwannee River, 50 mg/mL D$_2$O, pH 6.

ketonic groups are reduced with sodium borohydride to alcohols, about one-third of the protons in the aliphatic 2 peak are shifted to lower frequency.[11] Acid-, ester-, and aromatic-substituted aliphatics are all significant contributors to the remaining two-thirds of the aliphatic 2 peak.

The aliphatic 3 peak is associated with protons (predominantly methine[6]) attached to single C–O linkages of alcohols, ethers, and esters. Alcohols and ethers without adjacent electron-withdrawing groups (carbonyl and aromatics) occur between 3.3 and 4.1 ppm, whereas methine groups associated with acetal, ketal, α-ether and α-hydroxy acids are shifted to 4.1–4.9 ppm. Methine groups associated with the secondary alcohol portions of esters occur from 4.9 to 6.5 ppm depending on the degree of deshielding by adjacent electronegative functional groups. About 40% of the carboxyl groups in this fulvic acid are α-ether and α-ester acids.[12] Protons on the α-carbons of these acids contribute to this higher-frequency portion of the aliphatic 3 peak. Olefinic protons also potentially contribute to the aliphatic 3 peak, but the ester contribution seems to dominate in the 4.9–6.5 ppm region because proton intensity in this region disappears when structural esters are hydro-lyzed or converted to methyl esters.

The broad aromatic peak in Figure 16.3 gives a poorly resolved peak at 6.8 ppm, indicating aromatic C–O linkages, and a broad peak near 7.4 ppm that indicates aromatic protons on unsubstituted rings and aromatic protons that occur between electron-withdrawing (carbonyl groups) and electron-donating (phenol groups) substituents. Humic and fulvic acids with a greater degree of oxidation than this sample frequently give a broad peak near 8.0 ppm that is probably due to electron-withdrawing carboxyl and ketone groups.[5]

The detail in structural assignments in Figure 16.3 might be better exploited for quantitation by using spectral deconvolution programs rather than the simple integration used in this study. Groups that hydrolyze or enolize might be preserved in humic acids solubilized at high pH by determin-ing ^{1}H NMR spectra at low temperature to minimize these chemical changes.

Humic substances from different environments give a variety of spectral profiles in ^{1}H NMR spectrometry.[13] Humic substances from lakes and reser-voirs where autochthonous inputs dominate have many fewer aromatic pro-tons than do humic substances from rivers and swamps where allochthonous inputs dominate. Humic substances with minimal degradation have more protons associated with ketones in the aliphatic 2 peak, carbohydrates in the aliphatic 3 peak, and phenols in the 6.8 ppm region of the aromatic peak. As humification proceeds, ketones are oxidized and the aliphatic 2 peak decreases; carbohydrates degrade faster than aromatic and aliphatic, alicyclic structures and the aliphatic 3 peak decreases; and aromatic rings become carboxylated and aromatic protons, intensity shifts from 6.8 ppm down to 7.5–8.0 ppm.

Conclusions

The interpretations and assignments given to ^{1}H NMR spectra of fulvic acid from the Suwannee River in this chapter are more specific and definite than is generally possible because of the high degree of characterization of this fulvic acid. These interpretations might also apply to humic and fulvic acids from different environments. The full potential of information from ^{1}H NMR spectrometry of humic substances can be realized only when it is combined with ^{13}C NMR and infrared spectrometry, and with titrimetric, elemental analyses, and derivatization information that confirm tentative assignments for ^{1}H NMR spectra of humic substances.

References

1. Harvey, G. R., Boran, D. A., Chesal, L. A., and Tokar, J. M., The structure of marine fulvic and humic acids. *Mar. Chem.* 12, 119–132, 1983.

2. Hatcher, P. G., Maciel, G. E., and Dennis, L. W., Aliphatic structure of humic acids; a clue to their origin. *Org. Geochem.* 3, 43–48, 1981.

3. Noyes, T. I., and Leenheer, J. A., Proton nuclear-magnetic-resonance studies of fulvic acid from the Suwannee River. In *Humic Substances in the Suwannee River, Georgia: Interactions, Properties, and Proposed Structures*, Averett, R. C., Leenheer, J. A., McKnight, D. M., and Thorn, K. A. (eds.), US Geological Survey Open-File Rept. 87-557, Denver, CO, 1989, pp. 235–250.

4. Ruggiero, P., Interesse, F. S., and Sciacovelli, O., ^{1}H and ^{13}C-NMR studies on the importance of aromatic structures in fulvic and humic acids. *Geochem. Cosmochim. Acta* 43, 1771–1775, 1979.

5. Simons, W. W. (ed.), *The Sadtler Handbook of Proton NMR Spectra*, Sadtler Research Laboratories, Philadelphia, PA, 1978.

6. Thorn, K. A., Nuclear-magnetic-resonance spectrometry investigations of fulvic and humic acids from the Suwannee River. In *Humic Substances in the Suwannee River, Georgia: Interactions, Properties, and Proposed Structures*, Averett, R. C., Leenheer, J. A., McKnight, D. M., and Thorn, K. A. (eds.), US Geological Survey Open-File Rept. 87-557, Denver, CO, 1989, pp. 251–330.

7. Antweiler, R. C., The hydrolysis of Suwannee River fulvic acid. In *Organic Substances and Sediments in Water I: Humics and Soils*, Baker, R. A. (ed.), Lewis Publishers, Chelsea, MI, 1991, pp. 163–177.

8. Hayes, M. H. B., Extraction of humic substances from soils. In *Humic Substances in Soil, Sediment, and Water: Geochemistry, Isolation, and Characterization*, Aiken, G. R., McKnight, D. M., Wershaw, R. L., and MacCarthy, P. (eds.), Wiley, New York, NY, 1985, pp. 329–362.

9. Porter, L. K., Factors affecting the solubility and possible fractionation of organic colloids extracted from soil and leonardite with an acetone–H_2O–HCl solvent. *J. Agric. Food Chem.* 15, 807–811, 1967.

10. Averett, R. C., Leenheer, J. A., McKnight, D. M., and Thorn, K. A., *Humic Substances in the Suwannee River, Georgia: Interactions, Properties, and Proposed Structures*, US Geological Survey Open-File Rept. 87-557, Denver, CO, 1989.

11. Leenheer, J. A., Wilson, M. A., and Malcolm, R. L., Presence and potential significance of aromatic-ketone groups in aquatic humic substances. *Org. Geochem.* 11, 273–280, 1987.

12. Leenheer, J. A., Wershaw, R. L., and Reddy, M. M., Strong-acid, carboxyl-group structures in fulvic acid from the Suwannee River, Georgia, *Book of Abstracts, 205th ACS National Meeting, March 28–April 2 1993, Denver Colorado*, Environmental Abstract No. 99.

13. Leenheer, J. A., Chemistry of dissolved organic matter in rivers, lakes, and reservoirs. In *Environmental Chemistry of Lakes and Reservoirs*, Baker, L. A. (ed.), Adv. Chem. Ser. No. 237, American Chemical Society, Washington, DC, 1994, pp. 195–221.

17.

Preparation of Low-Carbon Sediments from the Mississippi River and Certain Tributaries for Solid-state CPMAS ^{13}C NMR Analysis

JERRY A. LEENHEER, GARY E. MACIEL, & T. I. NOYES

The nature of organic carbon in aquatic sediments and soils with low carbon contents and significant contents of paramagnetic elements such as iron and manganese is difficult to assess by solid-state, cross-polarization magic angle spinning (CPMAS) ^{13}C nuclear magnetic resonance (NMR) spectrometry because of the inherent low sensitivity of ^{13}C NMR analyses, and band broadening and sensitivity losses caused by paramagnetic elements. Other investigators have addressed this problem in the analysis of soils by enriching the organic carbon content by flotation, by magnetic separation of paramagnetic minerals, and by chemical reduction of iron by stannous chloride and sodium dithionite.[1] In this study, they found that satisfactory ^{13}C NMR spectra could be obtained if the C/Fe ratio was greater than 1 wt%. Each of the physical and chemical treatments used to increase the C/Fe ratio resulted in losses of organic matter and changes in the nature of organic matter through physical fractionation and chemical alteration. Suspended stream sediments frequently have equivalent contents of organic carbon and sesquioxide coatings with which the organic matter is associated. These sesquioxide coatings consist predominantly of iron and manganese oxyhydroxides[2,3] that cause problems with NMR analyses.

In this chapter we describe a method to enrich organic matter and remove iron and manganese from low-carbon sediments sampled from the Mississippi, Illinois, and Ohio Rivers with minimal loss and alteration of the organic matter. The second objective is to characterize the sedimentary organic matter by ^{13}C NMR using recent advances that increase instrument sensitivity.

Sediment Sampling and Treatment Procedures

Suspended and bed sediments were collected during a sampling cruise on the Mississippi River during May–June 1990.[4] Fine bed sediments were collected in depositional regions of the river or tributaries with a pipe dredge. Suspended silts were collected using a continuous-flow centrifuge operated on board the Research Vessel *Acadiana*.[5] Both bed sediments and suspended silts were freeze-dried prior to additional treatment procedures and NMR analyses.

A flow chart of selective mineral dissolution procedures is presented in Figure 17.1. The acid pyrophosphate treatment[6] was placed first in the sequence to remove calcium and magnesium minerals that would form insoluble oxalates in the following extraction. Acid pyrophosphate solution also removes some of the iron. The next step, use of 0.15 M oxalic acid at pH 3, was previously determined[7] to be the most efficient extraction method for removing amorphous aluminum, iron, and silicon from soils. Both the pyrophosphate and oxalate extractants also release adsorbed fulvic acid, which is recovered by adsorption chromatography on XAD-8 resin* The two hydrofluoric acid treatments were performed under dilute acid conditions at room temperature because a previous study[8] found that the high concentrations (2.88 M) and long reaction periods needed to completely remove silicates destroyed the carbohydrate portion of organic matter through acid hydrolysis. Dilute hydrofluoric acid destroys only hydrous silicates (clay minerals), but it does not attack silicon oxides (quartz).

Gentle mixing of the samples inside the dialysis bag was achieved by placing the bag inside a 1 L wide-mouth polyethylene bottle with the extracting solutions, and rotating the bottle attached to the rim of a bicycle wheel at six revolutions per minute. This rotation resulted in a gentle inversion of the sediment inside the dialysis bag so that the sediment mixed without physically abrading the bag. The cellulose acetate dialysis membrane was chemically stable to all the extracting solutions, and it greatly facilitated the separation of colloidal sediments from the extracting solutions without the usual requirement of filtration or centrifugation.

Organic carbon and nitrogen analyses of the sediment samples before and after treatment were performed by Huffman Laboratories, Inc., Golden, Colorado, USA. Organic carbon was determined as the difference between total carbon and carbonate carbon. Nitrogen was determined as total nitrogen, which included both ammonium and organic forms of nitrogen. The maximum relative standard deviation allowed by the laboratory for replicate sediment samples was 10% for organic carbon and 20% for nitrogen.

The effect of selective mineral dissolution treatments on suspended silts and bottom muds, on organic matter recovery, and on composition is presented in Table 17.1. The treatments removed an average of 59% of the weight of the suspended silt samples, and an average of 48% of the weight of the bottom mud samples. An average of 10% of the organic carbon was lost by these treatments; the anomalous increase in carbon after treatment for

---

*Any use of trade names in this manuscript is for descriptive purposes only and does not constitute endorsement by the US Geological Survey.

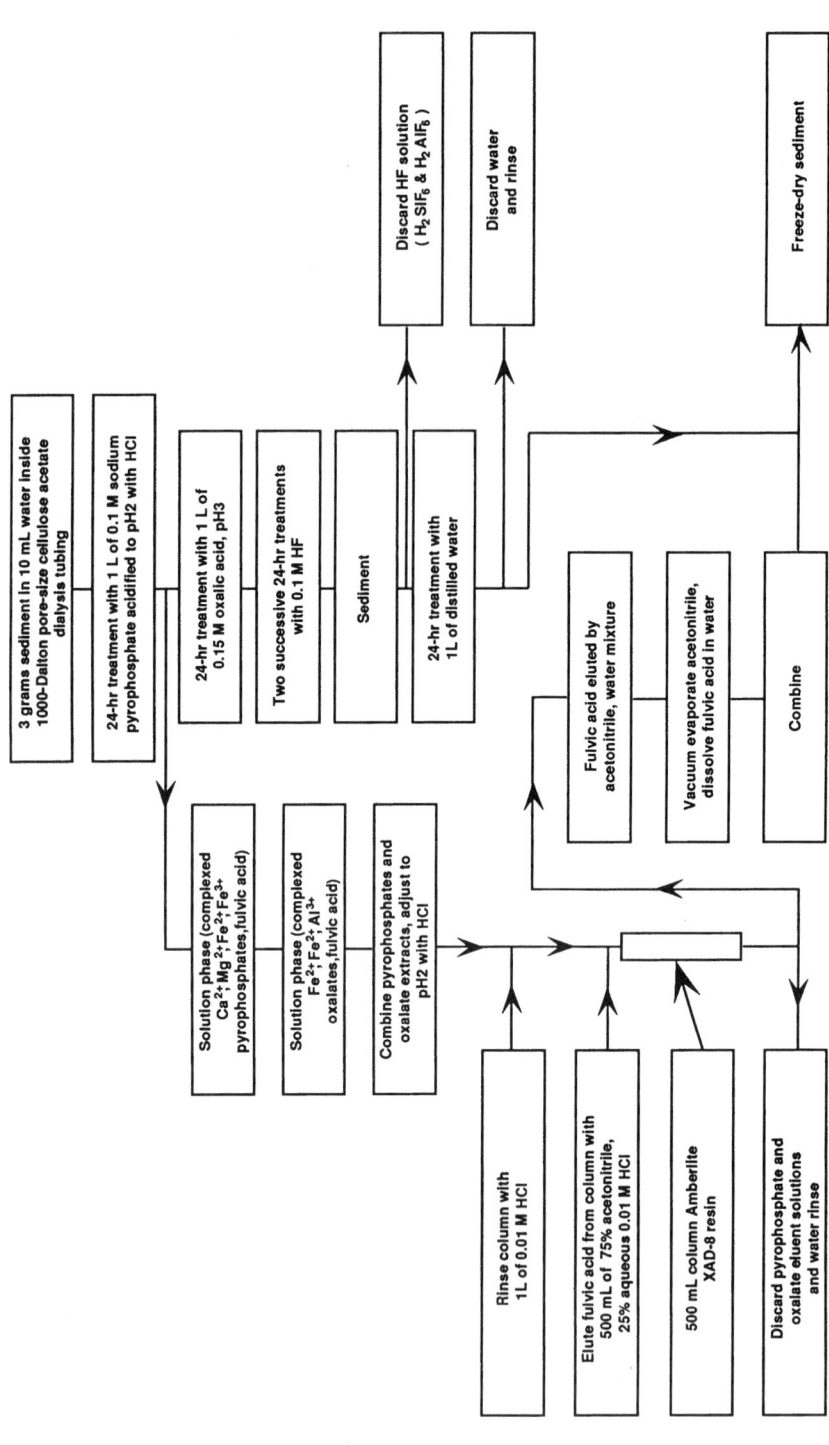

Figure 17.1 Flow chart of procedures used to prepare sediment samples for solid-state ^{13}C NMR spectral analyses.

Table 17.1 Effect of Selective Mineral Dissolution Treatments of Suspended Silts and Bottom Muds on Organic Matter Recoveries and Composition

Sampling Site and Sample	Carbon Content (%)		Nitrogen Content (%)		Weight Loss (%)	Organic Carbon Loss (%)	Atomic C/N	
	Before Treatment	After Treatment	Before Treatment	After Treatment			Before Treatment	After Treatment
Illinois River at Valley City, IL								
Suspended silt	3.2	7.3	0.44	0.85	60.8	11.7	8.6	10.0
Mississippi River below Grafton, IL								
Suspended silt	2.1	5.0	0.25	0.56	63.5	15.0	10.0	10.4
Bottom mud	1.3	2.7	0.14	0.26	45.9	−12.8	10.5	12.0
Mississippi River at St. Louis, MO								
Suspended silt	1.8	3.4	0.20	0.32	52.1	8.0	10.4	12.5
Mississippi River at Thebes, IL								
Suspended silt	1.8	3.6	0.21	0.36	57.6	15.2	10.1	11.8
Bottom mud	1.2	1.9	0.16	0.18	48.8	17.8	8.5	12.2
Ohio River at Olmsted, IL								
Suspended silt	2.1	4.5	0.24	0.46	60.2	13.5	10.1	11.5
Bottom mud	1.4	3.0	0.16	0.27	48.5	10.5	13.0	

the bottom mud sample below Grafton, Illinois, was included in the calcula-tions because of the inherent analytical variablity of the analyses and sub-sampling procedures. Increases in the C/N ratios indicated selective loss of organic nitrogen compounds. The loss may be caused by low-molecular-weight amines and amino acids that are released during the dilute acid treat-ments. Certain amines and amino acids are too small to be retained by the dialysis membrane, and they are not recovered with the fulvic acid fraction on Amberlite XAD-8 resin. The percentage of organic carbon was approxi-mately doubled by the treatments; this doubling is advantageous for ^{13}C NMR spectral analysis. Infrared studies of these sediments after treatment indicated the remaining mineral matter to be largely quartz and mineral fluorides, such as magnesium-rich clays converted to insoluble magnesium fluorides.

^{13}C Nuclear Magnetic Resonance Study

A recent advance in ^{13}C NMR spectrometry is the development of large-sample, magic angle spinners that provide enhanced signal to noise ratios in the spectra.[9] Solid-state ^{13}C NMR spectra of suspended silt samples from the Mississippi River at St. Louis, Missouri, were run on a Chemag-netics M-90S spectrometer using the large-sample spinners. The acquisition parameters used to generate the spectra follow:

(1) 90° pulse width for ^{1}H = 6.2 μs.
(2) Resonance frequency for ^{13}C = 22.6 MHz.
(2) Spectral bandwidth = 623.18 ppm.
(4) Total number of acquisition points = 1024.
(5) Number of data points after zero = 4096.
(6) Line broadening = 49.99 Hz.
(7) Cross-polarization contact time = 850 μs.
(8) Repetition delay = 700 ms.
(9) Number of scans = 3000.

The results on a treated and untreated silt sample from the Mississippi River at St. Louis, Missouri, are shown in Figure 17.2. The carbon distribu-tions for the treated and untreated silt samples are presented in Table 17.2. The carbon distributions based upon the various structural groups are almost identical as shown in Table 17.2, but the loss of sensitivity and band broad-ening problem caused by iron and manganese in the untreated sample (spec-trum B) has been improved by the treatment procedure (spectrum A).

The organic matter of this silt fraction appears to be a mixture of alipha-tic hydrocarbons (peak at 30 ppm), carbohydrates (peaks at 75 and 105 ppm), aromatic hydrocarbons (peak at 130 ppm), and carboxyl and amide func-tional groups (peak at 175 ppm). Low-molecular-weight nitrogen compounds (amines, amino acids, and peptides) which occur from 40 to 60 ppm may have been lost from the sample by the treatment process, as indicated by appear-

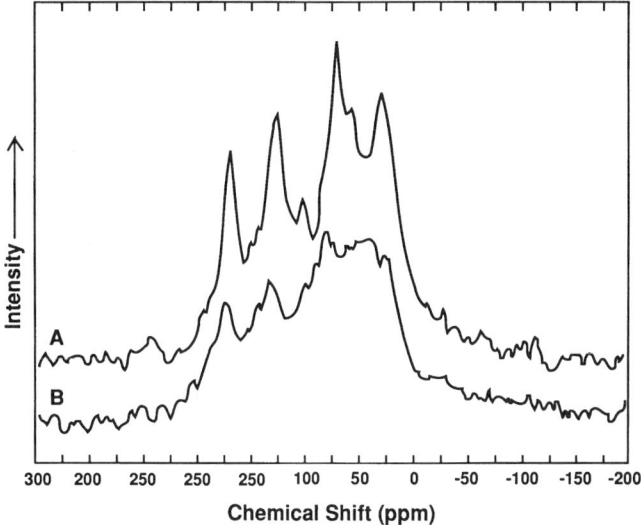

Figure 17.2   Solid-state ^{13}C NMR spectra of suspended silt samples from the Mississippi River at St. Louis in June, 1990: (A) treated for iron removal; (B) untreated for iron removal.

ance of a valley in this spectral region of the treated sample. This loss would not be indicated in Table 17.2 because it would be approximately equally distributed between the aliphatic and heteroaliphatic carbon regions. There appears to be no detectable destruction of carbohydrates by acid hydrolysis during the treatment procedure based upon the data of Table 17.2 and the anomeric carbon peak at 105 ppm in the ^{13}C NMR spectrum of the treated sample.

Infrared analyses of various suspended sediment size fractions isolated from the Mississippi River indicate, from observation of the amide II band,[10] that amides in proteins are a major contributor to the peak at 175 ppm observed in the ^{13}C NMR spectra of Figure 17.2. Therefore, the predominant nature of organic matter associated with the silt fraction is more like undegraded biopolymers containing carbohydrates, proteins, and lipids than like degraded humic substances that have greater acid and aromatic characteristics, although humic substances comprise a minor component of the silt coating.

Spin counting and contact time experiments were conducted to determine the quantitative accuracy of the peak areas in the ^{13}C NMR spectra. These experiments showed that all of the carbon in the sediment samples was not observed, and that peak areas should be interpreted with only general qualitative significance.

Solid-state ^{13}C NMR spectra were obtained on most of the treated samples listed in Table 17.1 using the standard small-sample spinners' because not

Table 17.2 Carbon Distributions[a] of Suspended Silt Samples, Treated and Untreated for Iron Removal, from the Mississippi River at St. Louis in June, 1990

Sample	Ketone 220–190 ppm	Carboxyl, ester, amide, quinone 160–165 ppm	Aromatic 165–110 ppm	Acetal/aromatic 110–90 ppm	Heteroaliphatic 90–50 ppm	Aliphatic 50–0 ppm
Treated for iron removal	3	10	23	7	30	27
Untreated for iron removal	3	11	23	6	29	28

[a] The numbers represent percentages of the total spectrum area.

enough sediment was collected or treated to use the large-sample spinner. The NMR conditions were almost the same as those listed previously except that number of scans was increased from 3000 to 100 000 to offset the sensitivity loss with the smaller spinner.

Solid-state ^{13}C NMR spectra of suspended silt and bottom mud collected from the Mississippi River at Thebes, Illinois, is shown in Figure 17.3. The treated suspended silt spectrum of Figure 17.3 was generally similar to the treated suspended silt spectrum of Figure 17.2, but the bottom mud sample of Figure 17.2 was depleted of carbohydrates (peak at 50–90 ppm) relative to the suspended silt spectra. This difference in organic matter composition may be related to hydrological fractionation processes or biological degradation processes, and may also affect contaminant binding, transport, and fate. The greater signal noise in the spectrum of the treated bottom mud sample is indicative of its low (1.9%) organic carbon content relative to the other samples whose organic carbon contents are listed in Table 17.1.

In summary, the mineral dissolution procedure removed paramagnetic metals and enriched carbon percentages in aquatic sediments so that satisfactory ^{13}C NMR spectra could be obtained. Furthermore, the acid treatments did not appear to degrade polysaccharides. The chief limitation of the procedure is that selective losses of low-molecular-weight, nitrogen-rich organic matter changed both the elemental composition of the organic matter and the ^{13}C NMR spectra. Therefore, a method for generating ^{13}C NMR spectra on low-carbon sediments without treatment to remove paramagnetic elements remains an objective for further research.

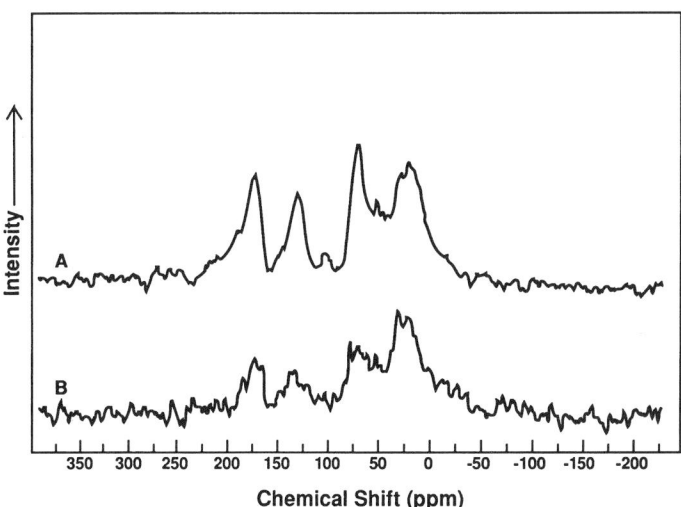

Figure 17.3   Solid-state ^{13}C NMR spectra of suspended silt (A) and bottom mud (B) samples collected from the Mississippi River at Thebes, Illinois, in June, 1990.

References

1. Arshad, M.A., Ripmeester, J. A., and Schnitzer, M., Attempts to improve solid state carbon-13 NMR spectra of whole mineral soils. *Can. J. Soil Sci.* 68, 593–602, 1988.
2. Gibbs, R. J., The geochemistry of the Amazon River system. I. The factors that control the salinity and the concentration and composition of the suspended solids. *Geol. Soc. Am. Bull.* 78, 1203–1232, 1967.
3. Jenne, E. A., Trace element sorption by sediments and soils—sites and processes. In *Symposium on Molybdenum in the Environment*, Chappel, W. R., and Peterson, K. K. (eds.), Marcel Dekker, New York, NY, 1977, pp. 425–553.
4. Moody, J. A., and Meade, R. H., *Hydrologic and Sedimentologic Data Collected During Three Cruises at Low Water on the Mississippi River and Some of its Tributaries, July 1987–June 1988*, US Geological Survey Open-File Rept. 91-485, Denver, CO, 1992.
5. Leenheer, J. A., Meade, R. H., Taylor, H. E., and Pereira, W. E., Sampling, fractionation, and dewatering of suspended sediment from the Mississippi River for geochemical and trace-contaminant analysis. In US *Geological Survey Toxic Substances Hydrology Program—Proc. Technical Meeting, Phoenix, Arizona, Sept. 26–30, 1988*, Mallard, G. E., and Ragone, S. E. (eds.), US Geol. Survey Water Resources Investigations Rept. 88-4220, 1989, pp. 501–512.
6. Gregor, J. E., and Powell, H. K. J., Acid pyrophosphate extraction of soil fulvic acids. *J. Soil Sci.* 37, 577–585, 1986.
7. Parfitt, R. L., Optimum conditions for extraction of aluminum, iron, and silicon from soils with acid oxalate. *Commun. Soil Sci. Plant Anal.* 20, 801–816, 1989.
8. Preston, C. M., Schnitzer, M., and Ripmeester, J. A., A spectroscopic and chemical investigation on the de-ashing of a humin. *Soil Sci. Soc. Am. J.* 53, 1444–1447, 1989.
9. Zhang, M., and Maciel, G. E., Enhanced signal-to-noise ratios in the nuclear magnetic resonance analysis of solids, using large-sample magic-angle spinners. *Anal. Chem.* 62, 633–638, 1990.
10. Leenheer, J. A., Organic substance structures that facilitate contaminant transport and transformations in aquatic sediments. In *Organic Substances and Sediments in Water I. Humics and Soils*, Baker, R. A. (ed.), Lewis Publishers, Chelsea, MI, 1991, pp. 3–22.
11. Jurkiewicz, A., Leenheer, J. A., and Maciel, G. E., Quantitativeness in the analysis of humic and sediment samples by solid-state ^{13}C NMR spectroscopy. *Book of Abstracts, 205th ACS National Meeting, March 28–April 2, 1993, Denver, Colorado*, Environmental Abstract No. 43.

18.

Research Needs for Environmental NMR

This chapter is the result of a panel discussion held at the end of the symposium "NMR Spectroscopy in Environmental Science and Technology" that was presented at the ACS National Meeting in Denver, Colorado, March 28–April 2, 1993. The intention of the panel discussion was to examine and make recommendations for the future of environmental NMR research. This chapter is a general synopsis of the answers and comments from the panelists and members of the audience to three posed questions. The six panelists were:

> Dr. Roger A. Minear (Moderator), University of Illinois, Urbana, IL
> Dr. H.-D. Lüdemann, Institut für Biophysik & Physikalische Biochemie, Regensburg, Germany
> Dr. Robert Wershaw, United States Geological Survey, Denver, CO
> Dr. Jerry A. Leenheer, United States Geological Survey, Denver, CO
> Dr. Gary Maciel, Colorado State University, Fort Collins, CO
> Dr. Leo Condron, Lincoln University, Canterbury, New Zealand

In What Environmental Research Areas will NMR Be Most Influential?

It was generally agreed that the area in which environmental NMR research will be the most influential is the examination of chemical and physical interactions between contaminants and the environmental matrix, especially for heterogeneous and complex matrices. This is because NMR can be used as an in-situ and non-invasive probe. One advantage of NMR for environmental studies is that it can specifically follow the chemistry occurring in complex environments and matrices. In addition, the wide range of NMR-accessible nuclei creates significant potential for research in this area.

A specific area where NMR could be useful is the examination of chemicals and their transformation in soils and sediments, both biotic and abiotic, without having to use extraction methods. This could provide information regarding precursors, reaction products, and changes occurring in soils, without jeopardizing sample integrity by extraction methods. Tracking reactions and reaction by-products in such matrices can be carried one step further by labeling compounds with NMR-sensitive nuclei and following the concurrent reactions. It will also be useful to use NMR in this fashion to examine the influence of the biota upon the reaction and the reaction products, which will in turn advance studies examining bioavailability and bioremediation processes.

Another area where NMR could be beneficial is in the examination of contaminant binding and transportation by different phases and minerals. Labeling studies with NMR-sensitive nuclei could shed new information about association processes by providing a molecular view of the interaction of pollutants and dissolved materials such as humic acids. By examining changes in the T_1 relaxation times, line broadening, coupling constants, and chemical shifts, NMR might provide a view of the type of association occurring: partitioning, noncovalent adsorption, and covalent bond formation.

Combining the topics mentioned above with various fractionation methods could provide information regarding chemical speciation in various components of environmental samples. Thus, new fractionation methods in conjunction with NMR studies will be important.

Finally, it was noted that NMR will be useful for the macroscopic examination of reactions, general structures, and fundamental characterization of whole samples. Solid-state NMR will be especially germane in this area. It was suggested that NMR be used in this fashion to study reactions at feasible concentrations and then the results could be extrapolated to low, environmentally realistic levels. This, of course, has the caveat that the results from the reaction run at higher concentration cannot usually be accurately extrapolated to lower concentrations because many environmental reactions may be concentration-dependent, e.g., aluminum hydrolysis.

What Are the Areas in which NMR Instrumentation and Techniques Require Further Research in Order for NMR to Become More Acceptable as a Routine Tool in Environmental Studies?

It was agreed by all that the major area of NMR instrumentation that requires enhancement is the sensitivity of the instrument. This can be achieved by increasing magnetic field strengths and building better probes. Advances in software as well as new pulse sequences have already enhanced, and will continue to improve, the detection capabilities of NMR spectrometers.

The areas of NMR technique that require further research are methods to deal with highly heterogeneous and complex matrices containing low concentrations of the compound being examined. Because the sample matrix is highly heterogeneous and complex, it was mentioned that an understanding of the basic parameters which affect NMR signals in such matrices need to be examined. A good approach might be to begin with a basic study of organic materials in several well-characterized environmental matrices.

Building upon this suggestion, it was mentioned that sample concentration and fractionation methodologies are important. Certain concentration and fractionation methods may change the sample; therefore, new, nonaltering methods are necessary if environmental samples are to be pretreated before NMR analysis. It was pointed out that even if concentration methods do not alter the sample, many environmental reactions are concentration dependent; hence, concentrated samples may behave differently.

To avoid concentration and fractionation artifacts and other problems usually encountered with environmental samples, e.g., low concentration, spiking environmental samples with compounds labeled with NMR-sensitive nuclei in specific, chemically active sites would be beneficial. This way, the chemical behavior of many compounds could be examined by NMR at or near typical concentration levels, and little or no sample preparation would be necessary.

What Are the General Difficulties That Environmental NMR Research Faces?

Six general difficulties were mentioned. The first three dealt with experimental details while the last three focused on the logistics of environmental NMR research, i.e., funding, access to NMR instrumentation, etc.

The first two difficulties mentioned were low sample concentration and the presence of paramagnetic materials in the sample. Both are the bane of any NMR spectroscopist; however, these are especially problematic for environmental scientists. Most environmentally pertinent compounds are present at concentrations far below NMR detection levels. Also, environmental samples, especially if they are from soils or sediments, usually contain some paramagnetic iron or manganese. Another difficulty occurs when using NMR to characterize trace amounts of a specific compound or compounds present in a highly heterogeneous matrix that contains similar nuclei to the compound. It is difficult to distinguish between NMR signals from the specific compound and from the matrix. For example, if trace amounts of a hydrophobic compound sorbed to humic material are being analyzed by ^{13}C NMR, then unless that hydrophobic compound is labeled with ^{13}C nuclei in specific sites, its signals will be just as intense as those of the humic material. Therefore, the ^{13}C NMR spectrum will contain signals from both the hydrophobic compound and the humic material, and it will be nearly impossible to identify the signals arising from the hydrophobic compound.

The last three difficulties mentioned dealt with the logistics involved when an environmental scientist uses NMR spectroscopy. The first problem listed was the prejudice of many "applied" researchers towards NMR as a tool. This prejudice primarily results from the low sensitivity of NMR, which many perceive as a barrier to obtaining useful information. Although NMR probably will never have the sensitivity that environmental analytical instrumentation such as gas chromatography/mass spectrometry has, yet it is useful in many aspects. It was stated, and agreed upon by many, that the environmental research community needs to advance the idea that NMR is indeed solving a wide range of important environmental problems.

Another logistical problem mentioned was concerned with the expense of obtaining, maintaining, and operating a NMR spectrometer. NMR spectrometers are expensive, and thus ownership is out of the question for many individual scientists. Pure research departments, such as chemistry departments, are better able (due to their size and to the large number of members who will use NMR) to pool their funds and purchase spectrometers. Another expense-related problem is that NMR spectroscopy is a complicated technique; therefore, environmental research projects usually require a spectroscopist as well as an environmental scientist. Because two researchers are required, more funding is necessary.

Finally, access to NMR spectrometers was mentioned as a difficulty. Because of their cost, NMR spectrometers are often not present in environmental laboratories. Even so, it was noted that environmental scientists may be able to find locations that allow access to instruments and that these opportunities should not be overlooked. It was suggested that creation of environmental science research centers that would house a wide variety of advanced instrumentation and provide a location where scientists could collaborate would help ease several of these problems.

Index

acetal, 301
acetaldehyde, 132, 134, 137
acetate, 166, 173
acetonitrile, 132, 134, 237
acetylation, 300
acid pyrophosphate treatment, 305
acoustic ringing, 125, 149
adenosine triphosphate, 234
adsorption, 126
    capacities of coal fly ash, 107t, 110t
    isotherms, 91, 116
    isotherms of hazardous organic, 102
    solid–gas equilibrium, 94, 103
    solid–liquid equilibrium, 94, 95, 107, 116
    surface, 237
aggregate, 175, 176, 222, 223, 231, 234, 235,
    236, 237, 238, 243, 244
alanine, 131, 132, 133f
alcohol, 300, 301
alcoholic hydroxyl, 300
aldehyde, 130
algal cells, 224, 227
algal growing, 231
aliphatic ketones, 300
alkaline bromination, 226, 234, 240, 241,
    243–244
allochthonous, 218
aluminosilicates, 97, 98
aluminum, 123, 125
    acidity of, 161

aluminum hydroxide, 162
    see also dissolved hydroxyaluminum
        (hydroxy-Al) species
    see also hydroxyaluminum (hydroxy-Al)
        products of aluminum hydrolysis
aqueous aluminum, 125, 161, 171f
chemistry of, 162
environmental effects of, 123, 161
health effects of, 161
hydrolysis of, 140, 164, 171, 314
    see also hydroxyaluminum (hydroxy-Al)
        products of aluminum hydrolysis
metal, 161
organo-aluminum, 126
precipitation of, 171, 172
speciation, 126
toxicity of, 162
tridecimer Al$_{13}$, 162, 172f
water treatment, 161
    see also dissolved hydroxyaluminum
        (hydroxy-Al) species
    see also hydroxyaluminum (hydroxy-Al)
        products of aluminum hydrolysis
aluminum (^{27}Al) NMR, 125, 147, 149, 162,
    165, 172
acoustic ringing, 149
aluminum chloride
    with acetic acid, 166f
    with citric acid, 170f
    with lactic acid, 168f